INTRODUCTION TO STATISTICS

INTRODUCTION TO STATISTICS

2nd EDITION

RONALD E. WALPOLE

Professor of Mathematics and Statistics
Roanoke College

MACMILLAN PUBLISHING CO., INC.
New York
COLLIER MACMILLAN PUBLISHERS
London

Macmillan Publishing Co., Inc.
866 Third Avenue, New York, New York 10022

Collier-Macmillan Canada, Ltd., Toronto, Ontario

Library of Congress Cataloging in Publication Data

Walpole, Ronald E.
  Introduction to statistics.
  Bibliography: p.
  1. Statistics. I. Title.
HA29.W3357 1974   519.5   72–12961
ISBN 0–02–424040–0

Printing: 3 4 5 6 7 8   Year: 5 6 7 8 9 0

DEDICATION

To Norma

PREFACE

Like the first edition, this book has been written to serve as a text for an introductory course in statistics for students majoring in any of the academic disciplines. Both the numerous illustrative examples and the exercises are selected from many different fields of application. A knowledge of high school algebra is sufficient to comprehend the basic concepts of descriptive and inductive statistics that are presented, although experience seems to indicate that a student benefits more from such a course when it has been preceded by a semester of college freshman mathematics. Consequently, the students enrolled in statistics at Roanoke College are, for the most part, second term freshmen and sophomores majoring in sociology, psychology, economics, business administration, and the sciences, who have completed a course in elementary functions or finite mathematics.

The field of modern statistics, with an increased emphasis on statistical inference, is based primarily on the theory of probability. An introduction to the basic concepts of probability theory using set notation is, therefore, presented in Chapter 2. The material in Chapters 3 and 4, introducing random variables and their mathematical expectations, has been completely revised. While retaining the discussion of discrete and continuous probability distributions and the various properties describing these distributions, a brief treatment of joint, marginal, and conditional distributions has been included. In keeping with the modern trend, the treatment of grouped sample data has been largely deleted. Relevant material on this subject retained from the first edition is included in Chapter 3.

An introduction to several discrete probability distributions is presented in Chapter 5 followed by a discussion of the normal distribution in Chapter 6. This then naturally leads to the treatment of sampling theory, estimation theory, and hypothesis testing in Chapters 7, 8, and 9, respectively. These three chapters have been revised in this edition so as to avoid much of the repetitiveness in both the theory and the examples that occurred in the original edition. Also, for greater emphasis, nonparametric tests now occupy separate sections in the chapter on hypothesis testing. Chapters 10 and 11 present an elementary discussion of regression theory, an expanded section on correlation analysis to include tests of significance, and an introduction to analysis of variance. Although some authors prefer to use a regression approach to the analysis of variance, no attempt has been made in this text to relate the two chapters. Either chapter may be considered without a knowledge of the other.

The text contains sufficient material to allow for flexibility in the length of the course and the selection of topics. A semester course, meeting three hours a week, should include all the material in Chapters 1 through 4; Sections 5.1, 5.2, and 5.3 of Chapters 5; Chapters 6 through 9, perhaps excluding Sections 9.9 and 9.10; and Sections 10.1, 10.2, 10.3, and 10.7 of Chapter 10.

I wish to acknowledge my appreciation to all those who assisted in the preparation of this textbook. I am particularly grateful to my wife for typing the original manuscript; to Miss Diane Milan, Miss Lynn Watson, and Miss Karen Carter for typing and proofreading the revised second edition; to David and Deborah Selman for generating data in several of the exercises; to Professor Suzanne Glass who read the first draft and tested the material in her classes; and to the teachers and reviewers of the first edition for their helpful suggestions in preparing this second edition of *Introduction to Statistics*.

I am indebted to the Literary Executor of the late Sir Ronald A. Fisher, F.R.S., Cambridge, and to Oliver & Boyd Ltd., Edinburgh, for their permission to reprint Table A.5 from their book *Statistical Methods for Research Workers*; to Professor E. S. Pearson and the Biometrika Trustees for permission to reprint in abridged form, as Tables A.6 and A.7, Tables 8 and 18 from *Biometrika Tables for Statisticians*, Vol. I; to D. Van Nostrand Company, Inc., for permission to reproduce in Table A.3 material from E. C. Molina's *Poisson's Exponential Binomial Limit*. I wish also to express my appreciation for permission to reproduce Table A.8 from the *Annals of Mathematical Statistics*, Table A.9 from the *Bulletin of the Educational Research at Indiana University*, Table A.10 from a publication by the American Cyanamid Company, and Table A.11 from *Biometrics*.

R.E.W.

CONTENTS

4 MATHEMATICAL EXPECTATION 55

5 SOME DISCRETE PROBABILITY DISTRIBUTIONS 77

6 NORMAL DISTRIBUTION 101

7 SAMPLING THEORY 121

8 ESTIMATION THEORY 153

INTRODUCTION TO STATISTICS

INTRODUCTION CHAPTER 1

1.1 Nature of Statistics

The statistician is basically concerned with the chance outcomes that occur in scientific investigations. He may be interested in the number of accidents occurring at a certain intersection, the outcome when a die is tossed, or the amount of residue deposited in a chemical experiment. Hence he is usually dealing with counts or numerical measurements. We shall refer to the recorded information in its original collected form as *raw data*. The statistician may be interested in various methods of describing large masses of raw data so as to yield otherwise obscure information, or he may be interested in making decisions and drawing conclusions about a large set of data about which he has only partial knowledge based on a smaller subset of the data.

The science of statistics deals with methods used in the collection, presentation, analysis, and interpretation of data. All problems involving the use of statistical methods may be categorized as belonging either to the field of *descriptive statistics* or *inductive statistics*. Any treatment of data leading to predictions or inferences concerning a larger group of data is known as *inductive statistics*. On the other hand, if our interest lies strictly with the data on hand and no attempt is made to generalize to a larger set of data, then we are in the field of *descriptive statistics*.

The distinction between descriptive and inductive statistics becomes quite clear if we consider a set of values representing the total precipitation in a certain part of the country during the month of July for the past 30 years. Any value describing data such as the average precipitation for July during the past 30 years or the driest July in the past 30 years is a value in the field of descriptive statistics. We are not attempting to say anything about the precipitation of any other years, except the 30 years from which the information was obtained. If the average precipitation for July in this area was 1.3 inches during the past 30 years and we make the statement that next July we can expect between 1.2 and 1.4 inches of rain, we are generalizing and thereby placing ourselves in the field of inductive statistics.

The generalizations of inductive statistics are subject to uncertainties, since we are dealing with only partial information obtained from a subset of the data of interest. To cope with uncertainties, an understanding of probability theory is essential. In this text we shall keep this to a minimum. Probability theory can be presented best by making use of set notation. Consequently we shall introduce sets and their properties early in Chapter 2. The mathematics required for a statistics course can be demanding if the proofs are to be rigorous. However, by omitting certain proofs that require a knowledge of calculus, one is able to learn a great deal about statistical procedures using only elementary concepts of high school algebra.

Every effort is made in this text to present material in such a manner that all areas of learning may benefit, whether it be the field of science, psychology, business, agriculture, or medicine. The basic techniques for collecting and analyzing data are the same no matter what the field of application may be. For example, the chemist runs an experiment using 3 variables and measures the amount of desired product. The results are then analyzed by statistical procedures. These same procedures are used to analyze the results obtained by measuring the yield of grain when 3 fertilizers are tested or to analyze the data representing the number of defectives produced by 3 similar machines. Many statistical methods that were derived primarily for agricultural applications have proved to be equally valuable for applications in other areas.

Statistics may be used to describe the center of a set of data and to provide a measure of variability or dispersion from this central value. The relationship between two variables can be measured by a numerical value called the *correlation coefficient*. Statistics enables one not only to predict and test hypotheses, but also to determine the accuracy of one's decisions. Every progressive industry today employs statisticians to direct their quality-control process and to assist in the establishment of good advertising and sales programs for their products. In business the statistician is responsible for the analysis of time series and the formation of index numbers.

Indeed, statistics is a very powerful tool if properly used. The abuse of statistical procedures can only lead to erroneous results. One should be careful to apply the correct and most efficient procedure for the given conditions to obtain maximum information from the data available.

The procedures used to analyze a set of data depend to a large degree on the method used to collect the information. For this reason it is desirable in any investigation to consult with the statistician from the time the project is planned until the final results are analyzed and interpreted.

1.2 History of Statistics

The field of descriptive statistics has a history that extends back to the beginning of mankind. In early biblical times statistics was used to provide information relative to taxes, wars, agricultural crops, and even athletic endeavors. Inductive statistics, depending largely on the theory of probability, has made its greatest strides since the sixteenth century. In fact, the present science of statistics, for the most part, is a result of active research by many people in various fields during the past 400 years.

Perhaps it was man's unquenchable thirst for gambling that led to the early development of probability theory. In an effort to increase their winnings, gamblers called upon the mathematician to provide optimum strategies for various games of chance. Some of the mathematicians providing the answers at this time were Pascal, Leibnitz, Fermat, and James Bernoulli.

In 1733 DeMoivre discovered the equation for the normal distribution upon which much of the theory of inductive statistics is based. This same bell-shaped distribution is often referred to as the *Gaussian distribution*, in honor of Gauss (1777–1855), who also derived its equation from a study of errors in repeated measurements of the same quantity. The work of Laplace, with applications of statistics to astronomy, was very prominent at this time.

During the nineteenth century a Belgian statistician, Adolph Quetelet (1796–1874), was very active in the application of statistical methods in the fields of education and sociology. Quetelet was one of the first statisticians to demonstrate that statistical techniques derived in one area of research are also applicable in most other areas.

Perhaps the greatest contributor to statistics for the social sciences was Sir Francis Galton (1822–1911). His most notable contributions were in the applications of statistics to problems encountered in the fields of heredity and eugenics. Galton is credited with the discovery of percentiles. A mathematician, Karl Pearson (1857–1936), worked with Galton to develop the theory of regression and correlation. Pearson went on to make many contributions to statistics and is responsible for much of the present theory of sampling.

At the beginning of the twentieth century, methods were developed by William S. Gosset for decision making based on small sets of data. Working for an Irish brewery that disallowed publications that might prove useful to its competitors, Gosset published his results under the name "Student." Further contributions on small-sample theory and also in experimental designs were advanced by Sir Ronald Fisher (1890–1962), regarded as this century's outstanding statistician to date.

The twentieth century has produced several noteworthy statisticians, most of whom are still very active in developing new theories and applications of statistics. During the past decade we have witnessed a definite shift from the so-called classical approach to estimation theory, based solely on objective information provided by the random sample, to the Bayesian approach, whereby sample information is combined with other available prior sub-jective information. Only since 1950 has statistics formed a separate program of study in many American colleges and universities. For the most part, graduate schools have awarded advanced degrees in statistics only since 1955. The availability of electronic computers is certainly a major factor in the modern development of statistics.

This, then, is the history and present status of the science of statistics. In the future we shall see many new theories developed. Productive research can be anticipated in the areas of mathematical statistics, probability, the theory of games, linear programming, stochastic processes, and experimental designs. Today the research worker considers statistics one of his most useful aids.

SETS AND PROBABILITY CHAPTER 2

2.1 Sets and Subsets

A fundamental concept in all branches of mathematics that is necessary for the study of probability and statistics is that of a *set*. We may think of a *set* as a well-defined collection of objects. For example, the rivers in Virginia, the letters in the alphabet, the members of the senate, and the monthly income from shoe sales all constitute sets. One can even think of a line segment as an infinite set of points. Each object in a set is called an *element* of the set or a *member* of the set.

Usually sets will be denoted by capital letters such as A, B, X, or Y, whereas small letters such as a, b, x, or y will be used to indicate the elements of a set. To have a well-defined set, we must have some means of indicating whether a given object is, or is not, a member of the set.

There are two ways to describe a set. First, if the set has a finite number of elements, we may *list* the members separated by commas and enclosed in braces. Thus the set A consisting of the numbers 2, 4, 6, and 8 may be written

$$A = \{2, 4, 6, 8\},$$

or the set B, of possible outcomes when a coin is tossed, may be written

$$B = \{H, T\},$$

where H and T correspond to "heads" and "tails," respectively.

Second, a set may be described by a *statement* or *rule*. For example, we may let C be the set of cities in the world with a population over 1 million. If x is an arbitrary element of C, we write

$$C = \{x \mid x \text{ is a city with a population over 1 million}\},$$

which reads "C is the set of all x such that x is a city with a population over 1 million." The vertical bar is read "such that." Similarly, if P is the set of points (x, y) on the boundary of a circle of radius 2, we write

$$P = \{(x, y) \mid x^2 + y^2 = 4\}.$$

Whether we describe the set by the rule method or by listing the elements will depend upon the specific problem at hand. It would be very difficult to list all elements in the set of people with blue eyes. On the other hand, there is no easy rule to specify the set

$$Y = \{\text{book, dog, coin, war}\}.$$

In set notation the symbol $\in$ means "is an element of" or "belongs to" and $\notin$ means "is not an element of" or "does not belong to." If x is an element of the set A and y is not, we write

$$x \in A \qquad \text{and} \qquad y \notin A.$$

Example 2.1 Let $A = \{2, 4, 6, 8\}$ and $B = \{x \mid x \text{ is an integer divisible by 3}\}$. Then $4 \in A$, $9 \in B$, $8 \notin B$, and $3 \notin A$.

DEFINITION   *Two sets are equal if they have exactly the same elements in them.*

If set A is equal or identical to set B, then every element that belongs to A also belongs to B, and every element that belongs to B also belongs to A. We denote this equality by writing $A = B$. However, if either of the sets A or B contains at least one element that is not common to both, the sets are said to be unequal, and then we write $A \neq B$.

Example 2.2 Let $A = \{1, 3, 5\}$, $B = \{3, 1, 5\}$, and $C = \{1, 3, 5, 7\}$. Then $A = B$, $A \neq C$, and $B \neq C$.

Note that a set does not change when the order of the elements is rearranged.

DEFINITION *The* null set *or* empty set *is a set that contains no elements. We denote this set by the symbol* $\emptyset$.

If we let A be the set of microscopic organisms detected by the naked eye in a biological experiment, then A must be the null set. Also if $B = \{x | x$ is a nonprime factor of 7$\}$, then B must be the null set, since the only possible factors of 7 are the prime numbers 1 and 7.

Let us consider the set $A = \{1, 2, 3, 4, 5\}$ and the set $B = \{2, 4\}$. We note that every element in the set B is also an element of the set A. The set B is said to be a *subset* of A. Symbolically we write this $B \subset A$, where $\subset$ means "is a subset of" or "is contained in."

DEFINITION *If every element of a set A is also an element of a set B, then A is called a* subset *of B.*

According to this definition every set is a subset of itself. Any subset of a set that is not the set itself is called a *proper subset* of the set. Therefore, B is a proper subset of A if $B \subset A$ and $B \neq A$.

Example 2.3 The set $B = \{2, 4\}$ is a proper subset of $A = \{1, 2, 3, 4, 5\}$. However, the set $C = \{3, 2, 5, 1, 4\}$ is a subset of A but not a proper subset, since $A = C$.

In many discussions all sets are subsets of one particular set. This set is called the *universal set* and is usually denoted by U. In a mathematical discussion the real numbers could be used as the universal set. The set of I.Q.s of all college students could be used as the universal set that has the I.Q.s of students of a certain college as a subset.

Example 2.4 All the subsets of the universal set $U = \{1, 2, 3\}$ are $\{1, 2, 3\}$, $\{1, 2\}, \{1, 3\}, \{2, 3\}, \{1\}, \{2\}, \{3\}$, and $\emptyset$.

Note that the number of subsets of a universal set containing three elements is $2^3 = 8$. In general a set with n elements has 2^n subsets.

The relationships between subsets and the corresponding universal set can be illustrated by means of *Venn diagrams*. In a Venn diagram we let the universal set be a rectangle and let subsets be circles drawn inside the rectangle. Thus in Figure 2.1 we see that A, B, and C are all subsets of the universal set U. It is also clear that $B \subset A$; B and C have no elements in common; A and C have at least one element in common.

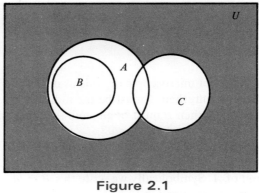

Figure 2.1
<u>Subsets of U.</u>

Sometimes it is convenient to shade various areas of the diagram as in Figure 2.2. In this case we take all the students of a certain college to be our universal set. The subset representing those students taking mathematics has been shaded by drawing straight lines in one direction and the subset representing those students studying history has been shaded by drawing lines in a different direction. The doubly shaded, or crosshatched, area represents the subset of students enrolled in both mathematics and history, and the unshaded part of the diagram corresponds to those students who are studying subjects other than mathematics or history.

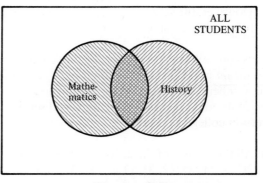

Figure 2.2
<u>Subsets indicated by shading.</u>

2.2 Set Operations

We now consider certain operations on sets that will result in the formation of new sets. These new sets will be subsets of the same universal set as the given sets.

DEFINITION *The* intersection *of two sets A and B is the set of elements that are common to A and B.*

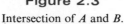

Symbolically, we write $A \cap B$ for the intersection of A and B. The elements in the set $A \cap B$ must be those and only those which belong to both A and B. These elements may either be *listed* or defined by the *rule method—$A \cap B = \{x \mid x \in A$ and $x \in B\}$*. In the Venn diagram in Figure 2.3 the shaded area corresponds to the intersection $A \cap B$.

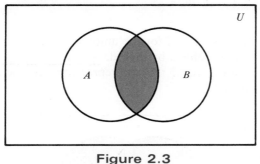

Figure 2.3

Intersection of A and B.

Example 2.5 Let $A = \{1, 2, 3, 4, 5\}$ and $B = \{2, 4, 6, 8\}$; then $A \cap B = \{2, 4\}$.

Example 2.6 If R is the set of all taxpayers and S is the set of all people over 65 years of age, then $R \cap S$ is the set of all taxpayers who are over 65 years of age.

Example 2.7 Let $P = \{a, e, i, o, u\}$ and $Q = \{r, s, t\}$; then $P \cap Q = \varnothing$. That is, P and Q have no elements in common.

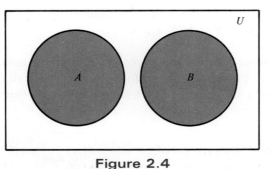

Figure 2.4

Disjoint sets.

DEFINITION *If $A \cap B = \emptyset$, then the sets A and B are* disjoint *; that is, A and B have no elements in common.*

Two disjoint sets A and B are illustrated in the Venn diagram in Figure 2.4. By shading the areas corresponding to the sets A and B, we find no doubly shaded area representing the set $A \cap B$. Hence $A \cap B$ is empty.

DEFINITION *The* union *of two sets A and B is the set of elements that belong to A or to B or to both.*

Symbolically we write $A \cup B$ for the union of A and B. The elements of $A \cup B$ may be listed or defined by the rule $A \cup B = \{x \mid x \in A \text{ or } x \in B\}$. In the Venn diagram in Figure 2.5 the area representing the elements of the set $A \cup B$ has been shaded.

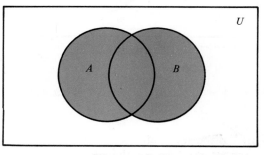

Figure 2.5

Union of A and B.

Example 2.8 Let $A = \{2, 3, 5, 8\}$ and $B = \{3, 6, 8\}$; then $A \cup B = \{2, 3, 5, 6, 8\}$.

Example 2.9 If $M = \{x \mid 3 < x < 9\}$ and $N = \{y \mid 5 < y < 12\}$, then $M \cup N = \{z \mid 3 < z < 12\}$.

Suppose that we consider the employees of some manufacturing firm as the universal set. Let all the smokers form a subset. Then all the nonsmokers form a set, also a subset of U, which is called the *complement* of the set of smokers.

DEFINITION *If A is a subset of the universal set U, then the* complement *of A with respect to U is the set of all elements of U that are not in A. We denote the complement of A by A′.*

The elements of A' may be listed or defined by the rule $A' = \{x|x \in U$ and $x \notin A\}$. In the Venn diagram in Figure 2.6 the area representing the elements of the set A' has been shaded.

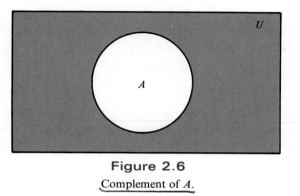

Figure 2.6

Complement of A.

Example 2.10 Let R be the set of red cards in an ordinary deck of 52 playing cards and let U be the entire deck. Then R' is the set of cards in the deck that are not red—the black cards.

Example 2.11 Consider the universal set $U = \{$book, dog, cigarette, coin, map, war$\}$. Let $A = \{$dog, war, book, cigarette$\}$. Then $A' = \{$coin, map$\}$.

Several results that follow from the above definitions which may easily be verified by means of Venn diagrams are

1. $A \cap \varnothing = \varnothing$.
2. $A \cup \varnothing = A$.
3. $A \cap A' = \varnothing$.
4. $A \cup A' = U$.
5. $U' = \varnothing$.
6. $\varnothing' = U$.
7. $(A')' = A$.

2.3 Sample Space

The scientist performs experiments to produce observations or measurements that will assist him in arriving at conclusions. If a chemist runs an analysis several times under the same conditions, in most cases he will obtain different measurements, indicating an element of chance in the experimental procedure. In statistics we use the word *experiment* to describe any process that generates raw data. In most cases the result of the experiment will depend on chance and therefore cannot be predicted with certainty.

An example of a statistical experiment might be the tossing of a coin. This experiment consists only of the 2 outcomes, heads or tails. Another experiment might be the drawing of 3 cards from an ordinary deck of playing cards and observing the suits. The opinions of voters concerning a new sales tax can also be considered as outcomes of an experiment.

DEFINITION *A set whose elements represent all possible outcomes of an experiment is called the sample space and is represented by S.*

The sample space is also sometimes called the *universal set*. In this book we shall use the notation S, rather than U, whenever we are dealing with a set whose members represent all possible outcomes of an experiment involving an element of chance.

DEFINITION *An element of a sample space is called a sample point.*

Example 2.12 Consider the experiment of tossing a die. If we are interested in the number that shows on the top face, then the sample space would be

$$S_1 = \{1, 2, 3, 4, 5, 6\}.$$

If we are interested only in whether the number is even or odd, then the sample space is simply

$$S_2 = \{\text{even, odd}\}.$$

This example illustrates the fact that more than one sample space can be used to describe the outcomes of an experiment. In this case S_1 provides more information than S_2. If we know which element in S_1 occurs, we can tell which outcome in S_2 occurs; however, a knowledge of what happens in S_2 in no way helps us to know which element in S_1 occurs. In general it is desirable to use a sample space that gives the most information concerning the outcomes of the experiment.

Example 2.13 Suppose three items are selected at random from a manufacturing process. Each item is inspected and classified defective or nondefective. The sample space providing the most information would be $S_1 = \{NNN, NDN, DNN, NND, DDN, DND, NDD, DDD\}$. A second sample space, although it provides less information, might be $S_2 = \{0, 1, 2, 3\}$, where the elements represent no defectives, one defective, two defectives, or three defectives in our random selection of three items.

In any given experiment we may be interested in the occurrence of certain events, rather than in the outcome of a specific element in the sample space. For instance, we might be interested in the event A that the outcome when a die is tossed is divisible by 3. This will occur if the outcome is an element of the subset $A = \{3, 6\}$ of the sample space S_1 in Example 2.12. As an additional illustration, we might be interested in the event B that the number of defectives is greater than 1 in Example 2.13. This will occur if the outcome is an element of the subset $B = \{DDN, DND, NDD, DDD\}$ of the sample space S_1.

To each event we assign a collection of sample points that constitutes a subset of the sample space. This subset represents all the elements for which the event is true.

DEFINITION   *An event is a subset of a sample space.*

Example 2.14 Given the subset $A = \{t | t < 5\}$ of the sample space $S = \{t | t \geq 0\}$, where t is the life in years of a certain electronic component, then A is the event that the component fails before the end of the fifth year.

This example illustrates the fact that corresponding to any subset we can state an event whose elements are the given subset. In practice we usually state the event first and then determine its set.

DEFINITION   *If an event is a set containing only one element of the sample space, then it is called a simple event. A compound event is one that can be expressed as the union of simple events.*

Example 2.15 The event of drawing a heart from a deck of 52 playing cards is the subset $A = \{$heart$\}$ of the sample space $S = \{$heart, spade, club, diamond$\}$. Therefore, A is a simple event. Now the event B of drawing a red card is a compound event, since $B = \{$heart $\cup$ diamond$\} = \{$heart, diamond$\}$.

Note that the union of simple events produces a compound event that is still a subset of the sample space. We should also note that if the 52 cards of the deck were the elements of the sample space rather than the 4 suits, then the event A of Example 2.15 would be a compound event.

2.4 Counting Sample Points

One of the problems that the statistician must consider and attempt to evaluate is the element of chance associated with the occurrence of certain events when an experiment is performed. These problems belong in the field of probability, a subject introduced in Section 2.5. In many cases we shall be able to solve a probability problem by counting the number of points in the sample space. A knowledge of the actual elements or a listing is not always required. The <u>fundamental principle of counting</u> is stated in the following theorem.

> THEOREM 2.1 *If an operation can be performed in n_1 ways, and if for each of these a second operation can be performed in n_2 ways, then the two operations can be performed together in $n_1 n_2$ ways.*

Example 2.16 How many sample points are in the sample space when a pair of dice is thrown once?

Solution. The first die can land in any 1 of 6 ways. For each of these 6 ways the second die can also land in 6 ways. Therefore, the pair of dice can land in $(6)(6) = 36$ ways. The student is asked to list these 36 elements in Exercise 12.

The above theorem may be extended to cover any number of events. The general case is stated in the following theorem.

> THEOREM 2.2 *If an operation can be performed in n_1 ways, and if for each of these a second operation can be performed in n_2 ways, and for each of the first two a third operation can be performed in n_3 ways, etc., then the sequence of k operations can be performed in $\underline{n_1 n_2 \cdots n_k}$ ways.*

Example 2.17 How many lunches are possible consisting of soup, a sandwich, dessert, and a drink if one can select from 4 soups, 3 kinds of sandwiches, 5 desserts, and 4 drinks?

Solution. The total number of lunches would be $(4)(3)(5)(4) = 240$.

Example 2.18 How many even three-digit numbers can be formed from the digits 1, 2, 5, 6, 9 if each digit can be used only once?

Solution. Since the number must be even, we have only 2 choices for the units position. For each of these we have 4 choices for the hundreds position and 3 choices for the tens position. Therefore we can form a total of $(2)(4)(3) = 24$ even three-digit numbers.

Frequently we are interested in a sample space that contains as elements all possible orders or arrangements of a group of objects. For example, we might want to know how many arrangements are possible for sitting 6 people around a table, or we might ask how many different orders are possible to draw 2 lottery tickets from a total of 20. The different arrangements are called *permutations*.

> DEFINITION *A* permutation *is an arrangement of all or part of a set of objects.*

Consider the three letters a, b, and c. The possible permutations are *abc*, *acb*, *bac*, *bca*, *cab*, and *cba*. Thus we see that there are 6 distinct arrangements. Using Theorem 2.2 we could have arrived at the answer without actually listing the different orders. There are 3 positions to be filled from the letters a, b, and c. Therefore, we have 3 choices for the first position, and 2 for the second, leaving only 1 choice for the last position, giving a total of $(3)(2)(1) = 6$ permutations. In general n distinct objects can be arranged in $n(n - 1)(n - 2)$ $\cdots (3)(2)(1)$ ways. We represent this product by the symbol $n!$, which is read "n factorial." Three objects can be arranged in $3! = (3)(2)(1) = 6$ ways. By definition $1! = 1$ and $0! = 1$.

> THEOREM 2.3 *The number of permutations of n distinct objects is n!.*

The number of permutations of the four letters a, b, c, and d will be $4! = 24$. Let us now consider the number of permutations that are possible by taking the 4 letters 2 at a time. These would be *ab, ac, ad, ba, ca, da, bc, cb, bd, db, cd,* and *dc*. Using Theorem 2.2 again, we have 2 positions to fill with 4 choices for the first and 3 choices for the second, a total of $(4)(3) = 12$ permutations. In general n distinct objects taken r at a time can be arranged in $n(n - 1)$ $(n - 2) \cdots (n - r + 1)$ ways. We represent this product by the symbol $_nP_r = n!/(n - r)!$.

> THEOREM 2.4 *The number of permutations of n distinct objects taken r at a time is*
>
> $$_nP_r = \frac{n!}{(n - r)!}.$$

Example 2.19   Two lottery tickets are drawn from 20 for first and second prizes. Find the number of sample points in the space S.

Solution.   The total number of sample points is

$$_{20}P_2 = \frac{20!}{18!} = (20)(19) = 380.$$

Example 2.20   How many ways can a basketball team schedule 3 exhibition games with 3 teams if they are all available on any of 5 possible dates?

Solution.   The total number of possible schedules is

$$_5P_3 = \frac{5!}{2!} = (5)(4)(3) = 60.$$

Permutations that occur by arranging objects in a circle are called *circular permutations*. Two circular permutations are not considered different, unless corresponding objects in the two arrangements are preceded or followed by a different object as we proceed in a clockwise direction. For example, if 4 people are playing bridge, we do not have a new permutation if they all move one position in a clockwise direction. By considering 1 person in a fixed position and arranging the other 3 in 3! ways, we find that there are 6 distinct arrangements for the bridge game.

THEOREM 2.5 *The number of permutations of n distinct objects arranged in a circle is* $(n - 1)!$.

So far we have considered permutations of distinct objects. That is, all the objects were completely different or distinguishable. Obviously, if the letters b and c are both equal to x, then the 6 permutations of the letters a, b, and c become axx, axx, xax, xax, xxa, and xxa, of which only 3 are distinct. Therefore, with 3 letters, 2 being the same, we have $3!/2! = 3$ distinct permutations. With the 4 letters a, b, c, and d we had 24 distinct permutations. If we let $a = b = x$ and $c = d = y$, we can only list the following: $xxyy$, $xyxy$, $yxxy$, $yyxx$, $xyyx$, and $yxyx$. Thus we have $4!/2!2! = 6$ distinct permutations.

THEOREM 2.6 *The number of distinct permutations of n things of which n_1 are of one kind, n_2 of a second kind, . . . , n_k of a kth kind is*

$$\frac{n!}{n_1! n_2! \cdots n_k!}.$$

Example 2.21 How many different ways can 3 red, 4 yellow, and 2 blue bulbs be arranged in a string of Christmas tree lights with 9 sockets?

Solution. The total number of distinct arrangements is

$$\frac{9!}{3!4!2!} = 1260.$$

Often we are concerned with the number of ways of partitioning a set of n objects into r subsets, called *cells*. A partition has been achieved if the intersection of every possible pair of the r subsets is the empty set $\emptyset$ and if the union of all subsets gives the original set. The order of the elements within a cell is of no importance. Consider the set $\{a, e, i, o, u\}$. The possible partitions into 2 cells, in which the first cell contains 4 elements and the second cell 1 element, are $\{(a, e, i, o), (u)\}$, $\{(a, i, o, u), (e)\}$, $\{(e, i, o, u), (a)\}$, $\{(a, e, o, u), (i)\}$, and $\{(a, e, i, u), (o)\}$. We see that there are 5 such ways to partition a set of 5 elements into 2 subsets or cells containing 4 elements in the first cell and 1 element in the second.

The number of partitions for this illustration is denoted by

$$\binom{5}{4, 1} = \frac{5!}{4!1!} = 5,$$

where the top number represents the total number of elements and the bottom numbers represent the number of elements going into each cell. We state this more generally in the following theorem.

THEOREM 2.7 *The number of ways of partitioning a set of n objects into r cells with n_1 elements in the first cell, n_2 elements in the second, and so on, is*

$$\binom{n}{n_1, n_2, \ldots, n_r} = \frac{n!}{n_1!n_2!\cdots n_r!},$$

where $n_1 + n_2 + \cdots + n_r = n$.

Example 2.22 How many ways can 7 people be assigned to 1 triple and 2 double rooms?

Solution. The total number of possible partitions would be

$$\binom{7}{3, 2, 2} = \frac{7!}{3!2!2!} = 210.$$

In several problems we are interested in the number of ways of *selecting* r objects from n without regard to order. These selections are called *combinations*. A combination is actually a partition with 2 cells, one cell containing the r objects selected and the other cell containing the $n - r$ objects that are left.

The number of such combinations, denoted by $\binom{n}{r, n - r}$, is usually shortened to $\binom{n}{r}$, since the number of elements in the second cell must be $n - r$.

THEOREM 2.8 *The number of combinations of n distinct objects taken r at a time is*

$$\binom{n}{r} = \frac{n!}{r!(n - r)!}.$$

Example 2.23 From 4 Republicans and 3 Democrats find the number of committees of 3 that can be formed with 2 Republicans and 1 Democrat.

Solution. The number of ways of selecting 2 Republicans from 4 is

$$\binom{4}{2} = \frac{4!}{2!2!} = 6.$$

The number of ways of selecting 1 Democrat from 3 is

$$\binom{3}{1} = \frac{3!}{1!2!} = 3.$$

Using Theorem 2.1, we find the number of committees that can be formed with 2 Republicans and 1 Democrat to be $(6)(3) = 18$.

2.5 Probability

The statistician is basically concerned with drawing conclusions or inferences from experiments involving uncertainties. For these conclusions and inferences to be reasonably accurate, an understanding of probability theory is essential.

What do we mean when we make the statements "John will probably win the tennis match," "I have a 50:50 chance of getting an even number when a die is tossed," "I am not likely to win at bingo tonight," or "Most of our graduating class will probably be married within 3 years"? In each case we are expressing an outcome of which we are not certain, but because of past information or from an understanding of the structure of the experiment, we have some degree of confidence in the validity of the statement.

The mathematical theory of probability for finite sample spaces provides a set of numbers, called *weights*, ranging from zero to 1, that provide a means of evaluating the likelihood of occurrence of events resulting from a statistical experiment. To every point in the sample space we assign a weight such that the sum of all the weights is 1. If we have reason to believe that a certain sample point is quite likely to occur when the experiment is conducted, the weight assigned should be close to 1. On the other hand, a weight closer to zero is assigned to a sample point that is not likely to occur. In many experiments, such as tossing a coin or a die, all the sample points have the same chance of occurring and are assigned equal weights. For points outside the sample space, that is, for simple events that cannot possibly occur, we assign a weight of zero.

To find the probability of any event A, we sum all the weights assigned to the sample points in A. This sum is called the *measure* of A or the probability of A and is denoted by $\Pr(A)$. Thus the measure of the set $\varnothing$ is zero and the measure of S is 1.

DEFINITION *The probability of any event A is the sum of the weights of all sample points in A. Therefore,*

$$0 \leq \Pr(A) \leq 1, \qquad \Pr(\varnothing) = 0, \qquad \Pr(S) = 1.$$

Example 2.24   A coin is tossed twice. What is the probability that at least 1 head occurs?

Solution.   The sample space for this experiment is $S = \{HH, HT, TH, TT\}$. If the coin is balanced, each of these outcomes would be equally likely to occur. Therefore, we assign a weight of w to each sample point. Then $4w = 1$ or $w = \frac{1}{4}$. If A represents the event of at least 1 head occurring, then $\Pr(A) = \frac{3}{4}$.

Example 2.25   A die is loaded in such a way that an even number is twice as likely to occur as an odd number. If E is the event that a number less than 4 occurs on a single toss of the die, find $\Pr(E)$.

Solution.   The sample space is $S = \{1, 2, 3, 4, 5, 6\}$. We assign a weight of w to each odd number and a weight of $2w$ to each even number. Since the sum of the weights must be 1, we have $9w = 1$ or $w = \frac{1}{9}$. Hence weights of $\frac{1}{9}$ and $\frac{2}{9}$ are assigned to each odd number and even number, respectively. Therefore,

$$\Pr(E) = \tfrac{1}{9} + \tfrac{2}{9} + \tfrac{1}{9} = \tfrac{4}{9}.$$

We can think of weights as being probabilities associated with simple events. If the experiment is of such a nature that we can assume equal weights for the sample points of S, then the probability of any event A is the ratio of the number of elements in A to the number of elements in S.

> **THEOREM** 2.9 *If an experiment can result in any one of N different equally likely outcomes, and if exactly n of these outcomes correspond to event A, then the probability of event A is*
>
> $$\Pr(A) = \frac{n}{N}.$$

Example 2.26 If a card is drawn from an ordinary deck, find the probability that it is a heart.

Solution. The number of possible outcomes is 52, of which 13 are hearts. Therefore, the probability of event A of getting a heart is $\Pr(A) = \frac{13}{52} = \frac{1}{4}$.

If the weights cannot be assumed equal, they must be assigned on the basis of prior knowledge or experimental evidence. For example, if a coin is not balanced, we could estimate the two weights by tossing the coin a large number of times and recording the outcomes. The true weights would be the fractions of heads and tails that occur in the long run. This method of arriving at weights is known as the *relative frequency* definition of probability.

To find a numerical value that represents adequately the probability of winning at tennis, we must depend on our past performance at the game as well as that of our opponent. Similarly, to find the probability that a horse will win a race, we must arrive at a weight based on the previous records of all the horses entered in the race.

2.6 Some Probability Laws

Often it is easier to calculate the probability of an event from known probabilities of other events. This may well be true if the event can be represented as the union of two other events or as the complement of an event. Several important laws that frequently simplify the computation of probabilities are listed below.

Before proceeding to these laws, we state the following definition.

> **DEFINITION** *Two events A and B are mutually exclusive if $A \cap B = \varnothing$.*

Using set language, we could say that two events are mutually exclusive if they are disjoint or if they have no points in common. In a practical situation we would be more likely to say that A and B are mutually exclusive if they cannot both occur at the same time.

Example 2.27 Suppose a die is tossed. Let A be the event that an even number turns up and let B be the event that an odd number shows. The intersection of the sets $A = \{2, 4, 6\}$ and $B = \{1, 3, 5\}$ is $A \cap B = \varnothing$, since they have no points in common. Therefore, A and B are mutually exclusive events. Since both an even and an odd number could not occur at the same time on a single toss of a die, we could have concluded that the events were mutually exclusive without finding their intersection.

> **THEOREM** 2.10 *If A and B are any two events, then*
> $$\Pr(A \cup B) = \Pr(A) + \Pr(B) - \Pr(A \cap B).$$

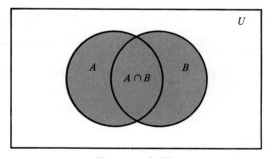

Figure 2.7

Proof. Consider the Venn diagram in Figure 2.7. The $\Pr(A \cup B)$ is the sum of the weights of the sample points in $A \cup B$. Now $\Pr(A) + \Pr(B)$ is the sum of all the weights in A plus the sum of all the weights in B. Therefore, we have added the weights in $A \cap B$ twice. Since these weights add up to give $\Pr(A \cap B)$, we must subtract this probability once to obtain the sum of the weights in $A \cup B$, which is $\Pr(A \cup B)$.

COROLLARY 1 *If A and B are mutually exclusive, then*
$$Pr(A \cup B) = Pr(A) + Pr(B).$$

The corollary is an immediate result of Theorem 2.10, since if A and B are mutually exclusive, $A \cap B = \varnothing$ and then $Pr(A \cap B) = Pr(\varnothing) = 0$. In general we write

COROLLARY 2 *If $A_1, A_2, A_3, \ldots, A_n$ are mutually exclusive, then*
$$Pr(A_1 \cup A_2 \cup \cdots \cup A_n) = Pr(A_1) + Pr(A_2) + \cdots + Pr(A_n).$$

Note that if $A_1, A_2, \ldots, A_n$ is a partition of a sample space S, then

$$\begin{aligned} Pr(A_1 \cup A_2 \cup \cdots \cup A_n) &= Pr(A_1) + Pr(A_2) + \cdots + Pr(A_n) \\ &= Pr(S) \\ &= 1. \end{aligned}$$

Example 2.28 The probability that a student passes mathematics is $\frac{2}{3}$, and the probability that he passes English is $\frac{4}{9}$. If the probability of passing at least one course is $\frac{4}{5}$, what is the probability that he will pass both courses?

Solution. If M is the event "passing mathematics" and E the event "passing English," then by transposing the terms in Theorem 2.10 we have

$$\begin{aligned} Pr(M \cap E) &= Pr(M) + Pr(E) - Pr(M \cup E) \\ &= \tfrac{2}{3} + \tfrac{4}{9} - \tfrac{4}{5} \\ &= \tfrac{14}{45}. \end{aligned}$$

Example 2.29 What is the probability of getting a total of 7 or 11 when a pair of dice is tossed?

Solution. Let A be the event that 7 occurs and B the event that 11 comes up. Now a total of 7 occurs for 6 of the 36 sample points and a total of 11 occurs for only 2 of the sample points. Since all sample points are equally likely, we have $Pr(A) = \frac{1}{6}$ and $Pr(B) = \frac{1}{18}$. The events A and B are mutually exclusive, since a total of 7 and 11 cannot both occur on the same toss. Therefore,

$$Pr(A \cup B) = Pr(A) + Pr(B)$$
$$= \tfrac{1}{6} + \tfrac{1}{18}$$
$$= \tfrac{2}{9}.$$

> **THEOREM 2.11** *If A and A′ are complementary events, then*
>
> $$Pr(A') = 1 - Pr(A).$$

Proof. Since $A \cup A' = S$ and the sets A and A' are disjoint, then

$$1 = Pr(S)$$
$$= Pr(A \cup A')$$
$$= Pr(A) + Pr(A').$$

Therefore,

$$Pr(A') = 1 - Pr(A).$$

Example 2.30 A coin is tossed 6 times in succession. What is the probability that at least 1 head occurs?

Solution. Let E be the event that at least 1 head occurs. The sample space S consists of $2^6 = 64$ sample points, since each toss can result in 2 outcomes. Now, $Pr(E) = 1 - Pr(E')$, where E' is the event that no head occurs. This can happen in only one way—when all tosses result in a tail. Therefore $Pr(E') = \tfrac{1}{64}$ and $Pr(E) = 1 - \tfrac{1}{64} = \tfrac{63}{64}$.

2.7 Conditional Probability

The probability of an event B occurring when it is known that some event A has occurred is called a *conditional probability* and is denoted by $Pr(B|A)$. The symbol $Pr(B|A)$ is usually read "the probability that B occurs given that A occurs" or simply "the probability of B, given A."

> **DEFINITION** *The* conditional probability *of B, given A, denoted by* $Pr(B|A)$, *is defined by the equation*
>
> $$Pr(B|A) = \frac{Pr(A \cap B)}{Pr(A)} \qquad \text{if } Pr(A) > 0.$$

Consider the event B of getting a 4 when a die is tossed. The die is constructed so that the even numbers are twice as likely to occur as the odd numbers. Based on the sample space $S = \{1, 2, 3, 4, 5, 6\}$, with weights of $\frac{1}{9}$ and $\frac{2}{9}$ assigned to the odd and even numbers, respectively, the probability of B occurring is $\frac{2}{9}$. Now suppose that it is known that the toss of the die resulted in a number greater than 3. We are now dealing with a reduced sample space $A = \{4, 5, 6\}$, which is a subset of S. To find the probability that B occurs, relative to the space A, we must first assign new weights to the elements of A proportional to their original weights such that their sum is 1. Assigning a weight of w to the odd number in A and a weight of $2w$ to each of the two even numbers, we have $5w = 1$ or $w = \frac{1}{5}$. Relative to the space A, we find

$$\Pr(B|A) = \tfrac{2}{5},$$

which can also be written

$$\Pr(B|A) = \tfrac{2}{5} = \frac{\frac{2}{9}}{\frac{5}{9}} = \frac{\Pr(A \cap B)}{\Pr(A)},$$

where $\Pr(A \cap B)$ and $\Pr(A)$ are found from the original sample space S. This example illustrates that events may have different probabilities when considered relative to different sample spaces.

As an additional illustration suppose that our sample space S is the population of adults in a small town who have completed the requirements for a college degree. We shall categorize them according to sex and employment status.

	Employed	Unemployed
Male	460	40
Female	140	260

One of these individuals is to be selected at random for a tour throughout the country to publicize the advantages of establishing new industries in the town. We shall be concerned with the following events:

M: a man is chosen,

E: the chosen one is employed.

Using the reduced sample space E, we find

$$\Pr(M|E) = \frac{460}{600} = \frac{23}{30}.$$

Let $n(A)$ denote the number of elements in any set A. Using this notation, we can write

$$\Pr(M|E) = \frac{n(E \cap M)}{n(E)} = \frac{n(E \cap M)/n(S)}{n(E)/n(S)} = \frac{\Pr(E \cap M)}{\Pr(E)},$$

where $\Pr(E \cap M)$ and $\Pr(E)$ are found from the original sample space S. To verify this result, note that

$$\Pr(E) = \frac{600}{900} = \frac{2}{3}$$

and

$$\Pr(E \cap M) = \frac{460}{900} = \frac{23}{45}.$$

Hence

$$\Pr(M|E) = \frac{\frac{23}{45}}{\frac{2}{3}} = \frac{23}{30},$$

as before.

Multiplying the formula in the preceding definition for conditional probability by $\Pr(A)$, we obtain the following important *multiplication theorem*:

THEOREM 2.12 *If in an experiment the events A and B can both occur, then*

$$\Pr(A \cap B) = \Pr(A)\Pr(B|A).$$

Thus the probability that both A and B occur is equal to the probability that A occurs multiplied by the probability that B occurs, given that A occurs.

To illustrate the use of Theorem 2.12, suppose that we have a fuse box containing 20 fuses, of which 5 are defective. If 2 fuses are selected at random and removed from the box in succession without replacing the first, what is the probability that both fuses are defective? To answer this question, we shall let A be the event that the first fuse is defective and B the event that the second fuse is defective; then we interpret $A \cap B$ as the event that A occurs, and then B occurs after A has occurred. The probability of first removing a defective fuse is $\frac{1}{4}$ and then the probability of removing a second defective fuse from the remaining 4 is $\frac{4}{19}$. Hence $\Pr(A \cap B) = (\frac{1}{4})(\frac{4}{19}) = \frac{1}{19}$.

Generalizing Theorem 2.12, we write:

THEOREM 2.13 *If in an experiment the events* $A_1, A_2, A_3, \ldots,$ *can occur, then*

$$\Pr(A_1 \cap A_2 \cap A_3 \cap \cdots) = \Pr(A_1)\Pr(A_2|A_1)\Pr(A_3|A_1 \cap A_2) \cdots.$$

If in the above illustration the first fuse is replaced and the fuses thoroughly rearranged before the second is removed, then the probability of a defective fuse on the second selection is still $\frac{1}{4}$, that is, $\Pr(B|A) = \Pr(B)$. When this is true, the events A and B are said to be *independent*.

DEFINITION *The events A and B are* independent *if and only if*

$$\Pr(A \cap B) = \Pr(A)\Pr(B).$$

Example 2.31 A pair of dice is thrown twice. What is the probability of getting totals of 7 and 11?

Solution. Let $A_1, A_2, B_1,$ and B_2 be the respective independent events that a 7 occurs on the first throw, a 7 occurs on the second throw, an 11 occurs on the first throw, and an 11 occurs on the second throw. We are interested in the probability of the union of the mutually exclusive events $A_1 \cap B_2$ and $B_1 \cap A_2$. Therefore,

$$\Pr[(A_1 \cap B_2) \cup (B_1 \cap A_2)] = \Pr(A_1 \cap B_2) + \Pr(B_1 \cap A_2)$$
$$= \Pr(A_1)\Pr(B_2) + \Pr(B_1)\Pr(A_2)$$
$$= (\tfrac{1}{6})(\tfrac{1}{18}) + (\tfrac{1}{18})(\tfrac{1}{6})$$
$$= \tfrac{1}{54}.$$

2.8 Bayes' Rule

Let us return to the illustration of Section 2.7, where an individual is being selected at random from the adults of a small town to tour the country and publicize the advantages of establishing new industries in the town. At that time we had no difficulty in establishing the fact that $\Pr(E) = \frac{2}{3}$, where E is the event that the one chosen is employed. Suppose we are given the additional information that 36 of those employed and 12 of those unemployed are members of the Rotary Club. What is the probability that the individual selected is employed if it is known that the person belongs to the Rotary Club?

Let A be the event that the person selected is a member of the Rotary Club. The conditional probability that we seek is then given by

$$\Pr(E|A) = \frac{\Pr(E \cap A)}{\Pr(A)}.$$

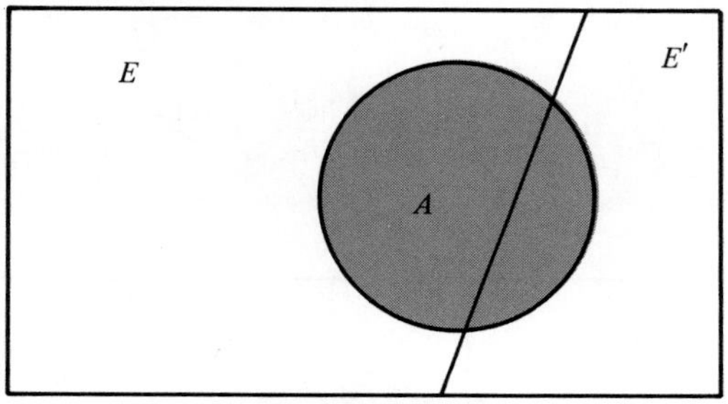

Figure 2.8

Venn diagram showing the events A, E, and E'.

Referring to Figure 2.8, we can write A as the union of the two mutually exclusive events $E \cap A$ and $E' \cap A$. Hence

$$A = (E \cap A) \cup (E' \cap A),$$

and by Corollary 1 of Theorem 2.10,

$$\Pr(A) = \Pr(E \cap A) + \Pr(E' \cap A).$$

We can now write

$$\Pr(E|A) = \frac{\Pr(E \cap A)}{\Pr(E \cap A) + \Pr(E' \cap A)}.$$

The data of Section 2.7, together with the additional information about the set A, enable one to compute

$$\Pr(E \cap A) = \frac{36}{900} = \frac{1}{25},$$

$$\Pr(E' \cap A) = \frac{12}{900} = \frac{1}{75}.$$

Hence

$$\Pr(E|A) = \frac{\frac{1}{25}}{\frac{1}{25} + \frac{1}{75}} = \frac{3}{4}.$$

A generalization of the foregoing procedure leads to the following theorem, called *Bayes' rule*.

THEOREM 2.14 (BAYES' RULE) *Let $\{B_1, B_2, \ldots, B_n\}$ be a set of events forming a partition of the sample space S, where $\Pr(B_i) \neq 0$, for $i = 1, 2, \ldots, n$. Let A be any event of S such that $\Pr(A) \neq 0$. Then, for $k = 1, 2, \ldots, n$,*

$$\Pr(B_k|A) = \frac{\Pr(B_k \cap A)}{\Pr(B_1 \cap A) + \Pr(B_2 \cap A) + \cdots + \Pr(B_n \cap A)}.$$

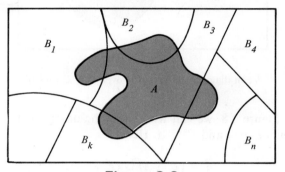

Figure 2.9

Partitioning of the sample space S.

Proof. Consider the Venn diagram in Figure 2.9. The event A is seen to be the union of the mutually exclusive events $B_1 \cap A, B_2 \cap A, \ldots, B_n \cap A$; that is,

$$A = (B_1 \cap A) \cup (B_2 \cap A) \cup \cdots \cup (B_n \cap A).$$

Using Corollary 2 of Theorem 2.10, we have

$$\Pr(A) = \Pr(B_1 \cap A) + \Pr(B_2 \cap A) + \cdots + \Pr(B_n \cap A).$$

By the definition of conditional probability,

$$\Pr(B_k|A) = \frac{\Pr(B_k \cap A)}{\Pr(A)}$$

$$= \frac{\Pr(B_k \cap A)}{\Pr(B_1 \cap A) + \Pr(B_2 \cap A) + \cdots + \Pr(B_n \cap A)},$$

which completes the proof.

Example 2.32 Three members of a private country club have been nominated for the office of president. The probability that Mr. Adams will be elected is 0.3, the probability that Mr. Brown will be elected is 0.5, and the probability that Mr. Cooper will be elected is 0.2. Should Mr. Adams be elected, the probability for an increase in membership fees is 0.8. Should Mr. Brown or Mr. Cooper be elected, the corresponding probabilities for an increase in fees are 0.1 and 0.4. If someone is considering joining the club but delays his decision for several weeks only to find out that the fees have been increased, what is the probability that Mr. Cooper was elected president of the club?

Solution. We consider the following events:

A: The person elected increased fees,

B_1: Mr. Adams is elected,

B_2: Mr. Brown is elected,

B_3: Mr. Cooper is elected.

Using Bayes' rule, we write

$$\Pr(B_3|A) = \frac{\Pr(B_3 \cap A)}{\Pr(B_1 \cap A) + \Pr(B_2 \cap A) + \Pr(B_3 \cap A)}.$$

Now

$$\Pr(B_1 \cap A) = \Pr(B_1)\Pr(A|B_1) = (0.3)(0.8) = 0.24,$$
$$\Pr(B_2 \cap A) = \Pr(B_2)\Pr(A|B_2) = (0.5)(0.1) = 0.05,$$
$$\Pr(B_3 \cap A) = \Pr(B_3)\Pr(A|B_3) = (0.2)(0.4) = 0.08.$$

Hence

$$\Pr(B_3|A) = \frac{0.08}{0.24 + 0.05 + 0.08} = \frac{8}{37} \simeq 23\%$$

In view of the fact that fees have increased, this result suggests that Mr. Cooper is probably not the president of the club.

EXERCISES

1. List the elements of each of the following sets:

(a) The set of integers between 1 and 50 divisible by 7.

(b) The set $A = \{x|x^2 + x - 6 = 0\}$.

(c) The set of outcomes when a die and a coin are tossed simultaneously.

    (d) The set $B = \{x|x \text{ is a continent}\}$.

    (e) The set $C = \{x|2x - 4 = 0 \text{ and } x > 5\}$.

2. Let $E = \{3, 5, 6, 8, 9\}$. Which of the following statements are true and which are false?

    (a) $5 \in E$.   (b) $7 \in E$.   (c) $9 \notin E$.

    (d) $\{5, 8, 3\} \subset E$.   (e) $\{6, 8, 3, 4\} \not\subset E$.   (f) $\varnothing \subset E$.

3. Which of the following sets are equal?

    (a) $A = \{1, 3\}$.

    (b) $B = \{x|x \text{ is a number on a die}\}$.

    (c) $C = \{x|x^2 - 4x + 3 = 0\}$.

    (d) $D = \{x|x \text{ is the number of heads when 6 coins are tossed}\}$.

4. How many subsets can be formed from the set $M = \{p, q, r, s\}$? List all the subsets that have exactly 3 elements.

5. List all the proper subsets of the universal set $U = \{\text{air, land, sea}\}$.

6. Construct a Venn diagram to illustrate the following subsets of all college students: all juniors, all mathematics majors, all women.

7. Construct a Venn diagram to illustrate the following subsets of $U = \{\text{copper, sodium, nitrogen, potassium, uranium, oxygen, zinc}\}$:

    $A = \{\text{copper, sodium, zinc}\}$.

    $B = \{\text{sodium, nitrogen, potassium}\}$.

    $C = \{\text{oxygen}\}$.

8. If $U = \{0, 1, 2, 3, 4, 5, 6, 7, 8, 9\}$ and $A = \{0, 2, 4, 6, 8\}$, $B = \{1, 3, 5, 7, 9\}$, $C = \{2, 3, 4, 5\}$, $D = \{1, 6, 7\}$, list the elements in the following sets:

    (a) $A \cup C$.   (b) $A \cap B$.   (c) C'.

    (d) $(C' \cap D) \cup B$.   (e) $(U \cap C)'$.   (f) $A \cap C \cap D'$.

9. Referring to Exercise 7, list the elements in the following sets:

    (a) A'.   (b) $A \cup C$.   (c) $(A \cap B') \cup C'$.

    (d) $(B' \cap C')$.   (e) $A \cap B \cap C$.   (f) $(A' \cup B') \cap (A' \cap C)$.

10. If $P = \{x|1 < x < 9\}$ and $Q = \{y|y < 5\}$, find $P \cup Q$ and $P \cap Q$.

11. Let A, B, and C be subsets of the universal set U. Using Venn diagrams, shade the areas representing the sets

    (a) $A \cap B'$.   (b) $(A \cup B)'$.   (c) $(A \cap \dot{C}) \cup B$.

12. An experiment involves tossing a pair of dice, 1 green and 1 red, and recording the numbers that come up.

    (a) List the elements of the sample space S.

    (b) List the elements of S corresponding to event A that the sum is less than 5.

    (c) List the elements of S corresponding to event B that a 6 occurs on either die.

    (d) List the elements of S corresponding to event C that a 2 comes up on the green die.

13. An experiment consists of flipping a coin and then flipping it a second time if a head occurs. If a tail occurs on the first flip, then a die is tossed once.

    (a) List the elements of the sample space S.

    (b) List the elements of S corresponding to event A that a number less than 4 occurred on the die.

    (c) List the elements of S corresponding to event B that 2 tails occurred.

14. An experiment consists of asking 3 women at random if they wash their dishes with brand X detergent.
  (a) List the elements of a sample space S using the letter Y for "yes" and N for "no."
  (b) List the elements of S corresponding to event E that at least 2 of the women use brand X.
  (c) Define an event that has as its elements the points $\{YYY, NYY, YYN, NYN\}$.

15. The résumés of 2 male applicants for a college teaching position in psychology are placed in the same file as the résumés of 2 female applicants. Two positions become available and the first, at the rank of assistant professor, is filled by selecting one of the 4 applicants at random. The second position, at the rank of instructor, is then filled by selecting at random one of the remaining 3 applicants.
  (a) List the elements of the sample space S.
  (b) List the elements of S corresponding to event A that the position of assistant professor is filled by a male applicant.
  (c) List the elements of S corresponding to event B that exactly one of the 2 positions was filled by a male applicant.
  (d) List the elements of S corresponding to event C that neither position was filled by a male applicant.

16. (a) How many ways can 5 people be lined up to get on a bus?
  (b) If a certain 2 persons refuse to follow each other, how many ways are possible?

17. A college freshman must take a science course, a social studies course, and a mathematics course. If he may select any of 3 sciences, any of 4 social studies, and any of 2 mathematics courses, how many ways can he arrange his program?

18. In how many different ways can an 8-question true–false examination be answered?

19. How many distinct permutations can be made from the letters of the word "statistics"?

20. How many ways can the 5 starting positions on a basketball team be filled with 9 men who can play any of the positions?

21. (a) How many three-digit numbers can be formed from the digits 0, 1, 2, 3, 4, 5 if each digit can be used only once?
  (b) How many of these are odd numbers?
  (c) How many are greater than 330?

22. A contractor wishes to build 5 houses, each different in design. In how many ways can he place these homes on a street if 3 lots are on one side of the street and 2 lots are on the opposite side?

23. In how many ways can 4 boys and 3 girls sit in a row if the boys and girls must alternate?

24. Four married couples have bought 8 seats in a row for a concert. In how many different ways can they be seated:
  (a) With no restrictions?
  (b) If each couple is to sit together?
  (c) If all the men sit together to the right of all the women?

25. In how many ways can 6 trees be planted in a circle?

26. In how many ways can 2 oaks, 3 pines, and 2 maples be arranged in a straight line if one does not distinguish between trees of the same kind?

27. A college plays 8 football games during a season. In how many ways can the team end the season with 4 wins, 3 losses, and 1 tie?

28. Ten people are going on a skiing trip in 3 cars that will hold 2, 4, and 5 passengers, respectively. How many ways is it possible to transport the 10 people to the ski lodge?

29. From a group of 5 men and 3 women how many committees of size 3 are possible:
    (a) With no restrictions?
    (b) With 2 men and 1 woman?
    (c) With 1 man and 2 women if a certain woman must be on the committee?

30. How many bridge hands are possible containing 5 spades, 3 diamonds, 3 clubs, and 2 hearts?

31. From 3 red, 4 green, and 5 yellow apples, how many selections consisting of 6 apples are possible if 2 of each color are to be selected?

32. A shipment of 10 television sets contains 3 defective sets. In how many ways can a hotel purchase 4 of these sets and receive at least 2 of the defective sets?

33. Three men are seeking public office. Candidates A and B are given about the same chance of winning but candidate C is given twice the chance of either A or B. What is the probability that C wins? What is the probability that A does not win?

34. Find the probability of event A in Exercise 12.

35. A box contains 500 envelopes of which 50 contain \$100 in cash, 100 contain \$25, and 350 contain \$10. An envelope may be purchased for \$25. What is the sample space for the different amounts of money? Assign weights to the sample points and then find the probability that the first envelope purchased contains less than \$100.

36. A four-sided die with sides numbered 1, 2, 3, and 4 is so constructed that the 4 occurs twice as often as the 2 and the odd numbers occur three times as often as the 4. Find the weights for the sample space $S = \{1, 2, 3, 4\}$. What is the probability that a perfect square occurs when the die is tossed once?

37. A pair of dice is tossed. What is the probability of getting a total of 5? At most a total of 4?

38. In a poker hand consisting of 5 cards, what is the probability of holding (a) 2 aces and 2 kings, (b) 5 spades?

39. If 3 books are picked at random from a shelf containing 4 novels, 3 books of poems and a dictionary, what is the probability that (a) the dictionary is selected, (b) 2 novels and 1 book of poems are selected?

40. Two cards are drawn in succession from a deck without replacement. What is the probability that both cards are greater than 2 and less than 9?

41. Which pair of events in Exercise 12 is mutually exclusive?

42. If A and B are mutually exclusive events and $\Pr(A) = 0.4$ and $\Pr(B) = 0.5$, find
    (a) $\Pr(A \cup B)$.   (b) $\Pr(A')$.   (c) $\Pr(A' \cap B)$.

43. If A, B, and C are mutually exclusive events and $\Pr(A) = 0.2$, $\Pr(B) = 0.1$, and $\Pr(C) = 0.4$, find
    (a) $\Pr(A \cup B \cup C)$.   (b) $\Pr[A' \cap (B \cup C)]$.   (c) $\Pr(B \cup C')'$.

44. A town has 2 fire engines operating independently. The probability that a specific fire engine is available when needed is 0.99.

(a) What is the probability that neither is available when needed?

(b) What is the probability that a fire engine is available when needed?

45. In a certain federal prison it is known that $\frac{2}{3}$ of the inmates are under 25 years of age. It is also known that $\frac{3}{5}$ of the inmates are male and that $\frac{5}{8}$ of the inmates are female or 25 years of age or older. What is the probability that a prisoner selected at random from this prison is female and at least 25 years old?

46. A pair of dice is thrown. If it is known that one die shows a 4, what is the probability that:

(a) The other die shows a 5?

(b) The total of both dice is greater than 7?

47. The probability that an automobile being filled with gasoline will also need an oil change is 0.25; the probability that it needs a new oil filter is 0.40; and the probability that both the oil and filter need changing is 0.18. If the oil had to be changed, what is the probability that a new oil filter is needed? $\quad\to P(\text{filter} \mid \text{oil})$

Handwritten notes in right margin:
$\to$ Direct application of definition Pg 23

$P(B|A) = \dfrac{P(B \cap A)}{P(A)}$

48. In a high school graduating class of 100 students, 42 studied mathematics, 68 studied psychology, 54 studied history, 22 studied both mathematics and history, 25 studied both mathematics and psychology, 7 studied history and neither mathematics nor psychology, 10 studied all three subjects, and 8 did not take any of the three. If a student is selected at random, find:

(a) The probability that he takes history and psychology but not mathematics.

(b) The probability that a person enrolled in history takes all three subjects.

(c) The probability that he takes mathematics only.

49. One bag contains 4 white balls and 3 black balls, and a second bag contains 3 white balls and 5 black balls. One ball is drawn from the first bag and is placed unseen in the second bag. What is the probability that a ball now drawn from the second bag is black?

50. The probability that a married man watches a certain television show is 0.4 and the probability that a married woman watches the show is 0.5. The probability that a man watches the show, given that his wife does, is 0.7. Find:

(a) The probability that a married couple watches the show.

(b) The probability that a wife watches the show given that her husband does.

(c) The probability that at least 1 person of a married couple will watch the show.

Handwritten notes in right margin:
.875

$P(m \cap w) = P(m) \cdot P(w/m)$
.35 = .4 ✗

$P(w \cap m) = P(w) \cdot P(m/w)$
.35 = (.5)(.7)

$P(m \cup m) = P(m) + P(w) - P(m \cap w)$
.5 + .4 - .35
= .55

51. If the probability that Tom will be alive in 20 years is 0.6 and the probability that Jim will be alive in 20 years is 0.9, what is the probability that neither will be alive in 20 years?

52. A basketball player sinks 50% of his shots. What is the probability that he makes exactly 3 of his next 4 shots?

53. A coin is biased so that a head is twice as likely to occur as a tail. If the coin is tossed 3 times, what is the probability of getting exactly 2 tails?

54. From a box containing 5 black balls and 3 green balls, 3 balls are drawn in succession, each ball being replaced in the box before the next draw is made. What is the probability that all 3 are the same color? What is the probability that each color is represented?

55. A real estate man has 8 master keys to open several new homes. Only 1 master key will open any given house. If 40% of these homes are usually left unlocked, what is the probability that the real estate man can get into a specific home if he selects 3 master keys at random before leaving the office?

56. Suppose that colored balls are distributed in three indistinguishable boxes as follows:

	Box		
	1	2	3
Red	2	4	3
White	3	1	4
Blue	5	3	3

A box is selected at random from which a ball is selected at random and it is observed to be red. What is the probability that box 3 was selected?

57. A commuter owns 2 cars, 1 a compact and 1 a standard model. About $\frac{3}{4}$ of the time he uses the compact to travel to work, and about $\frac{1}{4}$ of the time the larger car is used. When he uses the compact car, he usually gets home by 5:30 P.M. about 75% of the time; if he uses the standard-sized car, he gets home by 5:30 P.M. about 60% of the time (but he enjoys the air conditioner in the larger car). If he gets home at 5:35 P.M., what is the probability that he used the compact car?

58. A truth serum given to a suspect is known to be 90% reliable when the person is guilty and 99% reliable when the person is innocent. In other words, 10% of the guilty are judged innocent by the serum and 1% of the innocent are judged guilty. If the suspect was selected from a group of suspects of which only 5% have ever committed a crime, and the serum indicates that he is guilty, what is the probability that he is innocent?

DISTRIBUTIONS OF RANDOM VARIABLES

3.1 Concept of a Random Variable

The term *statistical experiment* has been used to describe any process by which several chance observations are obtained. All possible outcomes of an experiment comprise a set that we have called the *sample space*. Often we are not interested in the details associated with each sample point but only in some numerical description of the outcome. For example, when one tosses a coin 3 times, there are 8 sample points in the sample space that give, in complete detail, the outcome of each toss. If one is concerned with the number of heads that fall, then a numerical value of 0, 1, 2, or 3 will be assigned to each sample point.

The numbers 0, 1, 2, and 3 are random quantities determined by the outcome of an experiment. They may be thought of as the values assumed by some *random variable X*, which in this case represents the number of heads when a coin is tossed 3 times.

> DEFINITION *A function whose value is a real number determined by each element in the sample space is called a* random variable.

We shall use a capital letter, say X, to denote a random variable and its corresponding small letter, x in this case, for one of its values. Each possible value x of X then represents an event that is a subset of the sample space.

Example 3.1 Two balls are drawn in succession without replacement from an urn containing 4 red balls and 3 black balls. The possible outcomes and the values y of the random variable Y, where Y is the number of red balls, are

Simple Event	y
RR	2
RB	1
BR	1
BB	0

Example 3.2 A hatcheck girl returns 3 hats at random to 3 customers who had previously checked them. If Smith, Jones, and Brown, in that order, receive one of the 3 hats, list the sample points for the possible orders of returning the hats and find the values m of the random variable M that represents the number of correct matches.

Solution. If S, J, and B stand for Smith's, Jones', and Brown's hats, respectively, then the possible arrangements in which the hats may be returned and the number of correct matches are

Simple Event	m
SJB	3
SBJ	1
JSB	1
JBS	0
BSJ	0
BJS	1

If a sample space contains a finite number of points, as in the two examples above, or an unending sequence with as many elements as there are whole numbers, such as in the case of a die being thrown until a 5 occurs, it is called a *discrete sample space*. A random variable defined over a discrete sample space is called a *discrete random variable*.

Also, if the elements of a sample space are infinite, or as many as the number of points on a line segment, such as all possible heights, weights, temperatures, or life periods, we say we have a *continuous sample space*, and a variable defined over this space is called a *continuous random variable*.

In most practical problems continuous random variables represent *measured* data and discrete random variables represent *count* data, such as the number of defectives in a sample of k items or the number of accidents per year.

3.2 Discrete Probability Distributions

A discrete random variable assumes each of its values with a certain probability. In the case of tossing a coin 3 times, the variable X, representing the number of heads, assumes the value 2 with probability $\frac{3}{8}$, since 3 of the 8 equally likely sample points result in 2 heads and 1 tail. Assuming equal probabilities for the simple events in Example 3.2, the probability that no man gets back his right hat, that is, the probability that M assumes the value 0, is $\frac{1}{3}$. The possible values m of M and their probabilities are given by

m	0	1	3
$\Pr(M = m)$	$\frac{1}{3}$	$\frac{1}{2}$	$\frac{1}{6}$

Note that the values of m exhaust all possible cases, and hence the probabilities add to 1.

Frequently it is convenient to represent all the probabilities of a random variable X by a formula. Such a formula would necessarily be a function of the numerical values x that we shall denote by $f(x)$, $g(x)$, $r(x)$, and so forth. Hence we write $f(x) = \Pr(X = x)$; that is, $f(3) = \Pr(X = 3)$. The set of ordered pairs $(x, f(x))$ is called the *probability function* or *probability distribution* of the discrete random variable X.

DEFINITION *A table or a formula listing all possible values that a discrete variable can take on, together with the associated probabilities, is called a* discrete probability distribution.

Example 3.3 Find the probability distribution of the sum of the numbers when a pair of dice is tossed.

Solution. Let X be a random variable whose values x are the possible totals. Then x can be any integer from 2 to 12. Two dice can fall in $(6)(6) = 36$ ways, each with probability $\frac{1}{36}$. The $\Pr(X = 3) = \frac{2}{36}$, since a total of 3 can occur in only 2 ways. Consideration of the other cases leads to the following probability distribution:

x	2	3	4	5	6	7	8	9	10	11	12
$\Pr(X = x)$	$\frac{1}{36}$	$\frac{2}{36}$	$\frac{3}{36}$	$\frac{4}{36}$	$\frac{5}{36}$	$\frac{6}{36}$	$\frac{5}{36}$	$\frac{4}{36}$	$\frac{3}{36}$	$\frac{2}{36}$	$\frac{1}{36}$

Example 3.4 Find a formula for the probability distribution of the number of heads when a coin is tossed 4 times.

Solution. Since there are $2^4 = 16$ points in the sample space representing equally likely outcomes, the denominator for all probabilities, and therefore for our function, will be 16. To obtain the number of ways of getting, say 3 heads, we need to consider the number of ways of partitioning 4 outcomes into 2 cells, with 3 heads assigned to one cell and a tail assigned to the other. This can be done in $\binom{4}{3} = 4$ ways. In general x heads and $4 - x$ tails can occur in $\binom{4}{x}$ ways, where x can be 0, 1, 2, 3, or 4. Thus the probability distribution $f(x) = \Pr(X = x)$ is

$$f(x) = \frac{\binom{4}{x}}{16}, \qquad x = 0, 1, 2, 3, 4.$$

It is often helpful to look at a probability distribution in graphic form. One might plot the points $(x, f(x))$ of Example 3.4 to obtain Figure 3.1. By joining the points to the x axis, either with a dashed or solid line, we obtain what is commonly called a *bar chart*. Figure 3.1 makes it very easy to see what values of X are most likely to occur, and it also indicates a perfectly symmetric situation in this case.

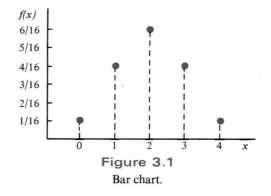

Figure 3.1

Bar chart.

Instead of plotting the points $(x, f(x))$, we more frequently construct rectangles, as in Figure 3.2. Here the rectangles are constructed so that their bases of equal width are centered at each value x and their heights are equal to the corresponding probabilities given by $f(x)$. The bases are constructed so as to leave no space between the rectangles. Figure 3.2 is called a *probability histogram*.

Since each base in Figure 3.2 has unit width, the $\Pr(X = x)$ is equal to the area of the rectangle centered at x. Even if the bases were not of unit width,

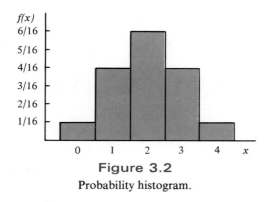

Figure 3.2
Probability histogram.

we could adjust the heights of the rectangles to give areas that would still equal the probabilities of X assuming any of its values x. This concept of using areas to represent probabilities is necessary for consideration of the probability distribution of a continuous random variable.

Certain probability distributions are applicable to more than one physical situation. The probability distribution of Example 3.4, for example, also applies to the random variable Y, where Y is the number of red cards that occurs when 4 cards are drawn at random from a deck in succession with each card replaced and the deck shuffled before the next drawing. Special discrete probability distributions that can be applied to many different experimental situations will be considered in Chapter 5.

3.3 Continuous Probability Distributions

A continuous random variable has a probability of zero of assuming exactly any of its values. Consequently its probability distribution cannot be given in tabular form. At first this may seem startling, but it becomes more plausible when we consider a particular example. Let us discuss a random variable whose values are the heights of all people over 21 years of age. Between any two values, say 63.5 and 64.5 inches, or even 63.99 and 64.01 inches, there are an infinite number of heights, one of which is 64 inches. The probability of selecting a person at random exactly 64 inches tall and not one of the infinitely large set of heights so close to 64 inches that you cannot humanly measure the difference is extremely remote, and thus we assign a probability of zero to the event. It follows that

$$\Pr(a < X \le b) = \Pr(a < X < b) + \Pr(X = b)$$
$$= \Pr(a < X < b).$$

That is, it does not matter whether we include an end point of the interval or not.

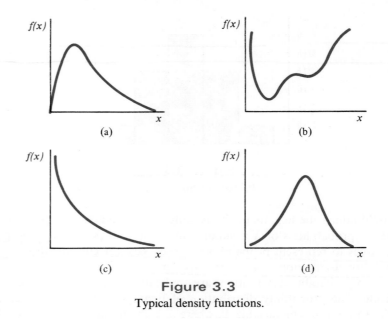

Figure 3.3
Typical density functions.

While the probability distribution of a continuous random variable cannot be presented in tabular form, it does have a formula. As before, we shall designate the probability distribution of X by the functional notation $f(x)$. In dealing with continuous variables, $f(x)$ is usually called the *density function*. Since X is defined over a continuous sample space, the graph of $f(x)$ will be continuous and may, for example, take one of the forms shown in Figure 3.3.

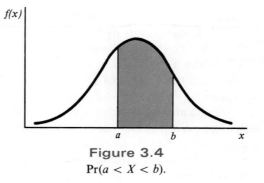

Figure 3.4
$\Pr(a < X < b)$.

A probability density function is constructed so that the area under its curve bounded by the x axis is equal to 1 when computed over the range of X for which $f(x)$ is defined. If $f(x)$ is represented in Figure 3.4, then the probability that X assumes a value between a and b is equal to the shaded area under the density function between the ordinates at $x = a$ and $x = b$.

DEFINITION *The function f(x) is called a* probability density function *for the continuous random variable X if the total area under its curve bounded by the x axis is equal to 1 and if the area under the curve between any two ordinates x = a and x = b gives the probability that X lies between a and b.*

The exact areas can be obtained only by methods of integral calculus beyond the scope of this book. Therefore we shall consider only those density functions that are used most frequently in conducting experiments and for which areas have been computed and put in tabular form. Because areas represent probabilities and probabilities are positive numerical values, the density function must be entirely above the x axis.

3.4 Empirical Distributions

Usually in an experiment involving a continuous random variable, the density function $f(x)$ is unknown and its equation is assumed. For the choice of $f(x)$ to be reasonably valid, good judgment based on all available information is needed in its selection. Statistical data, generated in large masses, can be very useful in studying the behavior of the distribution if presented in the form of a *frequency distribution*. Such an arrangement is obtained by grouping the data into classes and determining the number of measurements in each of the classes.

To illustrate the construction of a frequency distribution, consider the data of Table 3.1, which represents the lives of 40 similar car batteries recorded to the nearest tenth of a year. The batteries were guaranteed to last 3 years.

Table 3.1

Car Battery Lives

2.2	4.1	3.5	4.5	3.2	3.7	3.0	2.6
3.4	1.6	3.1	3.3	3.8	3.1	4.7	3.7
2.5	4.3	3.4	3.6	2.9	3.3	3.9	3.1
3.3	3.1	3.7	4.4	3.2	4.1	1.9	3.4
4.7	3.8	3.2	2.6	3.9	3.0	4.2	3.5

We must first decide on the number of classes into which the data are to be grouped. This is done arbitrarily, although we are guided by the amount of data available. Usually we choose between 5 and 20 class intervals. The smaller the number of data available, the smaller is our choice for the number of classes. For the data of Table 3.1 let us choose 7 class intervals. The *class width* must be large enough so that 7 class intervals accommodate all the data. To determine the approximate class width, we divide the difference between the largest and smallest measurements by the number of intervals. Therefore, in our example the class width can be no less than $(4.7 - 1.6)/7 = 0.443$. In practice it is desirable to choose equal class widths having the same number of significant places as the given data. Denoting this width by c, we choose $c = 0.5$. If we begin the lowest interval at 1.5, the second class would begin at 2.0, and so forth, to give the seven intervals 1.5–1.9, 2.0–2.4, 2.5–2.9, 3.0–3.4, 3.5–3.9, 4.0–4.4, and 4.5–4.9. The smallest and largest values that can fall into a given class are referred to as its *class limits*. For the interval 2.5–2.9, the smaller number, 2.5, is the *lower class limit* and the larger number, 2.9, is the *upper class limit*. The original data were recorded to the nearest tenth of a year, so that the 4 observations in the interval 2.5–2.9 are all the recorded measurements that are greater than 2.45 years and less than 2.95 years. The numbers 2.45 and 2.95 are called the *class boundaries* for the given interval. For the interval 2.5–2.9, the number 2.45 is called the *lower class boundary* and 2.95 is called the *upper class boundary*. However, 2.95 would also be the lower class boundary for the interval 3.0–3.4. Class boundaries should always be carried out to one more decimal place than the recorded observations. If an observation could have the same value as a class boundary, say 2.95, then it would be impossible to know whether it belonged to the class interval 2.45–2.95 or 2.95–3.45.

The frequency distribution for the data of Table 3.1, showing the midpoints of each class interval, is given in Table 3.2. Variations of Table 3.2 are obtained by listing the *relative frequencies* or *percentages* for each interval. The relative frequencies are simply the fraction or proportion of observations that fall in a given interval. A table listing relative frequencies is called a *relative frequency distribution*. If each relative frequency is multiplied by 100%, we have a *percentage distribution*.

Table 3.2

Frequency Distribution of Battery Lives

Class Interval	Class Boundaries	Class Midpoint	Frequency, f
1.5–1.9	1.45–1.95	1.7	2
2.0–2.4	1.95–2.45	2.2	1
2.5–2.9	2.45–2.95	2.7	4
3.0–3.4	2.95–3.45	3.2	15
3.5–3.9	3.45–3.95	3.7	10
4.0–4.4	3.95–4.45	4.2	5
4.5–4.9	4.45–4.95	4.7	3

The information provided by a frequency distribution in tabular form is easier to grasp if presented graphically. Using the midpoints of each interval and the corresponding frequencies, we construct a *frequency histogram*

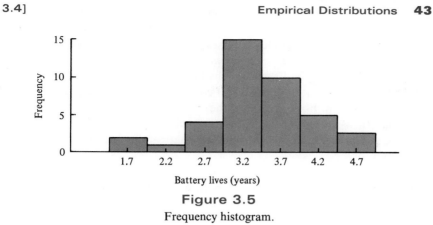

Figure 3.5

Frequency histogram.

(Figure 3.5) in exactly the same manner that we constructed the probability histogram of Section 3.2.

In Section 3.2 we suggested that the heights of the rectangles be adjusted so that the areas would represent probabilities. Once this is done the vertical axis may be omitted. If we wish to estimate the probability distribution $f(x)$ of a continuous random variable X by a smooth curve as in Figure 3.6, it is important that the rectangles of the frequency histogram be adjusted so that the total area is equal to 1.

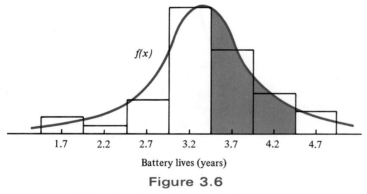

Figure 3.6

Estimating the probability density function.

The probability that a battery lasts between 3.45 and 4.45 years when selected at random from the infinite line of production of such batteries is given by the shaded area under the curve. Our estimated probability based on the recorded lives of the 40 batteries would be the sum of the areas contained in the rectangles between 3.45 and 4.45.

Although we have shown an estimate of the shape of $f(x)$ in Figure 3.6, we still have no knowledge of its formula or equation and therefore cannot find the area that has been shaded. To help understand the method of estimating the equation for $f(x)$, let us recall some elementary analytic geometry.

Parabolas, hyperbolas, circles, ellipses, and so forth, all have well-known forms of equations; in each case we would recognize their graphs. Thinking in reverse, if we only had their graphs but recognized their form, then it is not difficult to estimate the unknown constants or parameters and arrive at the exact equation. For example, if the curve appeared to have the form of a parabola, then we know that it has an equation of the form $f(x) = ax^2 + bx + x$, where a, b, and c are parameters that can be determined by various estimation procedures.

Many continuous distributions can be represented graphically by the characteristic bell-shaped curve of Figure 3.6. The equation of the probability density function $f(x)$ in this case is as well known as that of a parabola or circle and depends only on the determination of two parameters. Once these parameters are estimated from the data, we can write the estimated equation and then, using appropriate tables, find any probabilities we choose.

A distribution is said to be symmetrical if it can be folded along an axis so that the two sides coincide. A distribution that lacks symmetry with respect to a vertical axis is said to be *skewed*. The distribution illustrated in Figure 3.7(a) is said to be skewed to the right, since it has a long right tail and a much shorter left tail. In Figure 3.7(b) we see that the distribution is symmetrical, while in Figure 3.7(c) it is skewed to the left.

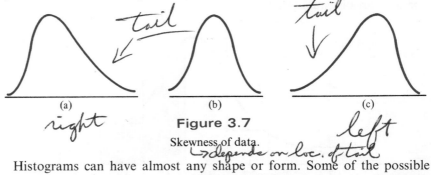

| (a) | (b) | (c) |

Figure 3.7

Skewness of data.

Histograms can have almost any shape or form. Some of the possible density functions that might arise were illustrated in Figure 3.3. In Chapters 6 and 7 we shall consider most of the important density functions that are used in experimental investigations.

In many situations we are concerned not with the number of observations in a given class but in the number that fall above or below a specified value. For example, in Table 3.2 the number of batteries lasting less than 3 years is 7. The total frequency of all values less than the upper class boundary of a given class interval is called the *cumulative frequency* up to and including that class. A table such as Table 3.3, showing the cumulative frequencies, is called a *cumulative frequency distribution*.

Two additional forms of Table 3.3 are possible using relative frequencies and percentages. Such distributions are called *relative cumulative frequency distributions* and *percentage cumulative distributions*. The percentage cumulative distribution enables one to read off the percentage of observations falling

<table>
<tr><th colspan="2">Table 3.3
RELATIVE
Cumulative Frequency
Distribution of Battery Lives</th><th colspan="2">Table 3.4
Percentage Cumulative
Distribution of Battery Lives</th></tr>
</table>

Class Boundaries	Cumulative Frequency	Class Boundaries	Cumulative %
Less than 1.45	0	Less than 1.45	0
Less than 1.95	2	Less than 1.95	5.0
Less than 2.45	3	Less than 2.45	7.5
Less than 2.95	7	Less than 2.95	17.5
Less than 3.45	22	Less than 3.45	55.0
Less than 3.95	32	Less than 3.95	80.0
Less than 4.45	37	Less than 4.45	92.5
Less than 4.95	40	Less than 4.95	100.0

below certain specified values. These values are called *percentiles*. In Table 3.4 we can see that 80% of the batteries last less than 3.95 years. Therefore, the 80th percentile is 3.95.

A line graph, called a *cumulative frequency polygon*, or *ogive*, is obtained by plotting the cumulative frequency less than any upper class boundary against the upper class boundary and joining all the consecutive points by straight lines. If relative cumulative frequencies or percentages are used, we call the graph a *relative frequency ogive* or *percentage ogive*.

The data of Table 3.4 have been used to construct the percentage ogive in Figure 3.8. In this case we have estimated the ogive for all values of X by drawing a smooth curve through the points.

Percentile points may be read quickly from the percentage ogive. In Figure 3.8 the dashed lines indicate that the twenty-fifth and seventieth percentiles are approximately 3.05 and 3.70, respectively. This means that 25% of all the batteries of this type are expected to last less than 3.05 years, while 70% of such batteries can be expected to last less than 3.70 years.

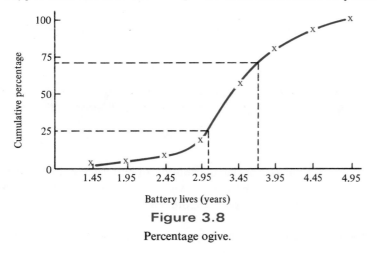

Battery lives (years)

Figure 3.8

Percentage ogive.

3.5 Percentiles, Deciles, and Quartiles

In Section 3.4 we defined a percentile to be a value below which a certain percentage of the observations falls. For any distribution there are 99 possible percentile values, which we shall denote by $P_1, P_2, \ldots, P_{99}$. Thus P_{18} is the value below which 18% of the observations fall and is called the *eighteenth percentile*.

> DEFINITION Percentiles *are values of the random variable X that divide a distribution into 100 equal parts. These values are denoted by* $P_1, P_2, \ldots,$ P_{99}.

Although one can always determine percentiles from a percentage ogive, as previously demonstrated, it may be advantageous and also less time-consuming to calculate a percentile directly from the frequency distribution. To illustrate the calculation of any percentile, we consider the following example.

Example 3.5 Find P_{85} for the distribution of battery lives in Table 3.2.

Solution. We are seeking the value below which $\frac{85}{100} \times 40 = 34$ observations fall. There are 32 observations falling below the class boundary 3.95. We still need 2 of the 5 observations assumed uniformly distributed between the class boundaries 3.95 and 4.45. Therefore we must go a distance $\frac{2}{5} \times 0.5 = 0.2$ beyond 3.95. Hence

$$P_{85} = 3.95 + 0.2$$
$$= 4.15 \text{ years.}$$

Therefore, we conclude that 85% of all batteries of this type will last less than 4.15 years.

> DEFINITION Deciles *are values of the random variable X that divide a distribution into 10 equal parts. These values are denoted by* $D_1, D_2, \ldots,$ D_9.

Example 3.6 Find D_7 for the distribution of battery lives in Table 3.2 and compare your answer with that obtained previously from the percentage ogive, Figure 3.8.

Solution. We need the value below which $\frac{70}{100} \times 40 = 28$ observations fall. There are 22 observations falling below 3.45. We still need 6 of the next 10 observations and therefore must go a distance $\frac{6}{10} \times 0.5 = 0.3$ beyond 3.45. Hence

$$D_7 = 3.45 + 0.3$$
$$= 3.75 \text{ years,}$$

compared with 3.70 years estimated from the ogive. We conclude that 70% of all batteries of this type will last less than 3.75 years.

> **DEFINITION** Quartiles *are values of the random variable X that divide a distribution into 4 equal parts. These values are denoted by* Q_1, Q_2, *and* Q_3.

Example 3.7 Find Q_1 for the distribution of battery lives in Table 3.2 and compare your answer with the result previously obtained from the percentage ogive, Figure 3.8.

Solution. We need the value below which $\frac{25}{100} \times 40 = 10$ observations fall. There are 7 observations falling below 2.95. We still need 3 of the next 15 observations and therefore must go a distance $\frac{3}{15} \times 0.5 = 0.1$ beyond 2.95. Hence

$$Q_1 = 2.95 + 0.1$$
$$= 3.05 \text{ years,}$$

which agrees exactly with our result obtained from the ogive.

The fiftieth percentile, fifth decile, and second quartile of a distribution are all equal to the same value, commonly referred to as the *median*. All the quartiles and deciles are percentiles. For example, the seventh decile is the seventieth percentile and the first quartile is the twenty-fifth percentile. Any percentile, decile, or quartile can be estimated from a percentage ogive.

3.6 Joint Probability Distributions

Our study of random variables and their probability distributions in the preceding sections was restricted to one-dimensional sample spaces, in that we recorded outcomes of an experiment assumed by a single random variable. There will be many situations, however, where we may find it desirable to record the simultaneous outcomes of several random variables. For example, we might measure the amount of precipitate P and the volume V of gas released from a controlled chemical experiment, giving rise to a two-dimensional sample space consisting of the outcomes (p, v); or one might be interested in the hardness H and tensile strength T of cold-drawn copper resulting in the outcomes (h, t). In a study to determine the likelihood of success in college, based upon high school data, one might use a three-dimensional sample space and record for each individual his aptitude test score, high school rank in class, and grade-point average at the end of the freshman year in college.

If X and Y are two random variables, the probability distribution for their simultaneous occurrence can be represented in functional notation by $f(x, y)$. It is customary to refer to $f(x, y)$ as the *joint probability distribution* of X and Y. Hence in the discrete case where a listing is possible, $f(x, y) = \Pr(X = x, Y = y)$; that is, $f(x, y)$ gives the probability that outcomes x and y occur at the same time. For example, if a television set is to be serviced and X represents the age of the set and Y represents the number of defective tubes in the set, then $f(5, 3)$ is the probability that the television set is 5 years old and needs 3 new tubes. When X and Y are continuous random variables, the *joint density function $f(x, y)$* is a surface lying above the xy-plane and the $\Pr[(X, Y) \in A]$, where A is any region in the xy-plane, is equal to the volume of the right cylinder bounded by the base A and the surface.

In keeping with the mathematical prerequisites of this text, we shall consider for the most part only joint probability functions of discrete random variables.

DEFINITION *A table or formula listing all possible values x and y of the* discrete *random variables X and Y, together with the associated probabilities f(x, y), is called a* joint probability distribution.

Example 3.8 Two refills for a ballpoint pen are selected at random from a box that contains 3 blue refills, 2 red refills, and 3 green refills. If X is the number of blue refills and Y is the number of red refills selected, find (a) the joint probability function $f(x, y)$ and (b) $\Pr[(X, Y) \in A]$, where A is the region $\{(x, y) | x + y \le 1\}$.

Solution. (a) The possible pairs of values (x, y) are $(0, 0)$, $(0, 1)$, $(1, 0)$, $(1, 1)$, $(0, 2)$, and $(2, 0)$. Now, $f(0, 1)$, for example, represents the probability that a red and a green refill are selected. The total number of equally likely ways of selecting any 2 refills from the 8 is $\binom{8}{2} = 28$. The number of ways of selecting 1 red from 3 red refills and 1 green from 3 green refills is $\binom{2}{1}\binom{3}{1} = 6$. Hence $f(0, 1) = \frac{6}{28} = \frac{3}{14}$. Similar calculations yield the probabilities for the other cases, which are presented in Table 3.5. Note that the probabilities sum to 1.

Table 3.5

Joint Probability Distribution
for Example 3.8

$f(x, y)$	0	x 1	2	Row Totals
0	$\frac{3}{28}$	$\frac{9}{28}$	$\frac{3}{28}$	$\frac{15}{28}$
y 1	$\frac{3}{14}$	$\frac{3}{14}$		$\frac{3}{7}$
2	$\frac{1}{28}$			$\frac{1}{28}$
Column Totals	$\frac{5}{14}$	$\frac{15}{28}$	$\frac{3}{28}$	1

In Chapter 3 it will become clear that the joint probability distribution of Table 3.5 can be represented by the formula

$$f(x, y) = \frac{\binom{3}{x}\binom{2}{y}\binom{3}{2 - x - y}}{\binom{8}{2}},$$

$$x = 0, 1, 2; y = 0, 1, 2; 0 \le x + y \le 2.$$

(b) $\Pr[(X, Y) \in A] = \Pr(X + Y \le 1)$

$$= f(0, 0) + f(0, 1) + f(1, 0)$$

$$= \tfrac{3}{28} + \tfrac{3}{14} + \tfrac{9}{28}$$

$$= \tfrac{9}{14}.$$

Given the joint probability distribution $f(x, y)$ of the discrete random variables X and Y, the one-dimensional probability distributions $g(x)$ of X alone and $h(y)$ of Y alone are given by the column and row totals in Table 3.5. We define $g(x)$ and $h(y)$ to be the *marginal distributions* of X and Y, respectively. Hence

$$g(0) = \Pr(X = 0) = \Pr(X = 0, Y = 0) + \Pr(X = 0, Y = 1)$$
$$+ \Pr(X = 0, Y = 2)$$
$$= f(0, 0) + f(0, 1) + f(0, 2)$$
$$= \tfrac{3}{28} + \tfrac{3}{14} + \tfrac{1}{28}$$
$$= \tfrac{5}{14},$$

$$g(1) = \Pr(X = 1) = \tfrac{15}{28},$$
$$g(2) = \Pr(X = 2) = \tfrac{3}{28}.$$

Similarly,

$$h(0) = \Pr(Y = 0) = \Pr(X = 0, Y = 0) + \Pr(X = 1, Y = 0)$$
$$+ \Pr(X = 2, Y = 0)$$
$$= f(0, 0) + f(1, 0) + f(2, 0)$$
$$= \tfrac{3}{28} + \tfrac{9}{28} + \tfrac{3}{28}$$
$$= \tfrac{15}{28},$$
$$h(1) = \Pr(Y = 1) = \tfrac{3}{7},$$
$$h(2) = \Pr(Y = 2) = \tfrac{1}{28}.$$

In tabular form the marginal distributions may be written as follows:

x	0	1	2	y	0	1	2
$g(x)$	$\tfrac{5}{14}$	$\tfrac{15}{28}$	$\tfrac{3}{28}$	$h(y)$	$\tfrac{15}{28}$	$\tfrac{3}{7}$	$\tfrac{1}{28}$

In Section 3.1 we stated that the value x of the random variable X represents an event that is a subset of the sample space. Using the definition of conditional probability as given in Chapter 2,

$$\Pr(B|A) = \frac{\Pr(A \cap B)}{\Pr(A)}, \qquad \Pr(A) = 0,$$

where A and B are now the events defined by $X = x$ and $Y = y$, respectively, then

$$\Pr(Y = y|X = x) = \frac{\Pr(X = x, Y = y)}{\Pr(X = x)}$$
$$= \frac{f(x, y)}{g(x)}, \qquad g(x) > 0,$$

when X and Y are discrete random variables.

It is not difficult to show that the function $f(x, y)/g(x)$, which is strictly a function of y with x fixed, satisfies all the conditions of a probability distribution. Writing this probability distribution as $f(y|x)$, we have

$$f(y|x) = \frac{f(x, y)}{g(x)}, \qquad g(x) > 0,$$

which is called the *conditional distribution* of the *discrete* random variable Y, given that $X = x$. Similarly we define $f(x|y)$ to be the conditional distribution of the *discrete* random variable X, given that $Y = y$, and write

$$f(x|y) = \frac{f(x, y)}{h(y)}, \qquad h(y) > 0.$$

Example 3.9 Referring to Example 3.8, find $f(x|1)$ and $\Pr(X = 0 | Y = 1)$.

Solution. First we find

$$h(1) = f(0, 1) + f(1, 1) + f(2, 1) = \tfrac{3}{14} + \tfrac{3}{14} + 0 = \tfrac{3}{7}.$$

Now,

$$f(x|1) = \frac{f(x, 1)}{h(1)} = \tfrac{7}{3} f(x, 1), \qquad x = 0, 1, 2.$$

Therefore,

$$f(0|1) = \tfrac{7}{3} f(0, 1) = (\tfrac{7}{3})(\tfrac{3}{14}) = \tfrac{1}{2},$$
$$f(1|1) = \tfrac{7}{3} f(1, 1) = (\tfrac{7}{3})(\tfrac{3}{14}) = \tfrac{1}{2},$$
$$f(2|1) = \tfrac{7}{3} f(2, 1) = (\tfrac{7}{3})(0) = 0,$$

and the conditional distribution of X, given that $Y = 1$, is

x	0	1	2
$f(x|1)$	$\tfrac{1}{2}$	$\tfrac{1}{2}$	0

Finally,

$$\Pr(X = 0 | Y = 1) = f(0|1) = \tfrac{1}{2}.$$

Suppose that the random variable X assumes the values 1, 2, 3, 4, 5, and 6 for the outcomes when a die is tossed, and Y assumes the values 0 and 1 for heads and tails, respectively, when a coin is tossed, then $f(3, 1)$ is the probability that both a 3 occurs on the die and a tail occurs on the coin if the die and coin are tossed together. In this experiment $f(3, 1) = \Pr(X = 3, Y = 1) = \Pr(X = 3)\Pr(Y = 1) = g(3)h(1)$, since obtaining a 3 on a die and a tail on a coin are independent events. Since this is true for all possible pairs (x, y), the random variables X and Y are said to be *statistically independent* and we write $f(x, y) = g(x)h(y)$.

> **DEFINITION** *Let X and Y be two random variables, discrete or continuous, with joint probability distribution $f(x, y)$ and marginal distributions $g(x)$ and $h(y)$, respectively. The random variables X and Y are said to be* statistically independent *if and only if*
>
> $$f(x, y) = g(x)h(y)$$
>
> *for all values of X and Y.*

The discrete random variables of Example 3.8 are not statistically in-dependent, as one can easily verify by considering the three probabilities $f(0, 1)$, $g(0)$, and $h(1)$. From Table 3.5 we find

$$f(0, 1) = \tfrac{3}{14},$$

$$g(0) = \sum_{y=0}^{2} f(0, y) = \tfrac{3}{28} + \tfrac{3}{14} + \tfrac{1}{28} = \tfrac{5}{14},$$

$$h(1) = \sum_{x=0}^{2} f(x, 1) = \tfrac{3}{14} + \tfrac{3}{14} + 0 = \tfrac{3}{7}.$$

Clearly,

$$f(0, 1) \neq g(0)h(1),$$

and therefore X and Y are not statistically independent.

EXERCISES

1. Classify the following random variables as discrete or continuous.
   X: The number of automobile accidents each year in Virginia.
   Y: The length of time to play 18 holes of golf.
   M: The amount of milk produced yearly by a particular cow.
   N: The number of eggs laid each month by 1 hen.
   P: The number of building permits issued each month in a certain city.
   Q: The weight of grain in pounds produced per acre.

2. From a box containing 4 black balls and 2 green balls, 3 balls are drawn in succession, each ball being replaced in the box before the next draw is made. Find the probability distribution for the number of green balls.

3. Find the probability distribution for the number of jazz records when 4 records are selected at random from a collection consisting of 5 jazz records, 2 classical records, and 3 polka records. Express your results by means of a formula.

4. Find a formula for the probability distribution of the random variable X representing the outcome when a single die is rolled once.

5. A shipment of 6 television sets contains 2 defective sets. A hotel makes a random purchase of 3 of the sets. If X is the number of defective sets purchased by the hotel, find the probability distribution of X. Express the results graphically as a probability histogram.

6. A coin is biased so that a head is twice as likely to occur as a tail. If the coin is tossed 3 times, find the probability distribution for the number of heads. Construct a probability histogram for the distribution.

7. A continuous random variable X that can assume values between $x = 1$ and $x = 3$ has a density function given by $f(x) = \tfrac{1}{2}$.
   (a) Show that the area under the curve is equal to 1.
   (b) Find $\Pr(2 < X < 2.5)$.
   (c) Find $\Pr(X \leq 1.6)$.

8. A continuous random variable X that can assume values between $x = 2$ and $x = 5$ has a density function given by $f(x) = 2(1 + x)/27$. Find (a) $\Pr(X < 4)$, (b) $\Pr(3 \leq X < 4)$.

9. The following scores represent the final examination grade for an elementary statistics course:

23	60	79	32	57	74	52	70	82	36
80	77	81	95	41	65	92	85	55	76
52	10	64	75	78	25	80	98	81	67
41	71	83	54	64	72	88	62	74	43
60	78	89	76	84	48	84	90	15	79
34	67	17	82	69	74	63	80	85	61

Using 10 intervals with the lowest starting at 9:

(a) Set up a frequency distribution.

(b) Construct a cumulative frequency distribution.

10. The following data represent the spendings in dollars and cents on extracurricular activities for a random sample of American college students during the first week of the second semester:

$16.91	$9.65	$22.68	$12.45	$18.24
11.79	6.48	12.93	7.25	13.02
15.90	8.10	3.25	9.00	9.90
12.87	17.50	10.05	27.43	16.01
6.63	4.95	14.73	8.59	6.50
20.35	8.84	13.45	18.75	24.10
13.57	9.18	5.74	9.50	7.14
10.41	12.80	32.09	6.74	11.38
17.95	7.25	4.32	19.25	8.31
6.50	13.80	9.78	6.29	14.59

Using 9 intervals with the lowest starting at $3.21:

(a) Set up a percentage distribution.

(b) Construct a percentage cumulative distribution.

11. The following data represent the number of hours per day from 6:00 A.M. to 6:00 P.M., measured to the nearest tenth, that a random sample of 30 Tennessee radio stations broadcast country music:

2.0	3.0	0.3	3.3	1.3	0.4
0.2	6.0	5.5	6.5	0.2	2.3
1.5	4.0	5.9	1.8	4.7	0.7
4.5	0.3	1.5	0.5	2.5	5.0
1.0	6.0	5.6	6.0	1.2	0.2

Using 6 intervals with the lowest starting at 0.1:

(a) Set up a relative frequency distribution.

(b) Construct a cumulative relative frequency distribution.

12. For the grouped data of Exercise 9:

(a) Construct a frequency histogram.

(b) Construct a smoothed cumulative frequency polygon.

(c) Estimate the number of people who made a score of at least 60 but less than 75.

(d) Discuss the skewness of the distribution.

13. For the grouped data of Exercise 10:
    (a) Construct a percentage histogram.
    (b) Construct a smoothed percentage ogive.
    (c) Estimate the thirtieth and seventy-fifth percentile from the ogive.
    (d) Discuss the skewness of the distribution.

14. For the grouped data of Exercise 11:
    (a) Construct a relative frequency histogram.
    (b) Construct a smoothed relative frequency ogive.
    (c) Estimate the value below which two thirds of the values fall.

15. Find P_{14}, Q_1, and D_7 for the grouped data of Exercise 9.

16. Find P_{56}, Q_3, and D_9 for the grouped data of Exercise 10.

17. Find P_{37}, Q_3, and D_2 for the grouped data of Exercise 11.

18. From a sack of fruit containing 3 oranges, 2 apples, and 3 bananas a random sample of 4 pieces of fruit is selected. If X is the number of oranges and Y is the number of apples in the sample, find:
    (a) The joint probability distribution of X and Y.
    (b) $\Pr[(X, Y) \in A]$, where A is the region $\{(x, y)|x + y \leq 2\}$.

19. Consider an experiment that consists of 2 rolls of a balanced die. If X is the number of 4s and Y is the number of 5s obtained in the 2 rolls of the die, find:
    (a) The joint probability distribution of X and Y.
    (b) $\Pr[(X, Y) \in A]$, where A is the region $\{(x, y)|2x + y < 3\}$.

20. Referring to Exercise 18, find:
    (a) $f(y|2)$.   (b) $\Pr(Y = 0|X = 2)$.

21. Suppose that X and Y have the following joint probability distribution:

y \ x	1	2	3
1	0	$\frac{1}{6}$	$\frac{1}{12}$
2	$\frac{1}{5}$	$\frac{1}{9}$	0
3	$\frac{2}{15}$	$\frac{1}{4}$	$\frac{1}{18}$

(a) Evaluate the marginal distribution of $g(x)$ and $h(y)$.
(b) Find $\Pr(Y = 3|X = 2)$.

22. Suppose that X and Y have the following joint probability function:

y \ x	2	4
1	0.10	0.15
3	0.20	0.30
5	0.10	0.15

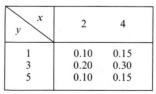

$$f(x,y) = g(x) f(y)$$

Find the marginal probability distributions and determine whether X and Y are independent.

23. Determine whether the two random variables of Exercise 21 are dependent or independent.

MATHEMATICAL EXPECTATION CHAPTER 4

4.1 Summation Notation

In statistics it is frequently necessary to work with sums of numerical values. For example, we may wish to compute the average cost of a certain brand of toothpaste sold at 10 different stores. Perhaps we would like to know the total number of heads that occur when 3 coins are tossed several times.

Consider a controlled diet experiment in which the decreases in weight over a 6-month period were 15, 10, 18, and 6 pounds, respectively. If we designate the first recorded value x_1, the second x_2, and so on, then we can write $x_1 = 15$, $x_2 = 10$, $x_3 = 18$, $x_4 = 6$.

Using the Greek letter Σ (capital sigma) to indicate "summation of," we can write the sum of the 4 weights as

$$\sum_{i=1}^{4} x_i,$$

where we read "summation of x_i, i going from 1 to 4." The numbers 1 and 4 are called the *lower* and *upper limits* of summation. Hence

$$\sum_{i=1}^{4} x_i = x_1 + x_2 + x_3 + x_4$$

$$= 15 + 10 + 18 + 6 = 49.$$

Also,

$$\sum_{i=2}^{3} x_i = x_2 + x_3 = 10 + 18 = 28.$$

55

In general the symbol $\sum\limits_{i=1}^{n}$ means that we replace i wherever it appears after the summation symbol by 1, then by 2, and so on up to n, and then add up the terms. Therefore we can write

$$\sum_{i=1}^{3} x_i^2 = x_1^2 + x_2^2 + x_3^2,$$

$$\sum_{j=2}^{5} x_j y_j = x_2 y_2 + x_3 y_3 + x_4 y_4 + x_5 y_5.$$

The subscript may be any letter, although i, j, and k seem to be preferred by statisticians. Obviously

$$\sum_{i=1}^{n} x_i = \sum_{j=1}^{n} x_j.$$

The lower limit of summation is not necessarily a subscript. For instance, the sum of the natural numbers from 1 to 9 may be written

$$\sum_{x=1}^{9} x = 1 + 2 + \cdots + 9 = 45.$$

When one is summing over all the values of x_i that are available, the limits of summation are often omitted and we simply write $\Sigma\, x_i$. If in the diet experiment only 4 people were involved, then $\Sigma\, x_i = x_1 + x_2 + x_3 + x_4$. In fact, some authors even drop the subscript and let $\Sigma\, x$ represent the sum of all available data.

Example 4.1 If $x_1 = 3$, $x_2 = 5$, and $x_3 = 7$, find

(a) $\sum x_i$.   (b) $\sum\limits_{i=1}^{3} 2x_i^2$.   (c) $\sum\limits_{i=2}^{3} (x_i - i)$.

Solution. (a) $\sum x_i = x_1 + x_2 + x_3 = 3 + 5 + 7 = 15$.

(b) $\sum\limits_{i=1}^{3} 2x_i^2 = 2x_1^2 + 2x_2^2 + 2x_3^2 = 18 + 50 + 98 = 166$.

(c) $\sum\limits_{i=2}^{3} (x_i - i) = (x_2 - 2) + (x_3 - 3) = 3 + 4 = 7$.

Example 4.2 Given $x_1 = 2$, $x_2 = -3$, $x_3 = 1$, $y_1 = 4$, $y_2 = 2$, and $y_3 = 5$, evaluate

(a) $\sum\limits_{i=1}^{3} x_i y_i$.   (b) $\left(\sum\limits_{i=2}^{3} x_i\right)\left(\sum\limits_{j=1}^{2} y_j^2\right)$.

Solution. (a) $\displaystyle\sum_{i=1}^{3} x_i y_i = x_1 y_1 + x_2 y_2 + x_3 y_3$

$$= (2)(4) + (-3)(2) + (1)(5) = 7.$$

(b) $\displaystyle\left(\sum_{i=2}^{3} x_i\right)\left(\sum_{j=1}^{2} y_j^2\right) = (x_2 + x_3)(y_1^2 + y_2^2)$

$$= (-2)(20) = -40.$$

Three theorems that provide basic rules in dealing with summation notation are given below.

THEOREM 4.1 *The summation of the sum of two or more variables is the sum of their summations. Thus*

$$\sum_{i=1}^{n} (x_i + y_i + z_i) = \sum_{i=1}^{n} x_i + \sum_{i=1}^{n} y_i + \sum_{i=1}^{n} z_i.$$

Proof. Expanding the left side and regrouping, we have

$$\sum_{i=1}^{n} (x_i + y_i + z_i)$$

$$= (x_1 + y_1 + z_1) + (x_2 + y_2 + z_2)$$
$$+ \cdots + (x_n + y_n + z_n)$$
$$= (x_1 + x_2 + \cdots + x_n) + (y_1 + y_2 + \cdots + y_n)$$
$$+ (z_1 + z_2 + \cdots + z_n)$$
$$= \sum_{i=1}^{n} x_i + \sum_{i=1}^{n} y_i + \sum_{i=1}^{n} z_i.$$

THEOREM 4.2 *If c is a constant, then*

$$\sum_{i=1}^{n} cx_i = c \sum_{i=1}^{n} x_i.$$

Proof. Expanding the left side and factoring, we get

$$\sum_{i=1}^{n} cx_i = cx_1 + cx_2 + \cdots + cx_n$$

$$= c(x_1 + x_2 + \cdots + x_n)$$

$$= c \sum_{i=1}^{n} x_i.$$

THEOREM 4.3 *If c is a constant, then*

$$\sum_{i=1}^{n} c = nc.$$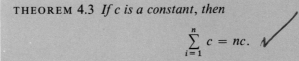

Proof. If in Theorem 4.2 all the x_i are equal to 1, then

$$\sum_{i=1}^{n} c = \underbrace{c + c + \cdots + c}_{n \text{ terms}} = nc.$$

The use of the above theorems in simplifying summation problems is illustrated in the following examples.

Example 4.3 If $x_1 = 2$, $x_2 = 4$, $y_1 = 3$, $y_2 = -1$, find the value of

$$\sum_{i=1}^{2} (3x_i - y_i + 4).$$

Solution.

$$\sum_{i=1}^{2} (3x_i - y_i + 4) = \sum_{i=1}^{2} 3x_i - \sum_{i=1}^{2} y_i + \sum_{i=1}^{2} 4$$

$$= 3 \sum_{i=1}^{2} x_i - \sum_{i=1}^{2} y_i + (2)(4)$$

$$= (3)(2 + 4) - (3 - 1) + 8$$

$$= 24.$$

Example 4.4 Simplify

$$\sum_{i=1}^{3} (x - i)^2.$$

Solution.

$$\sum_{i=1}^{3} (x - i)^2 = \sum_{i=1}^{3} (x^2 - 2xi + i^2)$$

$$= \sum_{i=1}^{3} x^2 - \sum_{i=1}^{3} 2xi + \sum_{i=1}^{3} i^2$$

$$= 3x^2 - 2x \sum_{i=1}^{3} i + \sum_{i=1}^{3} i^2$$

$$= 3x^2 - 2x(1 + 2 + 3) + (1 + 4 + 9)$$

$$= 3x^2 - 12x + 14.$$

In our treatment of joint probability distributions it is often more convenient to adopt a double-summation notation. The symbol $\sum\limits_{i=1}^{m} \sum\limits_{j=1}^{n}$ means that we first sum over the subscript j, using the theory for single summations, and then perform a second summation by allowing i to assume values from 1 to m. Hence

innermost out

$$\sum_{i=1}^{2} \sum_{j=2}^{4} x_{ij} = \sum_{i=1}^{2} (x_{i2} + x_{i3} + x_{i4})$$
$$= (x_{12} + x_{13} + x_{14}) + (x_{22} + x_{23} + x_{24})$$

and

$$\sum_{i=1}^{3} \sum_{j=1}^{2} f(x_i, y_j) = \sum_{i=1}^{3} [f(x_i, y_1) + f(x_i, y_2)]$$
$$= [f(x_1, y_1) + f(x_1, y_2)] + [f(x_2, y_1) + f(x_2, y_2)]$$
$$+ [f(x_3, y_1) + f(x_3, y_2)].$$

Some useful theorems that simplify the use of double-summation notation are found in Exercises 5 through 8.

4.2 Expected Value of a Random Variable

If two coins are tossed 16 times and X is the number of heads that occur per toss, then the values of X can be 0, 1, and 2. Suppose the experiment yields no heads, 1 head, and 2 heads, a total of 4, 7, and 5 times, respectively. The average number of heads per toss of the 2 coins is then

$$\frac{(0)(4) + (1)(7) + (2)(5)}{16} = (0)(\tfrac{4}{16}) + (1)(\tfrac{7}{16}) + (2)(\tfrac{5}{16})$$

$$= 1.06.$$

This is an average value and is not necessarily a possible outcome for the experiment. For instance, a salesman's average monthly income is not likely to be equal to any of his monthly paychecks.

The numbers $\tfrac{4}{16}$, $\tfrac{7}{16}$, and $\tfrac{5}{16}$ are the fractions of the total tosses resulting in 0, 1, and 2 heads, respectively. These fractions are also the relative frequencies for the different outcomes.

Let us now consider the problem of calculating the average number of heads per toss that we might expect in the long run. We denote this expected value or mathematical expectation by $E(X)$. From the relative frequency definition of probability we can, in the long run, expect no heads about $\tfrac{1}{4}$ of the time, 1 head about $\frac{1}{2}$ the time, and 2 heads about $\frac{1}{4}$ of the time. Therefore,

$$E(X) = (0)(\tfrac{1}{4}) + (1)(\tfrac{1}{2}) + (2)(\tfrac{1}{4}) = 1.$$

This means that a person who throws 2 coins over and over again will, on the average, get 1 head per toss.

The above illustration suggests that the average or expected value of any random variable may be obtained my multiplying each value of the random variable by its corresponding probability and summing the results. This is true only if the variable is discrete. Calculating the expected value of a continuous variable requires a knowledge of calculus and is therefore beyond the scope of this text.

DEFINITION *Let X be a discrete random variable with the following probability distribution:*

x	x_1	x_2 $\cdots$ x_n
$\Pr(X = x)$	$f(x_1)$	$f(x_2) \cdots f(x_n)$

The expected value *of X or the* mathematical expectation *of X is*

$$E(X) = \sum_{i=1}^{n} x_i f(x_i).$$

Example 4.5 Find the expected value of X, where X represents the outcome when a die is tossed.

Solution. Each of the numbers 1, 2, 3, 4, 5, and 6 occurs with probability $\frac{1}{6}$. Therefore,

$$E(X) = (1)(\tfrac{1}{6}) + (2)(\tfrac{1}{6}) + \cdots + (6)(\tfrac{1}{6}) = 3.5.$$

This means that a person will, on the average, roll 3.5.

Example 4.6 In a gambling game a man is paid $5 if he gets all heads or all tails when 3 coins are tossed, and he pays out $3 if either 1 or 2 heads show. What is his expected gain?

Solution. The random variable of interest is Y, the amount he can win. The possible values of Y are 5 and -3 with probabilities $\frac{1}{4}$ and $\frac{3}{4}$, respectively. Therefore,

$$E(Y) = (5)(\tfrac{1}{4}) + (-3)(\tfrac{3}{4}) = -1.$$

In this game the gambler will, on the average, lose $1 per toss.

A game is considered "fair" if the gambler will, on the average, come out even. Therefore an expected gain of zero defines a fair game.

Example 4.7 Find the expected number of boys on a committee of 3 selected at random from 4 boys and 3 girls.

Solution. Let X represent the number of boys on the committee. The probability distribution of X is given by

$$f(x) = \frac{\binom{4}{x}\binom{3}{3-x}}{\binom{7}{3}}, \qquad x = 0, 1, 2, 3.$$

A few simple calculations yield $f(0) = \frac{1}{35}$, $f(1) = \frac{12}{35}$, $f(2) = \frac{18}{35}$, and $f(3) = \frac{4}{35}$. Therefore,

$$E(X) = (0)(\tfrac{1}{35}) + (1)(\tfrac{12}{35}) + (2)(\tfrac{18}{35}) + (3)(\tfrac{4}{35}) = \tfrac{12}{7} = 1.7.$$

Thus if a committee of 3 is selected at random over and over again from 4 boys and 3 girls, it would contain on the average 1.7 boys.

Now let us consider a new random variable $g(X)$, which depends on X; that is, each value of $g(X)$ is determined by knowing the values of X. For instance, $g(X)$ might be X^2 or $3X - 1$, so that whenever X assumes the value 2, $g(X)$ assumes the value $g(2)$. In particular, if X is a discrete random variable with probability distribution $f(x)$, $x = -1, 0, 1, 2$, and $g(X) = X^2$, then

$$\Pr[g(X) = 0] = \Pr(X = 0) = f(0),$$
$$\Pr[g(X) = 1] = \Pr(X = -1) + \Pr(X = 1) = f(-1) + f(1),$$
$$\Pr[g(X) = 4] = \Pr(X = 2) = f(2),$$

so that the probability distribution of $g(X)$ may be written

$g(x)$	0	1	4
$\Pr[g(X) = g(x)]$	$f(0)$	$f(-1) + f(1)$	$f(2)$

By the definition of an expected value of a random variable we obtain

$$E[g(X)] = 0\,f(0) + 1[f(-1) + f(1)] + 4\,f(2)$$
$$= (-1)^2 f(-1) + (0)^2 f(0) + (1)^2 f(1) + (2)^2 f(2)$$
$$= \sum_{i=1}^{4} g(x_i)f(x_i),$$

where $x_1 = -1$, $x_2 = 0$, $x_3 = 1$, and $x_4 = 2$. This result is generalized in Theorem 4.4.

THEOREM 4.4 *Let X be a discrete random variable with the following probability distribution:*

x	x_1	x_2	$\cdots$	x_n
$\Pr(X = x)$	$f(x_1)$	$f(x_2)$	$\cdots$	$f(x_n)$

The expected value of a new random variable $g(X)$ is

$$E[g(X)] = \sum_{i=1}^{n} g(x_i)f(x_i).$$

Example 4.8 Let $Y = 2X - 1$, where X represents the outcome when a die is tossed. Find the expected value of Y.

Solution. The probability distribution for X and the corresponding values of Y are as follows:

x	1	2	3	4	5	6
y	1	3	5	7	9	11
$\Pr(X = x)$	$\frac{1}{6}$	$\frac{1}{6}$	$\frac{1}{6}$	$\frac{1}{6}$	$\frac{1}{6}$	$\frac{1}{6}$

Therefore,

$$E(Y) = (1)(\tfrac{1}{6}) + (3)(\tfrac{1}{6}) + \cdots + (11)(\tfrac{1}{6})$$
$$= \tfrac{36}{6} = 6.$$

Example 4.9 Let X be a random variable with probability distribution as follows:

x	0	1	2	3
$\Pr(X = x)$	$\frac{1}{3}$	$\frac{1}{2}$	0	$\frac{1}{6}$

Find $E[\{X - E(X)\}^2]$.

Solution. First we must find $E(X)$. Therefore,

$$E(X) = (0)(\tfrac{1}{3}) + (1)(\tfrac{1}{2}) + (2)(0) + (3)(\tfrac{1}{6}) = 1.$$

The problem now reduces to find $E[(X - 1)^2]$. We have

$$E[(X - 1)^2] = \sum_{x=0}^{3} (x - 1)^2 f(x)$$

$$= (-1)^2 f(0) + (0)^2 f(1) + (1)^2 f(2) + (2)^2 f(3)$$
$$= (1)(\tfrac{1}{3}) + (0)(\tfrac{1}{2}) + (1)(0) + (4)(\tfrac{1}{6})$$
$$= 1.$$

We shall now extend our concept of mathematical expectation to the case of two random variables X and Y with joint probability distribution $f(x, y)$.

DEFINITION *Let X and Y be random variables with joint probability distribution $f(x, y)$, where $x = x_1, x_2, \ldots, x_m$ and $y = y_1, y_2, \ldots, y_n$. The expected value of the function $g(X, Y)$ is*

$$E[g(X, Y)] = \sum_{i=1}^{m} \sum_{j=1}^{n} g(x_i, y_j) f(x_i, y_j).$$

Example 4.10 Let X and Y be random variables with joint probability distribution given by Table 3.5. Find the expected value of $g(X, Y) = XY$.

Solution. By the above definition, we write

$$E(XY) = \sum_{x=0}^{2} \sum_{y=0}^{2} xy\, f(x, y)$$
$$= (0)(0)f(0, 0) + (0)(1)f(0, 1) + (0)(2)f(0, 2)$$
$$\quad + (1)(0)f(1, 0) + (1)(1)f(1, 1) + (2)(0)f(2, 0)$$
$$= f(1, 1) = \tfrac{3}{14}.$$

4.3 Laws of Expectation

We shall now develop some useful laws that will simplify the calculations of mathematical expectations. These laws or theorems will permit us to calculate expectations in terms of other known expectations or expectations that are easily computed. All the results below are valid for both discrete and continuous random variables. Proofs will be given only for the discrete case.

THEOREM 4.5 *If a and b are constant, then*
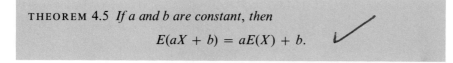

$$E(aX + b) = aE(X) + b.$$

Proof. By the definition of an expected value,

$$E(aX + b) = \sum_{i=1}^{n} (ax_i + b)f(x_i)$$

$$= (ax_1 + b)f(x_1) + (ax_2 + b)f(x_2) + \cdots$$
$$+ (ax_n + b)f(x_n)$$
$$= a[x_1 f(x_1) + x_2 f(x_2) + \cdots + x_n f(x_n)]$$
$$+ b[f(x_1) + f(x_2) + \cdots + f(x_n)]$$
$$= a \sum_{i=1}^{n} x_i f(x_i) + b \sum_{i=1}^{n} f(x_i).$$

The first sum on the right is $E(X)$ and the second sum equals 1. Therefore, we have

$$E(aX + b) = aE(X) + b.$$

COROLLARY 1 *Setting $a = 0$, we see that $E(b) = b$.*

COROLLARY 2 *Setting $b = 0$, we see that $E(aX) = aE(X)$.*

THEOREM 4.6 *The expected value of the sum or difference of two or more functions of a random variable X is the sum or difference of the expected values of the functions, that is,*

$$E[g(X) \pm h(X)] = E[g(X)] \pm E[h(X)].$$

Proof. By definition,

$$E[g(X) \pm h(X)]$$
$$= \sum_{i=1}^{n} [g(x_i) \pm h(x_i)]f(x_i)$$
$$= [g(x_1) \pm h(x_1)]f(x_1) + [g(x_2) \pm h(x_2)]f(x_2) + \cdots$$
$$+ [g(x_n) \pm h(x_n)]f(x_n)$$
$$= [g(x_1)f(x_1) + g(x_2)f(x_2) + \cdots + g(x_n)f(x_n)]$$
$$\pm [h(x_1)f(x_1) + h(x_2)f(x_2) + \cdots + h(x_n)f(x_n)]$$
$$= \sum_{i=1}^{n} g(x_i)f(x_i) \pm \sum_{i=1}^{n} h(x_i)f(x_i)$$
$$= E[g(X)] \pm E[h(X)].$$

Example 4.11 In Example 4.9 we could write

$$E[(X - 1)^2] = E(X^2 - 2X + 1)$$
$$= E(X^2) - 2E(X) + E(1).$$

We have shown that $E(X) = 1$, and from Corollary 1 of Theorem 4.5, we have $E(1) = 1$. Now

$$E(X^2) = \sum_{x=0}^{3} x^2 f(x) = (0)(\tfrac{1}{3}) + (1)(\tfrac{1}{2}) + (4)(0) + (9)(\tfrac{1}{6}) = 2.$$

Hence $E[(X - 1)^2] = 2 - (2)(1) + 1 = 1$, as before.

Suppose that we have two variables X and Y with joint probability distribution $f(x, y)$. Two additional laws that will be very useful in succeeding chapters involve the expected values of the sum, difference, and product of two random variables. First, however, let us prove a theorem on the expected value of the sum or difference of functions of the given variables. This, of course, is merely an extension of Theorem 4.6.

THEOREM 4.7 *The expected value of the sum or difference of two or more functions of the random variables X and Y is the sum or difference of the expected values of the functions, that is,*

$$E[g(X, Y) \pm h(X, Y)] = E[g(X, Y)] \pm E[h(X, Y)].$$

Proof. By definition,

$$E[g(X, Y) \pm h(X, Y)]$$
$$= \sum_{i=1}^{m} \sum_{j=1}^{n} [g(x_i, y_j) \pm h(x_i, y_j)] f(x_i, y_j)$$
$$= \sum_{i=1}^{m} \sum_{j=1}^{n} g(x_i, y_j) f(x_i, y_j) \pm \sum_{i=1}^{m} \sum_{j=1}^{n} h(x_i, y_j) f(x_i, y_j)$$
$$= E[g(X, Y)] \pm E[h(X, Y)].$$

COROLLARY *Setting g(X, Y) = X and h(X, Y) = Y, we see that*

$$E(X \pm Y) = E(X) \pm E(Y).$$

If X represents the daily production of an item from machine A and Y the daily production of the same kind of item from machine B, then $X + Y$ represents the total number of items produced daily from both machines. The corollary of Theorem 4.7 states that the average daily production for both machines is equal to the sum of the average daily production of each machine.

> **THEOREM 4.8** *Let X and Y be two independent random variables. Then*
>
> $$E(X, Y) = E(X)E(Y).$$

Proof. By definition,

$$E(XY) = \sum_{i=1}^{m} \sum_{j=1}^{n} x_i y_j f(x_i, y_j).$$

Since X and Y are independent, we may write

$$f(x, y) = g(x)h(y),$$

where $g(x)$ and $h(y)$ are the marginal distributions of X and Y, respectively. Hence

$$\begin{aligned}
E(XY) &= \sum_{i=1}^{m} \sum_{j=1}^{n} x_i y_j g(x_i) h(y_j) \\
&= \left[\sum_{i=1}^{m} x_i g(x_i) \right] \left[\sum_{j=1}^{n} y_j h(y_j) \right] \quad \text{(by Exercise 8)} \\
&= E(X)E(Y).
\end{aligned}$$

This theorem can be illustrated by tossing a green die and a red die. Let the random variable X represent the outcome on the green die and the random variable Y represent the outcome on the red die. Then XY represents the product of the numbers that occur on the pair of dice. In the long run, the average of the products of the numbers is equal to the product of the average number that occurs on the green die and the average number that occurs on the red die.

4.4 Special Mathematical Expectations

By allowing $g(X)$ in Theorem 4.4 to assume certain special forms, we obtain some very useful mathematical expectations. In the case where $g(X) = X$, for example, we have the expected value of the random variable X itself. Because the expected value of a random variable X is somewhat special, in that it represents the *mean* of the random variable, we shall write it as μ_X or simply μ. Thus

$$\mu = E(X).$$

If $g(X) = (X - \mu)^2$, Theorem 4.4 yields an expected value, called the *variance of the random variable* X, which we donate by σ_X^2 or simply σ^2. Thus

$$\sigma^2 = E[(X - \mu)^2].$$

The variance, which is just the average of the squares of the deviations of the values of X from the mean, tells us something about the variability of the measurements. Clearly, if the measurements are all close to the mean, then the deviations from the mean will be small and consequently σ^2 will be small. On the other hand, one can expect a large value of σ^2 if the values of X are widely dispersed about the mean. The positive square root of the variance is a measure called the *standard deviation*.

An alternative and preferred formula for σ^2 is given in the following theorem.

THEOREM 4.9 *The variance of a random variable X is given by*
$$\sigma^2 = E(X^2) - \mu^2.$$

Proof.
$$\sigma^2 = E[(X - \mu)^2]$$
$$= E(X^2 - 2\mu X + \mu^2)$$
$$= E(X^2) - 2\mu E(X) + E(\mu^2)$$
$$= E(X^2) - \mu^2,$$

since $\mu = E(X)$ by definition and $E(\mu^2) = \mu^2$ by Corollary 1 of Theorem 4.5.

Example 4.12 Calculate the variance of X, where X is the number of boys on a committee of 3 selected at random from 4 boys and 3 girls.

Solution. In Example 4.7 we showed that $\mu = \frac{12}{7}$. Now
$$E(X^2) = (0)(\tfrac{1}{35}) + (1)(\tfrac{12}{35}) + (4)(\tfrac{18}{35}) + (9)(\tfrac{4}{35})$$
$$= \tfrac{24}{7}.$$

Therefore,
$$\sigma^2 = \tfrac{24}{7} - (\tfrac{12}{7})^2 = \tfrac{24}{49}.$$

If $g(X, Y) = (X - \mu_X)(Y - \mu_Y)$, where $\mu_X = E(X)$ and $\mu_Y = E(Y)$, then $E[g(X, Y)]$ is called the *covariance of X and Y, which we denote by* σ_{XY} or cov(X, Y). Therefore,
$$\sigma_{XY} = E[(X - \mu_X)(Y - \mu_Y)]$$
$$= \sum_{i=1}^{m} \sum_{j=1}^{n} (x_i - \mu_X)(y_j - \mu_Y)f(x_i, y_j).$$

The covariance will be positive when large values of X are associated with large values of Y and small values of X are associated with small values of Y.

If small values of X are associated with large values of Y, and vice versa, then the covariance will be negative. When X and Y are statistically independent, it can be shown that the covariance is zero (see the corollary of Theorem 4.10). The converse, however, is not generally true. Two variables may have zero covariance and still not be statistically independent.

The alternative and preferred formula for σ_{XY} is given in the following theorem:

THEOREM 4.10 *The covariance of two random variables X and Y with means μ_X and μ_Y, respectively, is given by*

$$\sigma_{XY} = E(XY) - \mu_X\mu_Y.$$

Proof.

$$\begin{aligned}
\sigma_{XY} &= E[(X - \mu_X)(Y - \mu_Y)] \\
&= E(XY - \mu_X Y - \mu_Y X + \mu_X\mu_Y) \\
&= E(XY) - \mu_X E(Y) - \mu_Y E(X) + E(\mu_X\mu_Y) \\
&= E(XY) - \mu_X\mu_Y,
\end{aligned}$$

since $\mu_X = E(X)$ and $\mu_Y = E(Y)$ by definition and $E(\mu_X\mu_Y) = \mu_X\mu_Y$ by Corollary 1 of Theorem 4.5.

COROLLARY *If X and Y are independent random variables, then $\sigma_{XY} = 0$.*

This result follows from Theorem 4.10 by noting that $E(XY) = E(X)E(Y) = \mu_X\mu_Y$ for independent variables.

Example 4.13 Referring to the joint probability distribution of Example 3.5 and to the computations of Example 4.10, we see that $E(XY) = \frac{3}{14}$. Now,

$$\begin{aligned}
\mu_X = E(X) &= \sum_{x=0}^{2} \sum_{y=0}^{2} xf(x, y) = \sum_{x=0}^{2} xg(x) \\
&= (0)(\tfrac{10}{28}) + (1)(\tfrac{15}{28}) + (2)(\tfrac{3}{28}) \\
&= \tfrac{3}{4},
\end{aligned}$$

and

$$\begin{aligned}
\mu_Y = E(Y) &= \sum_{x=0}^{2} \sum_{y=0}^{2} yf(x, y) = \sum_{y=0}^{2} yh(y) \\
&= (0)(\tfrac{15}{28}) + (1)(\tfrac{3}{7}) + (2)(\tfrac{1}{28}) \\
&= \tfrac{1}{2}.
\end{aligned}$$

Therefore,

$$\sigma_{XY} = E(XY) - \mu_X\mu_Y$$

$$= \tfrac{3}{14} - (\tfrac{3}{4})(\tfrac{1}{2})$$

$$= -\tfrac{9}{56}.$$

4.5 Properties of the Variance

We shall now prove three theorems that are useful in calculating variances or standard deviations. If we let $g(X)$ be a function of the random variable X, then the mean and variance of $g(X)$ will be denoted by $\mu_{g(X)}$ and $\sigma^2_{g(X)}$, respectively.

THEOREM 4.11 *Let X be a random variable with probability distribution $f(x)$. The variance of the function $g(X)$ is*

$$\sigma^2_{g(X)} = E[\{g(X) - \mu_{g(X)}\}^2].$$

just subst in form. bottom pg 66

Proof. Since $g(X)$ is a random variable, the result follows from the definition of the variance.

THEOREM 4.12 *If X is a random variable and b is a constant, then,*

$$\sigma^2_{X+b} = \sigma^2_X = \sigma^2.$$

Proof.
$$\sigma^2_{X+b} = E[\{(X + b) - \mu_{X+b}\}^2].$$

Now,

$$\mu_{X+b} = E(X + b) = E(X) + b = \mu + b,$$

by Theorem 4.5. Therefore,

$$\sigma^2_{X+b} = E[(X + b - \mu - b)^2]$$

$$= E[(X - \mu)^2]$$

$$= \sigma^2.$$

This theorem states that the variance is unchanged if a constant is added to or subtracted from a random variable. The addition or subtraction of a constant simply shifts the values of X to the right or to the left but does not change their variability.

THEOREM 4.13 *If X is a random variable and a is any constant, then*

$$\sigma_{aX}^2 = a^2\sigma_X^2 = a^2\sigma^2.$$

Proof. $\sigma_{aX}^2 = E[\{aX - \mu_{aX}\}^2].$

Now

$$\mu_{aX} = E(aX) = aE(X) = a\mu,$$

by Corollary 2 of Theorem 4.5. Therefore,

$$\sigma_{aX}^2 = E[(aX - a\mu)^2]$$
$$= a^2E[(X - \mu)^2]$$
$$= a^2\sigma^2.$$

Therefore, if a random variable is multiplied or divided by a constant, the variance is multiplied or divided by the square of the constant.

THEOREM 4.14 *If X and Y are random variables with joint probability distribution f(x, y), then*

$$\sigma_{aX+bY}^2 = a^2\sigma_X^2 + b^2\sigma_Y^2 + 2ab\sigma_{XY}.$$

Proof. $\sigma_{aX+bY}^2 = E[(aX + bY) - \mu_{aX+bY}]^2.$

Now,

$$\mu_{aX+bY} = E(aX + bY) = aE(X) + bE(Y) = a\mu_X + b\mu_Y,$$

by using Theorem 4.6 followed by Corollary 2 of Theorem 4.5. Therefore,

$$\sigma_{aX+bY}^2 = E\{[(aX + bY) - (a\mu_X + b\mu_Y)]^2\}$$
$$= E\{[a(X - \mu_X) + b(Y - \mu_Y)]^2\}$$
$$= a^2E[(X - \mu_X)^2] + b^2E[(Y - \mu_Y)^2]$$
$$\quad + 2abE[(X - \mu_X)(Y - \mu_Y)]$$
$$= a^2\sigma_X^2 + b^2\sigma_Y^2 + 2ab\sigma_{XY}.$$

COROLLARY 1 *If X and Y are independent random variables, then*

$$\sigma_{aX+bY}^2 = a^2\sigma_X^2 + b^2\sigma_Y^2.$$

COROLLARY 2 *If X and Y are independent random variables, then*

$$\sigma^2_{aX - bY} = a^2 \sigma^2_X + b^2 \sigma^2_Y.$$

Corollary 2 follows by writing $X - Y$ as $X + (-Y)$. Then $\sigma^2_{X-Y} = \sigma^2_X + \sigma^2_{(-Y)}$. From Theorem 4.13 we know that $\sigma^2_{(-Y)} = (-1)^2 \sigma^2_Y = \sigma^2_Y$. Therefore,

$$\sigma^2_{X-Y} = \sigma^2_X + \sigma^2_Y.$$

Example 4.14 If X and Y are independent random variables with variances $\sigma^2_X = 1$ and $\sigma^2_Y = 2$, find the variance of the random variable $Z = 3X - 2Y + 5$.

Solution. $\sigma^2_Z = \sigma^2_{3X - 2Y + 5}$

$\qquad\qquad = \sigma^2_{3X - 2Y} \qquad$ by Theorem 4.12

$\qquad\qquad = 9\sigma^2_X + 4\sigma^2_Y \qquad$ by Corollary 2 of Theorem 4.14

$\qquad\qquad = (9)(1) + (4)(2)$

$\qquad\qquad = 17.$

4.6 Chebyshev's Theorem

In Section 4.4 we stated that the variance of a random variable tells us something about the variability of the observations about the mean. If a random variable has a small variance or standard deviation, we would expect most of the values to be grouped around the mean. Therefore the probability that a random variable assumes a value within a certain interval about the mean is greater than for a similar random variable with a larger standard deviation. If we think of probability in terms of area, we would expect a continuous distribution with a small standard deviation to have most of its area close to μ, as in Figure 4.1(a). However, a large value of σ indicates a greater variability and therefore we would expect the area to be more spread out, as in Figure 4.1(b).

We can argue the same way for a discrete distribution. The area in the probability histogram in Figure 4.2(b) is spread out much more than that of Figure 4.2(a), indicating a more variable distribution of measurements or outcomes.

The Russian mathematician, Chebyshev, discovered that the fraction of the area between any two values symmetric about the mean is related to the standard deviation. Since the area under a probability distribution curve or in a probability histogram adds to 1, the area between any two numbers is

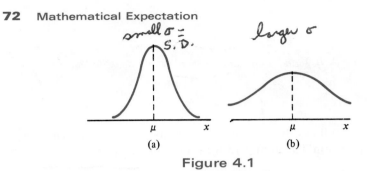

Figure 4.1

Variability of continuous observations about the mean.

the probability of the random variable assuming a value between these numbers.

The following theorem, due to Chebyshev, gives a conservative estimate of the probability of a random variable falling within k standard deviations of its mean. The proof is omitted.

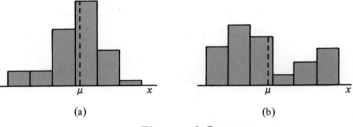

Figure 4.2

Variability of discrete observations about the mean.

CHEBYSHEV'S THEOREM. *The probability that any random variable X falls within k standard deviations of the mean is at least $1 - 1/k^2$. That is,*

$$\Pr(\mu - k\sigma < X < \mu + k\sigma) \geq 1 - \frac{1}{k^2}.$$

For $k = 2$ the theorem states that the random variable X has a probability of at least $1 - 1/2^2 = \frac{3}{4}$ of falling within 2 standard deviations of the mean, that is, $\frac{3}{4}$ or more of the observations of any distribution lie in the interval $\mu \pm 2\sigma$. Similarly, the theorem says that at least $\frac{8}{9}$ of the observations of any distribution fall in the interval $\mu \pm 3\sigma$.

Example 4.15 A random variable X has a mean $\mu = 8$, variance $\sigma^2 = 9$, and an unknown probability distribution. Find $\Pr(-4 < X < 20)$.

Solution. $\Pr(-4 < X < 20) = \Pr[8 - (4)(3) < X < 8 + (4)(3)]$.

Therefore, using Chebyshev's theorem with $k = 4$, we have

$$\Pr(-4 < X < 20) \geq \tfrac{15}{16}.$$

Chebyshev's theorem holds for any distribution of observations and, for this reason, the results are usually weak. The value given by the theorem is a lower bound only. That is, we know the probability of a random variable falling within 2 standard deviations of the mean can be *no less* than $\tfrac{3}{4}$, but we never know how much more it might actually be. Only when the probability distribution is known can we determine exact probabilities.

EXERCISES

1. Write each of the following expressions in full:

   (a) $\displaystyle\sum_{i=6}^{10} w_i^2$.   (b) $\displaystyle\sum_{h=2}^{4} (x_h + h)$.   (c) $\displaystyle\sum_{j=1}^{5} 3(v_j - 2)$.

2. Simplify, leaving your answer as a polynomial:

   (a) $\displaystyle\sum_{i=2}^{4} (2x + i)^2$.   (b) $\displaystyle\sum_{y=0}^{3} (x - y + 3)^3$.

3. If $x_1 = 4$, $x_2 = -3$, $x_3 = 6$, and $x_4 = -1$, evaluate the following:

   (a) $\displaystyle\sum_{i=1}^{4} x_i^2(x_i - 3)$.   (b) $\displaystyle\sum_{i=2}^{4} (x_i + 1)^2$.   (c) $\displaystyle\sum_{i=2}^{3} (x_i + 2)/x_i$.

4. Given $x_1 = -2$, $x_2 = 3$, $x_3 = 1$, $y_1 = 4$, $y_2 = 0$, and $y_3 = -5$, find the value of the following:

   (a) $\sum x_i y_i^2$.   (b) $\displaystyle\sum_{i=2}^{3} (2x_i + y_i - 3)$.   (c) $(\sum x^2)(\sum y)$.

5. Show that

   $$\sum_{i=1}^{m} \sum_{j=1}^{n} (x_{ij} + y_{ij} + z_{ij}) = \sum_{i=1}^{m} \sum_{j=1}^{n} x_{ij} + \sum_{i=1}^{m} \sum_{j=1}^{n} y_{ij} + \sum_{i=1}^{m} \sum_{j=1}^{n} z_{ij}.$$

6. Show that

   $$\sum_{i=1}^{m} \sum_{j=1}^{n} cx_{ij} = c \sum_{i=1}^{m} \sum_{j=1}^{n} x_{ij}.$$

7. Show that

   $$\sum_{i=1}^{m} \sum_{j=1}^{n} c = mnc.$$

8. Verify that

$$\sum_{i=1}^{m} \sum_{j=1}^{n} x_i y_j = \left(\sum_{i=1}^{m} x_i \right) \left(\sum_{j=1}^{n} y_j \right).$$

9. A shipment of 6 television sets contains 2 defectives. A hotel makes a random purchase of 3 of the sets. If X is the number of defective sets purchased by the hotel, find $E(X)$.

10. By investing in a particular stock, a man can make a profit in 1 year of $3000 with probability 0.3 or take a loss of $1000 with probability 0.7. What is his mathematical expectation?

11. In a gambling game a man is paid $2 if he draws a jack or queen and $5 if he draws a king or ace from an ordinary deck of 52 playing cards. If he draws any other card he loses. How much should he pay to play if the game is fair?

12. Find the expected number of jazz records when 4 records are selected at random from a collection consisting of 5 jazz records, 2 classical records, and 3 polka records.

13. The probability distribution of the discrete random variable X is

$$f(x) = \binom{3}{x} \left(\frac{1}{4}\right)^x \left(\frac{3}{4}\right)^{3-x}, \qquad x = 0, 1, 2, 3.$$

Find $E(X)$.

14. Let X represent the outcome when a balanced die is tossed. Find $E(Y)$, where $Y = 2X^2 - 5$.

15. A race-car driver wishes to insure his car for the racing season for $10,000. The insurance company estimates a total loss may occur with probability 0.002, a 50% loss with probability 0.01, and a 25% loss with probability 0.1. Ignoring all other partial losses, what premium should the insurance company charge each season to make a profit of $100?

16. Suppose X and Y have the following joint probability function:

y \ x	2	4
1	0.10	0.15
3	0.20	0.30
5	0.10	0.15

Find the expected value of $g(X, Y) = XY^2$.

17. Let X be a random variable with the following probability distribution.

x	-3	6	9
$\Pr(X = x)$	$\frac{1}{6}$	$\frac{1}{2}$	$\frac{1}{3}$

Find $E(X)$ and $E(X^2)$ and then, using the laws of expectation, evaluate
(a) $E[(2X + 1)^2]$.   (b) $E[\{X - E(X)\}^2]$.

18. From a sack of fruit containing 3 oranges, 2 apples, and 3 bananas, a random sample of 4 pieces of fruit is selected. If X is the number of oranges and Y is the number of apples in the sample, find $E(X^2 Y - 2XY)$.

19. Let X represent the number that occurs when a red die is tossed and Y the number that occurs when a green die is tossed. Find
    (a) $E(X + Y)$.   (b) $E(X - Y)$.   (c) $E(XY)$.

20. Referring to the random variables whose joint probability distribution is given in Exercise 16, find
    (a) $E(2X - 3Y)$.   (b) $E(XY)$.

21. Find the variance of the random variable X of Exercise 9.

22. Let X be a random variable with the following probability distribution:

x	-2	3	5
$\Pr(X = x)$	0.3	0.2	0.5

Find the standard deviation of X.

23. From a group of 5 men and 3 women a committee of size 3 is selected at random. If X represents the number of women on the committee, find the mean and variance of X.

24. Find the covariance of the random variables X and Y of Exercise 18.

25. Suppose X and Y have the following joint probability distribution:

also
$p(x,y)$

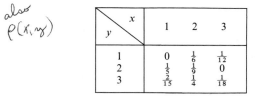

y \ x	1	2	3
1	0	$\frac{1}{6}$	$\frac{1}{12}$
2	$\frac{1}{5}$	$\frac{1}{9}$	0
3	$\frac{2}{15}$	$\frac{1}{4}$	$\frac{1}{18}$

Find the covariance of the random variables X and Y.

26. Show that $\operatorname{cov}(aX, bY) = ab \operatorname{cov}(X, Y)$.

27. Let X represent the number that occurs when a green die is tossed and Y the number that occurs when a red die is tossed. Find the variance of the random variable
    (a) $2X - Y$.   (b) $X + 3Y - 5$.

28. If X and Y are independent random variables with variances $\sigma_X^2 = 5$ and $\sigma_Y^2 = 3$, find the variance of the random variable $Z = -2X + 4Y - 3$.

29. Repeat Exercise 28 if X and Y are not independent and $\sigma_{XY} = 1$.

30. A random variable X has a mean $\mu = 12$, a variance $\sigma^2 = 9$, and an unknown probability distribution. Using Chebyshev's theorem, find
    (a) $\Pr(6 < X < 18)$.   (b) $\Pr(3 < X < 21)$.

31. If the distribution of I.Q.s of college students has a mean $\mu = 120$ and a standard deviation $\sigma = 8$, use Chebyshev's theorem to determine the interval containing at least $\frac{3}{4}$ of the scores. In what range can we be sure that no more than $\frac{1}{9}$ of the scores fall?

SOME DISCRETE
PROBABILITY DISTRIBUTIONS CHAPTER 5

5.1 Introduction

In Chapter 3 a discrete probability distribution was represented graphically, in tabular form, and if convenient by means of a formula. No matter what method of presentation is used, the behavior of a random variable is described. Many random variables associated with statistical experiments have similar properties and can be described by essentially the same probability distribution. For example, all random variables representing the number of successes in n independent trials of an experiment, where the probability of a success is constant for all n trials, have the same general type of behavior and therefore can be represented by a single formula. Thus if in firing a rifle at a target a direct hit is considered a success and the probability of a hit remains constant for successive firings, the formula for the distribution of hits in 5 firings of the rifle has the same structure as the formula for the distribution of 4s in 7 tosses of a die, where the occurrence of a 4 is considered a success.

Frequent reference will be made throughout the text to the parameters of a probability distribution. By this we shall mean the parameters of any random variable having that particular probability distribution. Therefore the mean or variance of a given probability distribution is defined to be the mean or variance of any random variable having that distribution.

Care should be exercised in choosing the probability distribution that correctly describes the observations being generated by the experiment. In this chapter we shall investigate several important discrete probability distributions that describe most random variables encountered in practice.

5.2 Uniform Distribution

The simplest of all discrete probability distributions is one in which the random variable assumes all its values with equal probability. Such a probability distribution is called the *uniform distribution*.

> UNIFORM DISTRIBUTION *If the random variable X assumes the values* $x_1, x_2, \ldots, x_k$, *with equal probability, then the discrete uniform distribution is given by*
>
> $$f(x; k) = \frac{1}{k}, \qquad x = x_1, x_2, \ldots, x_k.$$

We have used the notation $f(x; k)$ instead of $f(x)$ to indicate that the uniform distribution depends on the parameter k.

Example 5.1 When a die is tossed, each element of the sample space $S = \{1, 2, 3, 4, 5, 6\}$ occurs with probability $\frac{1}{6}$. Therefore we have a uniform distribution, with

$$f(x; 6) = \tfrac{1}{6}, \qquad x = 1, 2, 3, 4, 5, 6.$$

Example 5.2 Suppose that an employee is selected at random from a staff of 10 to supervise a certain project. Each employee has the same probability $\frac{1}{10}$ of being selected. If we assume that the employees have been numbered in some way from 1 to 10, the distribution is uniform with $f(x; 10) = \frac{1}{10}, x = 1, 2, \ldots, 10$.

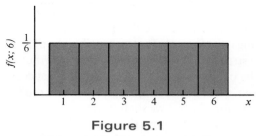

Figure 5.1

Histogram for the tossing of a die.

The graphic representation of the uniform distribution by means of a histogram always turns out to be a set of rectangles with equal heights. The histogram for Example 5.1 is shown in Figure 5.1.

It is not difficult to see that the distribution of all possible subsets of size n from a finite sample space is itself uniform. Suppose that the sample space consists of the 4 clergymen, A, B, C, and D, in a small town, from which 2 are to be chosen at random for a community Thanksgiving Day service. The total number of possible combinations is $\binom{4}{2} = 6$, which we could list as follows: AB, AC, AD, BC, BD, and CD. Since each has the same probability of being drawn, the distribution of subsets is uniform, with

$$f(x; 6) = \tfrac{1}{6}, \qquad x = 1, 2, \ldots, 6.$$

Therefore,

$$\Pr(B \text{ and } D \text{ are chosen}) = \Pr(X = 5) = f(5; 6) = \tfrac{1}{6}.$$

In general the value of k in the formula of the uniform distribution is given by $\binom{N}{n}$ when selecting a subset of size n from a finite sample space of size N.

Example 5.3 Find the uniform distribution for the subsets of months of size 3.

Solution. Since there are 12 possible months, we may choose 3 at random in $\binom{12}{3} = 220$ ways. Numbering these subsets from 1 to 220 the probability distribution is given by

$$f(x; 220) = \tfrac{1}{220}, \qquad x = 1, 2, \ldots, 220.$$

Thus the probability of choosing subset number 5 is

$$f(5; 220) = \tfrac{1}{220}.$$

5.3 Binomial Distribution

An experiment often consists of repeated trials, each with two possible outcomes, which may be labeled "success" or "failure." This is true in the flipping of a coin 5 times, where each trial may result in a head or a tail. We may choose to define either outcome as a success. It is also true if 5 cards are drawn in succession from an ordinary deck and each trial is labeled "success" or "failure," depending on whether the card is red or black. If each card is replaced and the deck shuffled before the next drawing, then the two experiments described above have similar properties, in that the repeated trials are independent and the probability of a success remains constant, $\tfrac{1}{2}$, from trial to trial. Experiments of this type are known as *binomial experiments*. Observe in the card-drawing example that the probabilities of a success for the repeated trials change if the cards are not replaced. That is, the probability of selecting a red card on the first draw is $\frac{1}{2}$, but on the second draw it is a conditional probability having a value of $\frac{26}{51}$ or $\frac{25}{51}$, depending on the color that occurred on the first draw. This then would no longer be considered a binomial experiment.

A binomial experiment is one that possesses the following properties:

1. The experiment consists of n repeated trials.
2. Each trial results in an outcome that may be classified as a success or a failure.
3. The probability of a success, denoted by p, remains constant from trial to trial.
4. The repeated trials are independent.

Consider the binomial experiment where a coin is tossed 3 times and a head is designated a success. The number of successes is a random variable X assuming integral values from 0 through 3. The 8 possible outcomes and the corresponding values of X are as follows:

Outcome	x
TTT	0
THT	1
TTH	1
HTT	1
THH	2
HTH	2
HHT	2
HHH	3

Since the trials are independent with constant probability of success equal to $\frac{1}{2}$, the $\Pr(HTH) = \Pr(H)\Pr(T)\Pr(H) = (\frac{1}{2})(\frac{1}{2})(\frac{1}{2}) = \frac{1}{8}$. Similarly, each of the other possible outcomes occur with probability $\frac{1}{8}$. The probability distribution of X is therefore given by

x	0	1	2	3
$\Pr(X = x)$	$\frac{1}{8}$	$\frac{3}{8}$	$\frac{3}{8}$	$\frac{1}{8}$

or by the formula

$$f(x) = \frac{\binom{3}{x}}{8}, \qquad x = 0, 1, 2, 3.$$

DEFINITION *The number X of successes in n trials of a binomial experiment is called a* **binomial random variable.**

The probability distribution of the binomial variable X is called the *binomial distribution* and will be denoted by $b(x; n, p)$, since its values depend on the number of trials and the probability of a success on a given trial. Thus, for the 3 tosses of a coin, the probability distribution should be rewritten

$$b(x; 3, \tfrac{1}{2}) = \frac{\binom{3}{x}}{8}, \qquad x = 0, 1, 2, 3.$$

Let us now generalize the above illustration to yield a formula for $b(x; n, p)$, that is, we wish to find a formula that gives the probability of x successes in n trials for a binomial experiment. First, consider the probability of x successes and $n - x$ failures in a specified order. Since the trials are independent, we can multiply all the probabilities corresponding to the different outcomes. Each success occurs with probability p and each failure with probability $q = 1 - p$. Therefore the probability for the specified order is $p^x q^{n-x}$. We must now determine the total number of sample points in the experiment that have x successes and $n - x$ failures. This number is equal to the number of partitions of n outcomes into two groups with x in one group and $n - x$ in the other and is given by $\binom{n}{x}$. Because these partitions are mutually exclusive, we add the probabilities of all the different partitions to obtain the general formula, or simply multiply $p^x q^{n-x}$ by $\binom{n}{x}$.

BINOMIAL DISTRIBUTION *If a binomial trial can result in a success with probability p and a failure with probability $q = 1 - p$, then the probability distribution of the binomial random variable X, the number of successes in n independent trials, is*

$$b(x; n, p) = \binom{n}{x} p^x q^{n-x}, \qquad x = 0, 1, 2, \ldots, n.$$

Note that when $n = 3$ and $p = \tfrac{1}{2}$,

$$b\left(x; 3, \frac{1}{2}\right) = \binom{3}{x}\left(\frac{1}{2}\right)^x \left(\frac{1}{2}\right)^{3-x} = \frac{\binom{3}{x}}{8},$$

which agrees with the answer above for the number of heads when three coins are tossed.

Example 5.4 Find the probability of obtaining exactly three 2s if an ordinary die is tossed 5 times.

Solution. The probability of a success on each of the 5 independent trials is $\frac{1}{6}$ and the probability of a failure is $\frac{5}{6}$. The outcome of 2 is considered a success here. Therefore,

$N = \#\text{ trials}$

$\#= \#\text{ of successes desired}$

$P = \text{prob success on a given trial}$

$q = \text{failure} = 1-P$

$\binom{m}{x} P^x q^{N-x}$

$b\left(3; 4, \frac{1}{6}\right) = \binom{5}{3}\left(\frac{1}{6}\right)^3\left(\frac{5}{6}\right)^2$

$$= \frac{5!}{3!2!} \cdot \frac{5^2}{6^5}$$

$$= 0.032.$$

Example 5.5 A basketball player hits on 75% of his shots from the free-throw line. What is the probability that he makes exactly 2 of his next 4 free shots?

Solution. Assuming the shots are independent and $p = \frac{3}{4}$ for each of the 4 shots, then

$$b\left(2; 4, \frac{3}{4}\right) = \binom{4}{2}\left(\frac{3}{4}\right)^2\left(\frac{1}{4}\right)^2$$

$$= \frac{4!}{2!2!} \cdot \frac{3^2}{4^4}$$

$$= 0.211.$$

The binomial distribution derives its name from the fact that the $(n + 1)$ terms in the binomial expansion of $(q + p)^n$ correspond to the values of $b(x; n, p)$ for $x = 0, 1, 2, \ldots, n$. That is,

$$(q + p)^n = \binom{n}{0} q^n + \binom{n}{1} pq^{n-1} + \binom{n}{2} p^2q^{n-2} + \cdots + \binom{n}{n} p^n$$

$$= b(0; n, p) + b(1; n, p) + b(2; n, p) + \cdots + b(n; n, p).$$

Since $p + q = 1$, we see that $\sum_{x=0}^{n} b(x; n, p) = 1$, a condition that must hold for any probability distribution.

Frequently we are interested in problems in which it is necessary to find $\Pr(X < r)$ or $\Pr(a \leq X \leq b)$. Fortunately binomial sums $\sum_{x=0}^{r} b(x; n, p)$ are available and are given in Table A.2 of the Appendix for samples of size $n = 5, 10, 15, 20$, and selected values of p from 0.1 to 0.9. We illustrate the use of Table A.2 with the following example:

Example 5.6 The probability that a patient recovers from a rare blood disease is 0.4. If 15 people are known to have contracted this disease, what is the probability that (a) at least 10 survive, (b) from 3 to 8 survive, and (c) exactly 5 survive?

Solution. (a) Let X be the number of people that survive. Then

$$\Pr(X \geq 10) = 1 - \Pr(X < 10)$$

$$= 1 - \sum_{x=0}^{9} b(x; 15, 0.4)$$

$$= 1 - 0.9662$$

$$= 0.0338.$$

(b) $\Pr(3 \leq X \leq 8) = \sum_{x=3}^{8} b(x; 15, 0.4)$

$$= \sum_{x=0}^{8} b(x; 15, 0.4) - \sum_{x=0}^{2} b(x; 15, 0.4)$$

$$= 0.9050 - 0.0271$$

$$= 0.8779.$$

(c) $\Pr(x = 5) = b(5; 15, 0.4)$

$$= \sum_{x=0}^{5} b(x; 15, 0.4) - \sum_{x=0}^{4} b(x; 15, 0.4)$$

$$= 0.4032 - 0.2173$$

$$= 0.1859.$$

THEOREM 5.1 *The mean and variance of the binomial distribution* $b(x; n, p)$ *are*

$$\mu = np \quad and \quad \sigma^2 = npq.$$

Proof. Let the outcome on the jth trial be represented by the random variable I_j, which assumes the values 0 and 1 with probabilities q and p, respectively. This is called an *indicator variable*, since $I_j = 0$ indicates a failure and $I_j = 1$ indicates a success.

Therefore, in a binomial experiment the number of successes can be written as the sum of the n independent indicator variables. Hence

$$X = I_1 + I_2 + \cdots + I_n.$$

The mean of any I_j is $E(I_j) = 0 \cdot q + 1 \cdot p = p$. Therefore, using the corollary of Theorem 4.7, the mean of the binomial distribution is

$$\mu = E(X) = E(I_1) + E(I_2) + \cdots + E(I_n)$$

$$= \underbrace{p + p + \cdots + p}_{n \text{ terms}}$$

$$= np.$$

The variance of any I_j is given by

$$\sigma_{I_j}^2 = E[(I_j - p)^2] = E(I_j^2) - p^2$$
$$= (0)^2 q + (1)^2 p - p^2$$
$$= p(1 - p) = pq.$$

Therefore, by Corollary 1 of Theorem 4.14, the variance of the binomial distribution is

$$\sigma_X^2 = \sigma_{I_1}^2 + \sigma_{I_2}^2 + \cdots + \sigma_{I_n}^2$$

$$= \underbrace{pq + pq + \cdots + pq}_{n \text{ terms}}$$

$$= npq.$$

Example 5.7 Using Chebyshev's theorem, find and interpret the interval $\mu \pm 2\sigma$ for Example 5.6.

Solution. Since Example 5.6 was a binomial experiment with $n = 15$ and $p = 0.4$, by Theorem 5.1 we have

$$\mu = (15)(0.4) = 6 \quad \text{and} \quad \sigma^2 = (15)(0.4)(0.6) = 3.6.$$

Taking the square root of 3.6, we find that $\sigma = 1.897$. Hence the required interval is $6 \pm (2)(1.897)$, or from 2.206 to 9.794. Chebyshev's theorem states that the recovery rate of 15 patients subjected to the given disease has a probability of at least $\frac{3}{4}$ of falling between 2.206 and 9.794.

The binomial experiment becomes a *multinomial experiment* if we let each trial have more than two possible outcomes. Hence the outcomes, when a pair of dice is tossed, might be recorded as a matching pair, a total of 7 or 11, or neither of these cases, and therefore constitutes a multinomial experiment. The drawing of a card from a deck *with replacement* is also a multinomial experiment if the 4 suits are the outcomes of interest.

In general, if a given trial can result in any one of k possible outcomes $E_1, E_2, \ldots, E_k$, with probabilities $p_1, p_2, \ldots, p_k$, then the *multinomial distribution* will give the probability that E_1 occurs x_1 times, E_2 occurs x_2 times, $\ldots$,

E_k occurs x_k times in n independent trials, where $x_1 + x_2 + \cdots + x_k = n$. We shall denote this joint probability distribution by $f(x_1, x_2, \ldots, x_k; p_1, p_2, \ldots, p_k, n)$. Clearly $p_1 + p_2 + \cdots + p_k = 1$, since the result of each trial must be one of the k possible outcomes.

To derive the general formula, we proceed as in the binomial case. Since the trials are independent, any specified order yielding x_1 outcomes for E_1, x_2 for $E_2, \ldots, x_k$ for E_k will occur with probability $p_1^{x_1} p_2^{x_2} \cdots p_k^{x_k}$. The total number of orders yielding similar outcomes for the n trials is equal to the number of partitions of n items into k groups with x_1 in the first group, x_2 in the second group, $\ldots, x_k$ in the kth group. This can be done in

$$\binom{n}{x_1, x_2, \ldots, x_k} = \frac{n!}{x_1! x_2! \cdots x_k!}$$

ways. Since the partitions are mutually exclusive and occur with equal probability, we obtain the multinomial distribution by multiplying the probability for a specified order by the total number of partitions.

MULTINOMIAL DISTRIBUTION *If a given trial can result in the k outcomes $E_1, E_2, \ldots, E_k$, with probabilities $p_1, p_2, \ldots, p_k$, then the probability distribution of the random variables $X_1, X_2, \ldots, X_k$, representing the number of occurrences for $E_1, E_2, \ldots, E_k$ in n independent trials, is*

$$f(x_1, x_2, \ldots, x_k; p_1, p_2, \ldots, p_k, n) = \binom{n}{x_1, x_2, \ldots, x_k} p_1^{x_1} p_2^{x_2} \cdots p_k^{x_k},$$

with $\displaystyle\sum_{i=1}^{k} x_i = n$ and $\displaystyle\sum_{i=1}^{k} p_i = 1$.

The multinomial distribution derives its name from the fact that the terms of the multinomial expansion of $(p_1 + p_2 + \cdots + p_k)^n$ correspond to all the possible values of $f(x_1, x_2, \ldots, x_k; p_1, p_2, \ldots, p_k, n)$.

Example 5.8 If a pair of dice is tossed 6 times, what is the probability of obtaining a total of 7 or 11 twice, a matching pair once, and any other combination 3 times?

Solution. We list the following possible events,

E_1: A total of 7 or 11 occurs,

E_2: A matching pair occurs,

E_3: Neither a pair nor a total of 7 or 11 occurs.

The corresponding probabilities for a given trial are $p_1 = \frac{2}{9}$,

$p_2 = \frac{1}{6}$, and $p_3 = \frac{11}{18}$. These values remain constant for all 6 trials. Using the multinomial distribution with $x_1 = 2$, $x_2 = 1$, and $x_3 = 3$, we obtain the required probability:

$$f\left(2, 1, 3; \frac{2}{9}, \frac{1}{6}, \frac{11}{18}, 6\right) = \binom{6}{2, 1, 3}\left(\frac{2}{9}\right)^2\left(\frac{1}{6}\right)^1\left(\frac{11}{18}\right)^3$$

$$= \frac{6!}{2!1!3!} \cdot \frac{2^2}{9^2} \cdot \frac{1}{6} \cdot \frac{11^3}{18^3}$$

$$= 0.1127.$$

5.4 Hypergeometric Distribution

In Section 5.3 we saw that the binomial distribution did not apply if we wished to find the probability of observing 3 red cards in 5 draws from an ordinary deck of 52 playing cards, unless each card is replaced and the deck reshuffled before the next drawing is made. To solve the problem of sampling without replacement, let us restate the problem. If 5 cards are drawn at random, we are interested in the probability of selecting 3 red cards from the 26 available and 2 black cards from the 26 black cards available in the deck. There are $\binom{26}{3}$ ways of selecting 3 red cards; for each of these ways we can choose 2 black cards in $\binom{26}{2}$ ways. Therefore the total number of ways to select 3 red and 2 black cards in 5 draws is the product $\binom{26}{3}\binom{26}{2}$. The total number of ways to select any 5 cards from the 52 that are available is $\binom{52}{5}$. Hence the probability of selecting 5 cards without replacement, of which 3 are red and 2 are black, is given by

$$\frac{\binom{26}{3}\binom{26}{2}}{\binom{52}{5}} = \frac{\dfrac{26!}{3!23!} \cdot \dfrac{26!}{2!24!}}{\dfrac{52!}{5!47!}} = 0.3251.$$

In general we are interested in the probability of selecting x successes from the k items labeled "success" and $n - x$ failures from the $N - k$ items labeled "failures," when a random sample of size n is selected from a finite population of size N. This is known as a *hypergeometric experiment*.

A hypergeometric experiment is one that possesses the following two properties:

A random sample of size n is selected from a population of N items.

2. k of the N items may be classified as successes and $N - k$ classified as failures.

> DEFINITION *The number X of successes in a hypergeometric experiment is called a* hypergeometric random variable.

The probability distribution of the hypergeometric variable X is called the *hypergeometric distribution* and will be denoted by $h(x; N, n, k)$, since its values depend on the number of successes k in the set N from which we select n items.

Example 5.9 A committee of size 5 is to be selected at random from 3 women and 5 men. Find the probability distribution for the number of women on the committee.

Solution. Let the random variable X be the number of women on the committee. The two properties of a hypergeometric experiment are satisfied. Hence

$$\Pr(X = 0) = h(0; 8, 5, 3) = \frac{\binom{3}{0}\binom{5}{5}}{\binom{8}{5}} = \frac{1}{56},$$

$$\Pr(X = 1) = h(1; 8, 5, 3) = \frac{\binom{3}{1}\binom{5}{4}}{\binom{8}{5}} = \frac{15}{56},$$

$$\Pr(X = 2) = h(2; 8, 5, 3) = \frac{\binom{3}{2}\binom{5}{3}}{\binom{8}{5}} = \frac{30}{56},$$

$$\Pr(X = 3) = h(3; 8, 5, 3) = \frac{\binom{3}{3}\binom{5}{2}}{\binom{8}{5}} = \frac{10}{56}.$$

In tabular form the hypergeometric distribution of X is as follows:

x	0	1	2	3
$\Pr(X = x)$	$\frac{1}{56}$	$\frac{15}{56}$	$\frac{30}{56}$	$\frac{10}{56}$

It is not difficult to see that the probability distribution can be given by the formula

$$h(x; 8, 5, 3) = \frac{\binom{3}{x}\binom{5}{5-x}}{\binom{8}{5}}, \qquad x = 0, 1, 2, 3.$$

Let us now generalize the above example to find a formula for $h(x; N, n, k)$. The total number of samples of size n chosen from N items is $\binom{N}{n}$. These samples are assumed to be equally likely. There are $\binom{k}{x}$ ways of selecting x successes from the k that are available, and for each of these ways we can choose the $n - x$ failures in $\binom{N-k}{n-x}$ ways. Thus the total number of favorable samples among the $\binom{N}{n}$ possible samples is given by $\binom{k}{x}\binom{N-k}{n-x}$. Hence we have the following definition.

HYPERGEOMETRIC DISTRIBUTION *If a population of size N contains k items labeled "success" and N − k items labeled "failure," then the probability distribution of the hypergeometric random variable X, the number of successes in a random sample of size n, is*

$$h(x; N, n, k) = \frac{\binom{k}{x}\binom{N-k}{n-x}}{\binom{N}{n}}, \qquad x = 0, 1, 2, \ldots, n.$$

Example 5.10 If 5 cards are dealt from a standard deck of 52 playing cards, what is the probability that 3 will be hearts?

Solution. Using the hypergeometric distribution with $n = 5$, $N = 52$, $k = 13$, and $x = 3$, we find the probability of receiving 3 hearts to be

$$h(3; 52, 5, 13) = \frac{\binom{13}{3}\binom{39}{2}}{\binom{52}{5}} = 0.0815.$$

To find the mean of the hypergeometric distribution, we can again write X as the sum of n indicator variables that now are no longer independent. Hence

$$X = I_1 + I_2 + \cdots + I_n,$$

where I_j assumes a value of 1 or 0, depending on whether we have a success or a failure on the jth draw. Now

$$E(I_1) = 0\Pr(I_1 = 0) + 1\Pr(I_1 = 1)$$

$$= 0\left(\frac{N-k}{N}\right) + 1\left(\frac{k}{N}\right)$$

$$= \frac{k}{N},$$

$$E(I_2) = 0[\Pr(I_1 = 0, I_2 = 0) + \Pr(I_1 = 1, I_2 = 0)]$$
$$\qquad + 1[\Pr(I_1 = 0, I_2 = 1) + \Pr(I_1 = 1, I_2 = 1)]$$

$$= 0\left(\frac{N-k}{N} \cdot \frac{N-k-1}{N-1} + \frac{k}{N} \cdot \frac{N-k}{N-1}\right)$$

$$\quad + 1\left(\frac{N-k}{N} \cdot \frac{k}{N-1} + \frac{k}{N} \cdot \frac{k-1}{N-1}\right)$$

$$= \frac{k(N-k+k-1)}{N(N-1)}$$

$$= \frac{k}{N},$$

$$\vdots$$

$$E(I_n) = \frac{k}{N}.$$

Therefore,

$$\mu = E(X) = \underbrace{\frac{k}{N} + \frac{k}{N} + \cdots + \frac{k}{N}}_{n \text{ terms}}$$

$$= \frac{nk}{N}.$$

Since the indicator variables are no longer independent, the problem of finding σ_X^2 becomes much more complex as a result of the various covariance terms. Therefore we omit the proof for the variance and merely include the result in the following theorem.

THEOREM 5.2 *The mean and variance of the hypergeometric distribution* $h(x; N, n, k)$ *are*

$$\mu = \frac{nk}{N},$$

$$\sigma^2 = \frac{N - n}{N - 1} \cdot n \cdot \frac{k}{N}\left(1 - \frac{k}{N}\right).$$

Example 5.11 Using Chebyshev's theorem, find and interpret the interval $\mu \pm 2\sigma$ for Example 5.10.

Solution. Since Example 5.10 was a hypergeometric experiment with $N = 52$, $n = 5$, and $k = 13$, then by Theorem 5.2 we have

$$\mu = \frac{(5)(13)}{52} = \frac{5}{4} = 1.25,$$

$$\sigma^2 = \left(\frac{52 - 5}{51}\right)(5)\left(\frac{13}{52}\right)\left(1 - \frac{13}{52}\right)$$

$$= 0.8640.$$

Taking the square root of 0.8640, we find $\sigma = 0.93$. Hence the required interval is $1.25 \pm (2)(0.93)$, or from -0.61 to 3.11. Chebyshev's theorem states that the number of hearts obtained when 5 cards are dealt from an ordinary deck of 52 playing cards has a probability of at least $\frac{3}{4}$ of falling between -0.61 and 3.11, that is, at least $\frac{3}{4}$ of the time the 5 cards include less than 4 hearts.

If n is small relative to N, the probability for each drawing will change only slightly. Hence we essentially have a binomial experiment and can approximate the hypergeometric distribution by using the binomial distribution with $p = k/N$. The mean and variance can also be approximated by the formulas

$$\mu = np = \frac{nk}{N},$$

$$\sigma^2 = npq = b \cdot \frac{k}{N}\left(1 - \frac{k}{N}\right).$$

Comparing these formulas with those of Theorem 5.2, we see that the mean is the same, while the variance differs by a correction factor of $(N - n)/)N - 1)$. This is negligible when n is small relative to N.

Example 5.12 The telephone company reports that among 5000 telephones installed in a new subdivision 4000 are nonblack. If 10 people are called at random, what is the probability that exactly 3 will be talking on black telephones?

Solution. Since the population size $N = 5000$ is large relative to the sample size $n = 10$, we shall approximate the desired probability by using the binomial distribution. The probability of calling someone with a black telephone is 0.2. Therefore the probability that exactly 3 people are called who have black telephones is

$$h(3; 5000, 10, 1000) \simeq b(3; 10, 0.2)$$

$$= \sum_{x=0}^{3} b(x; 10, 0.2) - \sum_{x=0}^{2} b(x; 10, 0.2)$$

$$= 0.8791 - 0.6778$$

$$= 0.2013.$$

The hypergeometric distribution can be extended to treat the case in which the population can be partitioned into k cells $A_1, A_2, \ldots, A_k$, with a_1 elements in the first cell, a_2 elements in the second cell, $\ldots$, a_k elements in the kth cell. We are now interested in the probability that a random sample of size n yields x_1 elements from A_1, x_2 elements from $A_2, \ldots$, and x_k elements from A_k. Let us represent this probability by $f(x_1, x_2, \ldots, x_k; a_1, a_2, \ldots, a_k, N, n)$.

To obtain a general formula, we note that the total number of samples that can be chosen of size n from a population of size N is still $\binom{N}{n}$. There are $\binom{a_1}{x_1}$ ways of selecting x_1 items from the items in A_1, and for each of these we can choose x_2 items from the items in A_2 in $\binom{a_2}{x_2}$ ways. Therefore we can select x_1 items from A_1 and x_2 items from A_2 in $\binom{a_1}{x_1}\binom{a_2}{x_2}$ ways. Continuing in this way, we can select all n items consisting of x_1 from A_1, x_2 from $A_2, \ldots$, and x_k from A_k in $\binom{a_1}{x_1}\binom{a_2}{x_2}\cdots\binom{a_k}{x_k}$ ways. The required probability distribution is now defined as follows.

EXTENSION OF THE HYPERGEOMETRIC DISTRIBUTION *If a population of size N can be partitioned into the k cells $A_1, A_2, \ldots, A_k$, with $a_1, a_2, \ldots, a_k$ elements, respectively, then the probability distribution of the random variables $X_1, X_2, \ldots, X_k$, representing the number of elements selected from $A_1, A_2, \ldots, A_k$ in a random sample of size n, is*

$$f(x_1, x_2, \ldots, x_k; a_1, a_2, \ldots, a_k, N, n) = \frac{\binom{a_1}{x_1}\binom{a_2}{x_2}\cdots\binom{a_k}{x_k}}{\binom{N}{n}},$$

with $\sum_{i=1}^{k} x_i = n$ *and* $\sum_{i=1}^{k} a_i = N.$

Example 5.13 A gardener wishes to landscape a piece of property by planting flowers across the front and back of the house. From a box containing 3 tulip bulbs, 4 daffodil bulbs, and 3 hyacinth bulbs he selects 5 at random to be planted at the front of the house, and the remaining 5 are planted at the rear of the house. What is the probability that 1 tulip plant, 2 daffodil plants, and 2 hyacinth plants bloom at the front of the house?

Solution. Using the extension of the hypergeometric distribution with $x_1 = 1, x_2 = 2, x_3 = 2, a_1 = 3, a_2 = 4, a_3 = 3, n = 10,$ and $n = 5$, we find that the desired probability is

$$f(1, 2, 2; 3, 4, 3, 10, 5) = \frac{\binom{3}{1}\binom{4}{2}\binom{3}{2}}{\binom{10}{5}} = \frac{3}{14}. \approx 21\%$$

5.5 Poisson Distribution

Experiments yielding numerical values of a random variable X, the number of successes occurring during a given time interval or in a specified region, are often called *Poisson experiments*. The given time interval may be of any length, such as a minute, a day, a week, a month, or even a year. Hence a Poisson experiment might generate observations for the random variable X representing the number of telephone calls per hour received by an office, the number of days school is closed due to snow during the winter, or the number of postponed games due to rain during a baseball season. The specified region could be a line segment, an area, a volume, or perhaps a piece of material. In this case X might represent the number of field mice per acre, the number of bacteria in a given culture, or the number of typing errors per page.

A Poisson experiment is one that possesses the following properties:

1. The average number of successes, μ, occurring in the given time interval or specified region is known.
2. The probability that a single success will occur during a very short time interval or in a small region is proportional to the length of the time interval or the size of the region and does not depend on the number of successes occurring outside this time interval or region.
3. The probability that more than one success will occur in such a short time interval or falling in such a small region is negligible.

DEFINITION *The number X of successes in a Poisson experiment is called a* Poisson random variable.

The probability distribution of the Poisson variable X is called the *Poisson distribution* and will be denoted by $p(x;\mu)$, since its values depend only on μ, the average number of successes occurring in the given time interval or specified region. The derivation of the formula for $p(x;\mu)$, based on the properties for a Poisson experiment listed above, is beyond the scope of this text. We list the result in the following definition.

POISSON DISTRIBUTION *The probability distribution of the Poisson random variable X, representing the number of successes occurring in a given time interval or specified region, is*

$$p(x;\mu) = \frac{e^{-\mu}\mu^x}{x!}, \qquad x = 0, 1, 2, \ldots,$$

where μ is the average number of successes occurring in the given time interval or specified region and $e = 2.71828\ldots$.

Table A.3 contains Poisson probability sums $\sum_{x=0}^{r} p(x;\mu)$ for a few selected values of μ ranging from 0.1 to 18. We illustrate the use of this table with the following two examples.

Example 5.14 The average number of days school is closed due to snow during the winter in a certain city in the eastern part of United States is 4. What is the probability that the schools in this city will close for 6 days during a winter?

Solution. Using the Poisson distribution with $x = 6$ and $\mu = 4$, we find from Table A.3 that

$$p(6;4) = \frac{e^{-4}4^6}{6!} = \sum_{x=0}^{6} p(x;4) - \sum_{x=0}^{5} p(x;4)$$

$$= 0.8893 - 0.7851 = 0.1042.$$

Example 5.15 The average number of field mice per acre in a 5-acre wheat field is estimated to be 10. Find the probability that a given acre contains more than 15 mice.

Solution. Let X be the number of field mice per acre. Then using Table A.3 we have

$$\Pr(X > 15) = 1 - \Pr(X \le 15)$$

$$= 1 - \sum_{x=0}^{15} p(x;10)$$

$$= 1 - 0.9513$$

$$= 0.0487.$$

The variance of the Poisson distribution can be shown to be equal to the mean. Thus in Example 5.14, where $\mu = 4$, we also have $\sigma^2 = 4$ and hence $\sigma = 2$. Using Chebyshev's theorem, we can state that our random variable has a probability of at least $\frac{3}{4}$ of falling in the interval $\mu \pm 2\sigma = 4 \pm (2)(2)$, or from 0 to 8. Therefore we conclude that at least $\frac{3}{4}$ of the time the schools of the given city will be closed anywhere from 0 to 8 days during the winter season.

The Poisson and binomial distributions have histograms with approximately the same shape when n is large and p is close to zero. Hence if these two conditions hold, the Poisson distribution, with $\mu = np$, can be used to approximate binomial probabilities. If p is close to 1, we can interchange what we have defined to be a success and a failure, thereby changing p to a value close to zero.

Example 5.16 Suppose that on the average 1 person in every 1000 is an alcoholic. Find the probability that a random sample of 8000 people will yield fewer than 7 alcoholics.

Solution. This is essentially a binomial experiment with $n = 8000$ and $p = 0.001$. Since p is very close to zero and n is quite large, we shall approximate with the Poisson distribution using $\mu = (8000)(0.001) = 8$. Hence if X represents the number of alcoholics, we have

$$\Pr(X < 7) = \sum_{x=0}^{6} b(x; 8000, 0.001)$$

$$\simeq \sum_{x=0}^{6} p(x; 8)$$

$$= 0.3134.$$

5.6 Negative Binomial Distribution

Let us consider an experiment in which the properties are the same as those listed for a binomial experiment, with the exception that the trials will be repeated until a *fixed* number of successes occur. Therefore, instead of finding the probability of x successes in n trials, where n is fixed, we are now interested in the probability that the kth success occurs on the xth trial. Experiments of this kind are called *negative binomial experiments*.

As an illustration consider a basketball player who hits on 60% of his shots from the floor. We are interested in finding the probability that his fifth basket occurs on his seventh shot. A shot is considered a success if it results in a basket. Designating a success by S and a failure by F, a possible order of achieving the desired result is *SFSSSFS*, which occurs with probability $(0.6)(0.4)(0.6)(0.6)(0.6)(0.4)(0.6) = (0.6)^5(0.4)^2$. We could list all possible orders by rearranging the Fs and Ss, except for the last outcome, which must be the fifth success. The total number of possible orders is equal to the number of partitions of the first 6 trials into two groups, with the 2 failures assigned to the one group and the 4 successes assigned to the other group. This can be done in

$$\binom{6}{4} = 15$$

mutually exclusive ways. Hence if X represents the outcome on which the fifth basket is made, then

$$\Pr(X = 7) = \binom{6}{4}(0.6)^5(0.4)^2 = 0.1866.$$

DEFINITION *The number X of trials to produce k successes in a negative binomial experiment is called a* negative binomial random variable.

The probability distribution of the negative binomial variable X is called the *negative binomial distribution* and will be denoted by $b^*(x; k, p)$, since its values depend on the number of successes desired and the probability of a success on a given trial. To obtain the general formula for $b^*(x; k, p)$, consider the probability of a success on the xth trial preceded by $k - 1$ successes and $x - k$ failures in some specified order. Since the trials are independent, we can multiply all the probabilities corresponding to each desired outcome. Each success occurs with probability p and each failure with probability $q = 1 - p$. Therefore the probability for the specified order, ending in a success, is $p^{k-1}q^{x-k}p = p^k q^{x-k}$. The total number of sample points in the experiment ending in a success, after the occurrence of $k - 1$ successes and $x - k$ failures in any order, is equal to the number of partitions of $x - 1$ trials into two groups with $k - 1$ successes corresponding to one group and $x - k$ failures corresponding to the other group. This number is given by the term $\binom{x - 1}{k - 1}$, all mutually exclusive and occurring with equal probability $p^k q^{x-k}$. We obtain the general formula by multiplying $p^k q^{x-k}$ by $\binom{x - 1}{k - 1}$.

NEGATIVE BINOMIAL DISTRIBUTION *If repeated independent trials can result in a success with probability p and a failure with probability* $q = 1 - p$, *then the probability distribution of the random variable X, the number of the trial on which the kth success occurs, is given by*

$$b^*(x; k, p) = \binom{x - 1}{k - 1} p^k q^{x-k}, \qquad x = k, k + 1, k + 2, \dots .$$

Example 5.17 Find the probability that a person tossing 3 coins will get either all heads or all tails for the second time on the fifth toss.

Solution. Using the negative binomial distribution with $x = 5, k = 2$, and $p = \frac{1}{4}$, we have

$$b^*\left(5; 2, \frac{1}{4}\right) = \binom{4}{1}\left(\frac{1}{4}\right)^2\left(\frac{3}{4}\right)^3$$

$$= \frac{4!}{1!\,3!} \cdot \frac{3^3}{4^5}$$

$$= \tfrac{27}{256}.$$

The negative binomial distribution derives its name from the fact that each term in the expansion of $p^k(1 - q)^{-k}$ corresponds to the values of $b^*(x; k, p)$ for $x = k, k + 1, k + 2, \dots .$

If we consider the special case of the negative binomial distribution where $k = 1$, we have a probability distribution for the number of trials required for a single success. An example would be the tossing of a coin until a head occurs. We might be interested in the probability that the first head occurs on the fourth toss. The negative binomial distribution reduces to the form $b^*(x; 1, p) = pq^{x-1}$, $x = 1, 2, 3, \ldots$. Since the successive terms constitute a geometric progression, it is customary to refer to this special case as the *geometric distribution* and denote it by $g(x; p)$.

> **GEOMETRIC DISTRIBUTION** *If repeated independent trials can result in a success with probability p and a failure with probability $q = 1 - p$, then the probability distribution of the random variable X, the number of the trial on which the first success occurs, is given by*
>
> $$g(x; p) = pq^{x-1}, \qquad x = 1, 2, 3, \ldots .$$

Example 5.18 Find the probability that a person flipping a balanced coin requires 4 tosses to get a head.

Solution. Using the geometric distribution with $x = 4$ and $p = \frac{1}{2}$, we have

$$g(4; \tfrac{1}{2}) = \tfrac{1}{2}(\tfrac{1}{2})^3 = \tfrac{1}{16}.$$

EXERCISES

1. Find a formula for the distribution of the random variable X representing the number on a tag drawn at random from a box containing 10 tags numbered 1 to 10. What is the probability that the number drawn is less than 4?

2. A roulette wheel is divided into 25 sectors of equal area numbered from 1 to 25. Find a formula for the probability distribution of X, the number that occurs when the wheel is spun.

3. Find the uniform distribution for the random samples of committees of size 3 chosen from 6 students.

4. A baseball player's batting average is 0.250. What is the probability that he gets exactly 1 hit in his next 4 times at bat?

5. If we define the random variable X to be equal to the number of heads that occur when a balanced coin is flipped once, find the probability distribution of X. What two well-known distributions describe the values of X?

6. A multiple-choice quiz has 15 questions, each with 4 possible answers of which only 1 is the correct answer. What is the probability that sheer guesswork yields from 5 to 10 correct answers?

7. The probability that a patient recovers from a delicate heart operation is 0.9. What is the probability that exactly 5 of the next 7 patients having this operation survive?

8. A pheasant hunter brings down 75% of the birds he shoots at. What is the probability that at least 3 of the next 5 pheasants shot at will escape?

9. A survey of the residents in a United States city showed that 20% preferred a white telephone over any other color available. What is the probability that more than one half of the next 20 telephones installed in this city will be white?

10. One quarter of the female freshmen entering a Virginia college are out-of-state students. If the students are assigned at random to the dormitories, 3 to a room, what is the probability that in one room at most 2 of the 3 roommates are out-of-state students?

11. If X represents the number of pheasants that escape in Exercise 8 when 5 pheasants are shot at, find the probability distribution of X. Using Chebyshev's theorem, find and interpret the interval $\mu \pm 2\sigma$.

12. Suppose that airplane engines operate independently in flight and fail with probability $q = \frac{1}{5}$. Assuming that a plane makes a safe flight if at least one half of its engines run, determine whether a 4-engine plane or a 2-engine plane has the highest probability for a successful flight.

13. Repeat Exercise 12 for $q = \frac{1}{2}$ and $q = \frac{1}{3}$.

14. In Exercise 6, how many correct answers would you expect based on sheer guesswork? Using Chebyshev's theorem, find and interpret the interval $\mu \pm 2\sigma$.

15. If 64 coins are tossed a large number of times, how many heads can we expect on the average per toss? Using Chebyshev's theorem, between what two values would you expect the number of heads to fall at least $\frac{3}{4}$ of the time?

16. A card is drawn from a well-shuffled deck of 52 playing cards, the result recorded, and the card replaced. If the experiment is repeated 5 times, what is the probability of obtaining 2 spades and 1 heart?

17. The surface of a circular dart board has a small center circle called the bulls-eye and 20 pie-shaped regions numbered from 1 to 20. Each of the pie-shaped regions is further divided into 3 parts such that a person throwing a dart that lands on a specified number scores the value of the number, double the number, or triple the number, depending on which of the 3 parts the dart falls. If a person hits the bulls-eye with probability 0.01, hits a double with probability 0.10, a triple with probability 0.05, and misses the dart board with probability 0.02, what is the probability that 7 throws will result in no bulls-eyes, no triples, a double twice, and a complete miss once?

18. Find the probability of obtaining 2 ones, 1 two, 1 three, 2 fours, 3 fives, and 1 six in 10 rolls of a balanced die.

19. According to the theory of genetics a certain cross of guinea pigs will result in red, black, and white offspring in the ratio 8:4:4. Find the probability that among 8 such offspring 5 will be red, 2 black, and 1 white.

20. If 6 cards are dealt from an ordinary deck of 52 playing cards, what is the probability that (a) exactly 2 of them will be face cards, (b) at least 1 of them will be a queen?

21. A homeowner plants 6 bulbs selected at random from a box containing 5 tulip bulbs and 4 daffodil bulbs. What is the probability that he planted 2 daffodil bulbs and 4 tulip bulbs?

22. A random committee of size 3 is selected from 4 men and 2 women. Write a formula for the probability distribution of the random variable X representing the number of men on the committee. Find the $\Pr(2 \leq X \leq 3)$.

[handwritten: $N = 6$, $k = men$, $m = 3$]

[handwritten: Hyper]

23. From a lot of 10 missiles 4 are selected at random and fired. If the lot contains 3 defective missiles that will not fire, what is the probability that (a) all 4 will fire, (b) at most 2 will not fire?

24. In Exercise 23 how many defective missiles might we expect to be included among the 4 that are selected? Use Chebyshev's theorem to describe the variability of the number of defective missiles included when 4 are selected from several lots each of size 10 containing 3 defective missiles.

25. If a person is dealt 13 cards from an ordinary deck of 52 playing cards several times, how many hearts per hand can he expect? Between what two values would you expect the number of hearts to fall at least 75% of the time?

[handwritten: $N = 52$, $n = 13$, $k = hearts$]

[handwritten: hyper]

26. It is estimated that 4000 of the 10,000 voting residents of a town are against a new sales tax. If 15 eligible voters are selected at random and asked their opinion, what is the probability that at most 7 favor the new tax?

27. An annexation suit is being considered against a county subdivision of 1200 residents by a neighboring city. If the occupants of one half the residences object to being annexed, what is the probability that in a random sample of 10 at least 3 favor the annexation suit?

28. Find the probability of being dealt a bridge hand of 13 cards containing 5 spades, 2 hearts, 3 diamonds, and 3 clubs.

29. A foreign student club lists as its members 2 Canadians, 3 Japanese, 5 Italians, and 2 Germans. Find the probability that all nationalities are represented if a committee of 4 is selected at random.

30. An urn contains 3 green balls, 2 blue balls, and 4 red balls. In a random sample of 5 balls, find the probability that both blue balls and at least 1 red ball are selected.

31. On the average a certain intersection results in 3 traffic accidents per month. What is the probability that exactly 5 accidents will occur at this intersection in any given month?

32. A secretary makes 2 errors per page on the average. What is the probability that she makes (a) 4 or more errors on the next page she types, (b) no errors?

33. A certain area of the eastern United States is, on the average, hit by 6 hurricanes a year. Find the probability that in a given year (a) fewer than 4 hurricanes will hit this area, (b) anywhere from 6 to 8 hurricanes will hit the area.

34. A restaurant prepares a tossed salad containing on the average 5 vegetables. What is the probability that on a given day the salad contains more than 5 vegetables?

35. The probability that a person dies from a certain respiratory infection is 0.002. Find the probability that fewer than 5 of the next 2000 so infected will die.

36. Suppose that on the average 1 person in 1000 makes a numerical error in preparing his income tax return. If 10,000 forms are selected at random and examined, find the probability that 6, 7, or 8 of the forms will be in error.

37. Using Chebyshev's theorem, and and interpret the interval $\mu \pm 2\sigma$ for Exercise 35.

38. The probability that a person will install a black telephone in a residence is estimated to be 0.3. Find the probability that the tenth phone installed in a new subdivision is the fifth black phone.

39. A scientist inoculates several mice, one at a time, with a disease germ until he finds 2 that have contracted the disease. If the possibility of contracting the disease is $\frac{1}{6}$, what is the probability that 8 mice are required?

40. Find the probability that a person flipping a coin gets the third head on the seventh flip.

41. Three people toss a coin and the odd man pays for the coffee. If the coins all turn up the same, they are tossed again. Find the probability that fewer than 4 tosses are needed.

42. The probability that a student pilot passes the written test for his private pilot's license is 0.7. Find the probability that a person passes the test **(a)** on the third try, **(b)** before the fourth try.

NORMAL DISTRIBUTION CHAPTER **6**

6.1 Normal Curve

The most important continuous probability distribution in the entire field of statistics is the *normal distribution*. Its graph, called the *normal curve*, is the bell-shaped curve of Figure 6.1 that describes so many sets of data that occur in nature, industry, and research. In 1733 DeMoivre developed the mathematical equation of the normal curve. This provided a basis upon which much of the theory of inductive statistics is based. The normal distribution is often referred to as the *Gaussian distribution* in honor of Gauss (1777–1855), who also derived its equation from a study of errors in repeated measurements of the same quantity.

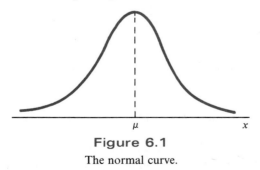

Figure 6.1
The normal curve.

A random variable X having the bell-shaped distribution of Figure 6. 1 is called a *normal random variable*. The mathematical equation for the probability distribution of the continuous normal variable depends upon the two parameters μ and σ, its mean and standard deviation. Hence we shall denote the density function of X by $n(x; \mu, \sigma)$.

NORMAL CURVE *If X is a normal random variable with mean μ and variance σ^2, then the equation of the normal curve is*

$$n(x; \mu, \sigma) = \frac{1}{\sqrt{2\pi}\sigma} e^{-\frac{1}{2}\left(\frac{x-\mu}{\sigma}\right)^2}, \qquad -\infty < x < \infty,$$

where $\pi = 3.14159\ldots$ and $e = 2.71828\ldots$.

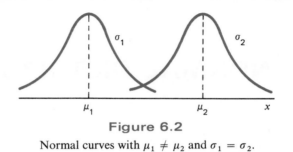

Figure 6.2

Normal curves with $\mu_1 \neq \mu_2$ and $\sigma_1 = \sigma_2$.

Once μ and σ are specified, the normal curve is completely determined. For example, if $\mu = 50$ and $\sigma = 5$, then the ordinates of $n(x; 50, 5)$ can easily be computed for various values of x and the curve drawn. In Figure 6.2 we have sketched two normal curves having the same standard deviation but different means. The two curves are identical in form but are centered at different positions along the horizontal axis.

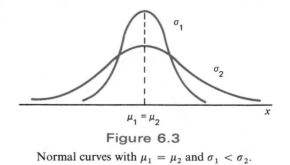

Figure 6.3

Normal curves with $\mu_1 = \mu_2$ and $\sigma_1 < \sigma_2$.

In Figure 6.3 we have sketched two normal curves with the same mean but different standard deviations. This time we see that the two curves are centered at exactly the same position on the horizontal axis, but the curve with the larger standard deviation is lower and spreads out farther. Remember that the area under a probability curve must be equal to 1, and therefore the more variable the set of observations, the lower and wider the corresponding curve will be.

Figure 6.4 shows the results of sketching two normal curves having different means and different standard deviations. Clearly they are centered at different positions on the horizontal axis and their shapes reflect the two different values of σ.

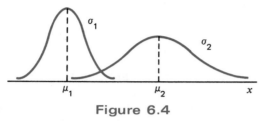

Figure 6.4

Normal curves with $\mu_1 \neq \mu_2$ and $\sigma_1 < \sigma_2$.

From an inspection of Figures 6.1 through 6.4 we list the following properties of the normal curve:

1. The *mode*, which is the point on the horizontal axis where the curve is a maximum, occurs at $x = \mu$.
2. The curve is symmetric about a vertical axis through the mean μ.
3. The normal curve approaches the horizontal axis asymptotically as we proceed in either direction away from the mean.
4. The total area under the curve and above the horizontal axis is equal to 1.

Many random variables have probability distributions that can be described adequately by the normal curve once μ and σ^2 are specified. In this chapter we shall assume these two parameters are known, perhaps from previous investigations. Later, in Chapter 7, we shall consider methods of estimating μ and σ^2 from the available experimental data.

6.2 Areas Under the Normal Curve

The curve of any continuous probability distribution or density function is constructed so that the area under the curve bounded by the two ordinates $x = x_1$ and $x = x_2$ equals the probability that the random variable X assumes a value between $x = x_1$ and $x = x_2$. Thus, for the normal curve in Figure 6.5, the $\Pr(x_1 < X < x_2)$ is represented by the area of the shaded region.

In Figures 6.2, 6.3, and 6.4 we saw how the normal curve is dependent upon the mean and the standard deviation of the distribution under investigation. The area under the curve between any two ordinates must then also depend upon the values of μ and σ. This is evident in Figure 6.6, where we

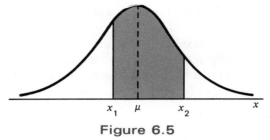

Figure 6.5

$\Pr(x_1 < X < x_2) =$ area of the shaded region.

have shaded regions corresponding to $\Pr(x_1 < X < x_2)$ for two curves with different means and variances. The $\Pr(x_1 < X < x_2)$, where X is the random variable describing distribution I, is indicated by the shaded region where the lines slope up to the right. If X is the random variable describing distribution II, then the $\Pr(x_1 < X < x_2)$ is given by the shaded region where the lines slope down to the right. Obviously the two shaded regions are different in size; therefore the probability associated with each distribution will be different.

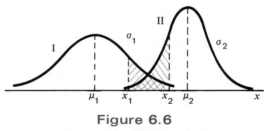

Figure 6.6

$\Pr(x_1 < X < x_2)$ for different normal curves.

It would be a hopeless task to attempt to set up separate tables of normal curve areas for every conceivable value of μ and σ. Yet we must use tables if we hope to avoid the use of integral calculus. Fortunately we are able to transform all the observations of any normal random variable X to a new set of observations of a normal random variable Z with mean zero and variance 1. This can be done by means of the transformation

$$Z = \frac{X - \mu}{\sigma}.$$

The mean of Z is zero, since

$$E(Z) = \frac{1}{\sigma} E(X - \mu) = \frac{1}{\sigma} (\mu - \mu) = 0,$$

and the variance is

$$\sigma_Z^2 = \sigma_{(X-\mu)/\sigma}^2 = \sigma_{X/\sigma}^2 = \frac{1}{\sigma^2}\sigma_X^2 = \frac{\sigma^2}{\sigma^2} = 1.$$

DEFINITION *The distribution of a normal random variable with mean zero and standard deviation equal to 1 is called a* standard normal *distribution.*

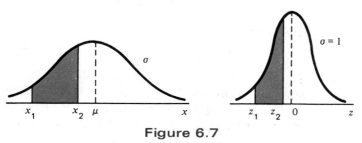

Figure 6.7

The original and transformed normal populations.

Whenever X is between the values $x = x_1$ and $x = x_2$, the random variable Z will fall between the corresponding values $z_1 = (x_1 - \mu)/\sigma$ and $z_2 = (x_2 - \mu)/\sigma$. The original and transformed distributions are illustrated in Figure 6.7. Since all the values of X falling between x_1 and x_2 have corresponding z values between z_1 and z_2, the area under the X curve between the ordinates $x = x_1$ and $x = x_2$ in Figure 6.7 equals the area under the Z curve between the transformed ordinates $z = z_1$ and $z = z_2$. Hence we have

$$\Pr(x_1 < X < x_2) = \Pr(z_1 < Z < z_2).$$

We have now reduced the required number of tables of normal-curve areas to one—that of the standard normal distribution. Table A.4 gives the area under the standard normal curve corresponding to $\Pr(Z < z)$ for values of z from -3.4 to 3.4. To illustrate the use of this table, let us find the probability that Z is less than 1.74. First we locate a value of z equal to 1.7 in the left column and then move across the row to the column under 0.04, where we read 0.9591. Therefore $\Pr(Z < 1.74) = 0.9591$.

Example 6.1 Given a normal distribution with $\mu = 50$ and $\sigma = 10$, find the probability that X assumes a value between 45 and 62.

Solution. The z values corresponding to $x_1 = 45$ and $x_2 = 62$ are

$$z_1 = \frac{45 - 50}{10} = -0.5,$$

$$z_2 = \frac{62 - 50}{10} = 1.2.$$

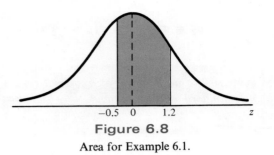

Figure 6.8

Area for Example 6.1.

Therefore,

$$\Pr(45 < X < 62) = \Pr(-0.5 < Z < 1.2).$$

The $\Pr(-0.5 < Z < 1.2)$ is given by the area of the shaded region in Figure 6.8. This area may be found by subtracting the area to the left of the ordinate $z = -0.5$ from the entire area to the left of $z = 1.2$. Using Table A.4, we have

$$\Pr(45 < X < 62) = \Pr(-0.5 < Z < 1.2)$$
$$= \Pr(Z < 1.2) - \Pr(Z < -0.5)$$
$$= 0.8849 - 0.3085$$
$$= 0.5764.$$

According to Chebyshev's theorem the probability that a random variable assumes a value within 2 standard deviations of the mean is at least $\frac{3}{4}$. If the random variable has a normal distribution, the z values corresponding to $x_1 = \mu - 2\sigma$ and $x_2 = \mu + 2\sigma$ are easily computed to be

$$z_1 = \frac{(\mu - 2\sigma) - \mu}{\sigma} = -2$$

and

$$z_2 = \frac{(\mu + 2\sigma) - \mu}{\sigma} = 2.$$

Hence

$$\Pr(\mu - 2\sigma < X < \mu + 2\sigma) = \Pr(-2 < Z < 2)$$
$$= \Pr(Z < 2) - \Pr(Z < -2)$$
$$= 0.9772 - 0.0228$$
$$= 0.9544,$$

which is a much stronger statement than that given by Chebyshev's theorem.

Some of the many problems in which the normal distribution is applicable are treated in the following examples. The use of the normal curve to approximate binomial probabilities will be considered in Section 6.3.

Example 6.2 A certain type of storage battery lasts on the average 3.0 years, with a standard deviation of 0.5 year. Assuming that the battery lives are normally distributed, find the probability that a given battery will last less than 2.3 years.

Solution. First construct a diagram such as Figure 6.9, showing the given distribution of battery lives and the desired area. To find the $\Pr(X < 2.3)$, we need to evaluate the area under the normal curve to the left of 2.3. This is accomplished by finding the area to the left of the corresponding z value. Hence we find

$$z = \frac{2.3 - 3}{0.5} = -1.4,$$

and then, using Table A.4, we have

$$\Pr(X < 2.3) = \Pr(Z < -1.4)$$
$$= 0.0808.$$

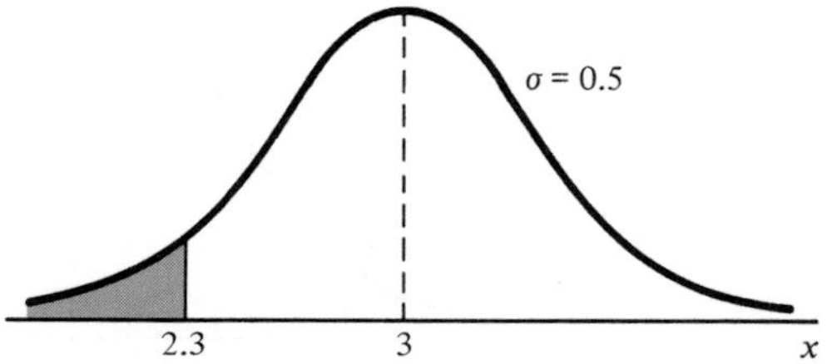

Figure 6.9
Area for Example 6.2.

(handwritten margin notes)
need
1. X normally distributed
2. μ
3. σ
4. Ques. Find Prob between x_1 and x_2

solve
1. $Z = \frac{X - \mu}{\sigma}$
   get $z_1 + z_2$
3. check table A.4

$X = \sigma Z + \mu$

Example 6.3 An electrical firm manufactures light bulbs that have a length of life that is normally distributed with mean equal to 800 hours and a standard deviation of 40 hours. Find the probability that a bulb burns between 778 and 834 hours.

Solution. The distribution of light bulbs is illustrated in Figure 6.10. The z values corresponding to $x_1 = 778$ and $x_2 = 834$ are

$$z_1 = \frac{778 - 800}{40} = -0.55,$$

$$z_2 = \frac{834 - 800}{40} = 0.85.$$

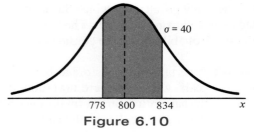

Figure 6.10

Area for Example 6.3.

Hence

$$\Pr(778 < X < 834) = \Pr(-0.55 < Z < 0.85)$$
$$= \Pr(Z < 0.85) - \Pr(Z < -0.55)$$
$$= 0.8023 - 0.2912$$
$$= 0.5111.$$

Example 6.4 On an examination the average grade was 74 and the standard deviation was 7. If 12% of the class are given As, and the grades are curved to follow a normal distribution, what is the lowest possible A and the highest possible B?

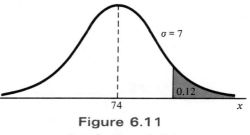

Figure 6.11

Area for Example 6.4.

Solution. The preceding two examples were solved by going first from a value of x to a z value and then computing the desired area. In this problem we reverse the process and begin with a known area or probability, find the z value, and then determine x from the formula $x = \sigma z + \mu$. An area of 0.12, corresponding to the fraction of students receiving As, is shaded in Figure 6.11. We require a z value such that $\Pr(Z > z) = 0.12$ or $\Pr(Z < z) = 0.88$. From Table A.4 we have $\Pr(Z < 1.175) = 0.88$, so that the desired z value is 1.175. Hence

$$x = (7)(1.175) + 74$$
$$= 82.225.$$

Therefore the lowest A is 83 and the highest B is 82.

Example 6.5 If the average height of miniature poodles is 12 inches, with a standard deviation of 1.8 inches, what percentage of miniature poodles exceeds 14 inches in height, assuming that the heights follow a normal distribution and can be measured to any desired degree of accuracy?

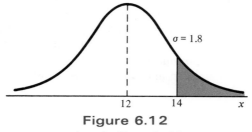

Figure 6.12
Area for Example 6.5.

Solution. A percentage is found by multiplying the relative frequency by 100%. Since the relative frequency for an interval is equal to the probability of falling in the interval, we must find the area to the right of $x = 14$ in Figure 6.12. This can be done by transforming $x = 14$ to the corresponding z value, obtaining the area to the left of z from Table A.4, and then subtracting this area from 1. We find

$$z = \frac{14 - 12}{1.8} = 1.11.$$

Hence

$$\Pr(X > 14) = \Pr(Z > 1.11)$$
$$= 1 - \Pr(Z < 1.11)$$
$$= 1 - 0.8665$$
$$= 0.1335.$$

Therefore 13.35% of miniature poodles exceed 14 inches in height.

Example 6.6 Find the percentage of miniature poodles exceeding 14 inches in Example 6.5 if the heights are all measured to the nearest inch.

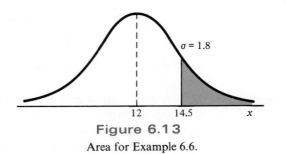

Figure 6.13

Area for Example 6.6.

Solution. This problem differs from Example 6.5 in that we now assign a measurement of 14 inches to all poodles whose heights are greater than 13.5 inches and less than 14.5 inches. We are actually approximating a discrete distribution by means of a continuous normal distribution. The required area is the region shaded to the right of 14.5 in Figure 6.13. We now find

$$z = \frac{14.5 - 12}{1.8} = 1.39.$$

Hence

$$\Pr(X > 14.5) = \Pr(Z > 1.39)$$
$$= 1 - \Pr(Z < 1.39)$$
$$= 1 - 0.9177$$
$$= 0.0823.$$

Therefore 8.23% of miniature poodles exceed 14 inches in height when measured to the nearest inch. The difference of 5.12% between this answer and that of Example 6.5 represents all those poodles having a height greater than 14 and less than 14.5 inches that are now recorded as being 14 inches tall.

Example 6.7 The quality grade-point averages of 300 college freshmen follow approximately a normal distribution with a mean of 2.1 and a standard deviation of 1.2. How many of these freshmen would you expect to have a score between 2.5 and 3.5 inclusive if the point averages are computed to the nearest tenth?

Solution. Since the scores are recorded to the nearest tenth, we require the area between $x_1 = 2.45$ and $x_2 = 3.55$, as indicated in Figure 6.14. The corresponding z values are

$$z_1 = \frac{2.45 - 2.1}{1.2} = 0.29,$$

$$z_2 = \frac{3.55 - 2.1}{1.2} = 1.21.$$

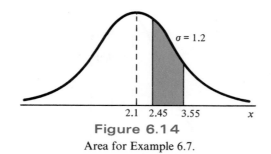

Figure 6.14
Area for Example 6.7.

Therefore,

$$Pr(2.45 < X < 3.55) = Pr(0.29 < Z < 1.21)$$
$$= Pr(Z < 1.21) - Pr(Z < 0.29)$$
$$= 0.8869 - 0.6141$$
$$= 0.2728.$$

Hence 27.28%, or approximately 82 of the 300 freshmen, should have a score between 2.5 and 3.5 inclusive.

6.3 Normal Approximation to the Binomial

Probabilities associated with binomial experiments are readily obtainable from the formula $b(x; n, p)$ of the binomial distribution or from Table A.2 when n is small. If n is not listed in any available table, we must compute the binomial probabilities by approximation procedures. In Section 5.5 we illustrated how the Poisson distribution can be used to approximate binomial probabilities when n is large and p is very close to zero or 1. Both the binomial and Poisson distributions are discrete. The first application of a continuous probability distribution to approximate probabilities over a discrete sample space was demonstrated in Section 6.2, Examples 6.6 and 6.7, where the normal curve was used. We shall now state a theorem that allows us to use areas under the normal curve to approximate binomial probabilities when n is sufficiently large.

> **THEOREM 6.1** *If X is a binomial random variable with mean $\mu = np$ and variance $\sigma^2 = npq$, then the limiting form of the distribution of*
>
> $$Z = \frac{X - np}{\sqrt{npq}},$$
>
> *as $n \to \infty$, is the standardized normal distribution $n(z; 0, 1)$.*

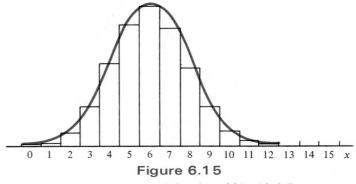

Figure 6.15

Normal-curve approximation of $b(x; 15, 0.4)$.

It turns out that the proper normal distribution provides a very accurate approximation to the binomial distribution when n is large and p is close to $\frac{1}{2}$. In fact, even when n is small and p is not extremely close to zero or 1, the approximation is fairly good.

To investigate the normal approximation to the binomial distribution, we first draw the histogram for $b(x; 15, 0.4)$ and then superimpose the particular normal curve having the same mean and variance as the binomial variable X. Hence we draw a normal curve with

$$\mu = np = (15)(0.4) = 6,$$
$$\sigma^2 = npq = (15)(0.4)(0.6) = 3.6.$$

The histogram of $b(x; 15, 0.4)$ and the corresponding superimposed normal curve, which is completely determined by its mean and variance, are illustrated in Figure 6.15.

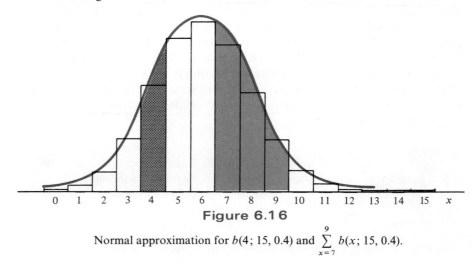

Figure 6.16

Normal approximation for $b(4; 15, 0.4)$ and $\sum_{x=7}^{9} b(x; 15, 0.4)$.

The exact probability of the binomial random variable X assuming a given value x is equal to the area of the rectangle whose base is centered at x. For example the exact probability that X assumes the value 4 is equal to the area of the rectangle with base centered at $x = 4$. Using the formula for the binomial distribution, we find this area to be

$$b(4; 15, 0.4) = 0.1268.$$

This same probability is approximately equal to the area of the shaded region under the normal curve between the two ordinates $x_1 = 3.5$ and $x_2 = 4.5$ in Figure 6.16. Converting to z values we have

$$z_1 = \frac{3.5 - 6}{1.9} = -1.316,$$

$$z_2 = \frac{4.5 - 6}{1.9} = -0.789.$$

If X is a binomial random variable and Z a standard normal variable, then

$$\begin{aligned}
\Pr(X = 4) &= b(4; 15, 0.4) \\
&\simeq \Pr(-1.316 < Z < -0.789) \\
&= \Pr(Z < -0.789) - \Pr(Z < -1.316) \\
&= 0.2151 - 0.0941 \\
&= 0.1210.
\end{aligned}$$

This agrees very closely with the exact value of 0.1268.

The normal approximation is most useful in calculating binomial sums for large values of n, which, without tables of binomial sums, is an impossible task. Referring to Figure 6.16, we might be interested in the probability that X assumes a value from 7 to 9 inclusive. The exact probability is given by

$$\begin{aligned}
\Pr(7 \leq X \leq 9) &= \sum_{x=7}^{9} b(x; 15, 0.4) \\
&= \sum_{x=0}^{9} b(x; 15, 0.4) - \sum_{x=0}^{6} b(x; 15, 0.4) \\
&= 0.9662 - 0.6098 \\
&= 0.3564,
\end{aligned}$$

which is equal to the sum of the areas of the rectangles with bases centered at $x = 7, 8$, and 9. For the normal approximation we find the area of the shaded region under the curve between the ordinates $x_1 = 6.5$ and $x_2 = 9.5$ in Figure 6.16. The corresponding z values are

$$z_1 = \frac{6.5 - 6}{1.9} = 0.263,$$

$$z_2 = \frac{9.5 - 6}{1.9} = 1.842.$$

Now

$$\Pr(7 \leq X \leq 9) \simeq \Pr(0.263 < Z < 1.842)$$
$$= \Pr(Z < 1.842) - \Pr(Z < 0.263)$$
$$= 0.9673 - 0.6037$$
$$= 0.3636.$$

Once again the normal-curve approximation provides a value that agrees very closely with the exact value of 0.3564. The degree of accuracy, which depends on how well the curve fits the histogram, will increase as n increases. This is particularly true when p is not very close to $\frac{1}{2}$ and the histogram is no longer symmetric. Figures 6.17 and 6.18 show the histograms for $b(x; 6, 0.2)$ and $b(x; 15, 0.2)$, respectively. It is evident that a normal curve would fit the histogram when $n = 15$ considerably better than when $n = 6$.

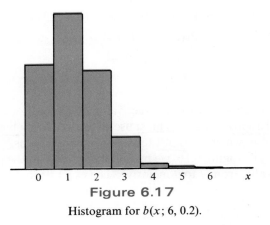

Figure 6.17

Histogram for $b(x; 6, 0.2)$.

In summary we use the normal approximation to evaluate binomial probabilities whenever p is not close to zero or 1. The approximation is excellent when n is large and fairly good for small values of n if p is reasonably close to $\frac{1}{2}$. One possible guide to determine when the normal approximation may be used is provided by calculating np and nq. If both np and nq are greater than 5, the approximation will be good.

Example 6.8 A basketball player hits on 60% of his shots from the floor. What is the probability that he makes less than one half of his next 100 shots?

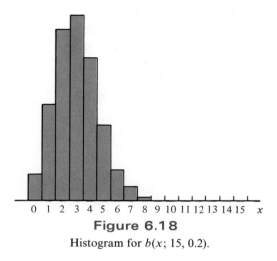

Figure 6.18

Histogram for $b(x; 15, 0.2)$.

Solution. Assuming that the shots are independent and $p = 0.6$ for each shot, we have a binomial experiment. Since the number of trials is large, we should obtain fairly accurate results using the normal-curve approximation with

$$\mu = np = (100)(0.6) = 60,$$

$$\sigma = \sqrt{npq} = \sqrt{(100)(0.6)(0.4)} = 4.9.$$

To obtain the desired probability, we have to find the area to the left of $x = 49.5$. The z value corresponding to 49.5. is

$$z = \frac{49.5 - 60}{4.9} = -2.143,$$

and the probability of making fewer than 50 baskets in the next 100 shots is given by the area of the shaded region in Figure 6.19. Hence if X represents the number of baskets, then

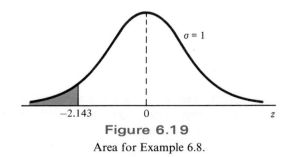

Figure 6.19

Area for Example 6.8.

$$\Pr(X < 50) = \sum_{x=0}^{49} b(x; 100, 0.6)$$

$$\simeq \Pr(Z < -2.143)$$

$$= 0.0160.$$

Example 6.9 A multiple-choice quiz has 200 questions, each with 4 possible answers, of which only 1 is the correct answer. What is the probability that sheer guesswork yields from 25 to 30 correct answers for 80 of the 200 problems about which the student has no knowledge?

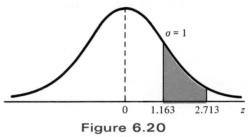

Figure 6.20
Area for Example 6.9.

Solution. The probability of a correct answer for each of the 80 questions is $p = \frac{1}{4}$. If X represents the number of correct answers due to guesswork, then

$$\Pr(25 \leq X \leq 30) = \sum_{x=25}^{30} b(x; 80, \tfrac{1}{4}).$$

Using the normal-curve approximation with

$$\mu = np = (80)(\tfrac{1}{4}) = 20,$$

$$\sigma = \sqrt{npq} = \sqrt{(80)(\tfrac{1}{4})(\tfrac{3}{4})} = 3.87,$$

we need the area between $x_1 = 24.5$ and $x_2 = 30.5$. The corresponding z values are

$$z_1 = \frac{24.5 - 20}{3.87} = 1.163,$$

$$z_2 = \frac{30.5 - 20}{3.87} = 2.713.$$

The probability of correctly guessing from 25 to 30 questions is given by the area of the shaded region in Figure 6.20. From Table A.4 we find

$$\Pr(25 \leq X \leq 30) = \sum_{x=25}^{30} b(x; 80, \tfrac{1}{4})$$
$$\simeq \Pr(1.163 < Z < 2.713)$$
$$= \Pr(Z < 2.713) - \Pr(Z < 1.163)$$
$$= 0.9967 - 0.8776$$
$$= 0.1191.$$

EXERCISES

1. Given a normal distribution with $\mu = 40$ and $\sigma = 6$, find:
  (a) The area below 32.
  (b) The area above 27.
  (c) The area between 42 and 51.
  (d) The point that has 45% of the area below it.
  (e) The point that has 13% of the area above it.

2. Given a normal distribution with $\mu = 200$ and $\sigma^2 = 100$, find:
  (a) The area below 214.
  (b) The area above 179.
  (c) The area between 188 and 206.
  (d) The point that has 80% of the area below it.
  (e) The two points containing the middle 75% of the area.

3. Given the normally distributed variable X with mean 18 and standard deviation 2.5, find:
  (a) $\Pr(X < 15)$.
  (b) The value of k such that $\Pr(X < k) = 0.2578$.
  (c) $\Pr(17 < X < 21)$.
  (d) The value of k such that $\Pr(X > k) = 0.1539$.

4. A soft-drink machine is regulated so that it discharges an average of 7 ounces per cup. If the amount of drink is normally distributed with standard deviation equal to 0.5 ounce:
  (a) What fraction of the cups will contain more than 7.8 ounces?
  (b) What is the probability that a cup contains between 6.7 and 7.3 ounces?
  (c) How many cups will likely overflow if 8 ounce cups are used for the next 1000 drinks?
  (d) Below what value do we get the smallest 25% of the drinks?

5. The finished inside diameter of a piston ring is normally distributed with a mean of 4.00 inches and a standard deviation of 0.01 inch.
  (a) What proportion of rings will have inside diameters exceeding 4.025 inches?
  (b) What is the probability that a piston ring will have an inside diameter between 3.99 and 4.01 inches?
  (c) Below what value of inside diameter will 15% of the piston rings fall?

6. A lawyer commutes daily from his suburban home to his midtown office. On the average the trip one way takes 24 minutes, with a standard deviation of 3.8 minutes.

Assume the distribution of trip times to be normally distributed.

(a) What is the probability that a trip will take at least $\frac{1}{2}$ hour?

(b) If the office opens at 9:00 A.M. and he leaves his house at 8:45 A.M. daily, what percentage of the time is he late for work?

(c) If he leaves the house at 8:35 A.M. and coffee is served at the office from 8:50 A.M. until 9:00 A.M., what is the probability that he misses coffee?

(d) Find the length of time above which we find the slowest 15% of the trips.

7. If a set of grades on a statistics examination are approximately normally distributed with a mean of 74 and a standard deviation of 7.9, find:

(a) The lowest passing grade if the lowest 10% of the students are given Fs.

(b) The highest B if the top 5% of the students are given As.

8. The heights of 1000 students are normally distributed with a mean of 68.5 inches and a standard deviation of 2.7 inches. Assuming that the heights are recorded to the nearest half inch, how many of these students would you expect to have heights

(a) Less than 63.0 inches?

(b) Between 67.5 and 71.0 inches inclusive?

(c) Equal to 69.0 inches?

(d) Greater than or equal to 74.0 inches?

9. In a mathematics examination the average grade was 82 and the standard deviation was 5. All students with grades from 88 to 94 received a grade of B. If the grades are approximately normally distributed and 8 students received a B grade, how many students took the examination?

10. A company pays its employees an average wage of $3.25 an hour with a standard deviation of 60 cents. If the wages are approximately normally distributed:

(a) What percentage of the workers receive wages between $2.75 and $3.69 an hour inclusive?

(b) The highest 5% of the hourly wages are greater than what amount?

11. The weights of a large number of miniature poodles are approximately normally distributed with a mean of 18 pounds and a standard deviation of 2 pounds. If measurements are recorded to the nearest pound, find the fraction of these poodles with weights:

(a) Over 21 pounds.

(b) At most 19 pounds.

(c) Between 16 and 20 pounds inclusive.

12. The tensile strength of a certain metal component is normally distributed with a mean of 10,000 pounds per square inch and a standard deviation of 100 pounds per square inch. Measurements are recorded to the nearest 50 pounds per square inch.

(a) What proportion of these components exceed 10,150 pounds per square inch in tensile strength?

(b) If specifications require that all components have tensile strength between 9,800 and 10,200 pounds per square inch inclusive, what proportion of pieces would we expect to scrap?

13. If a set of observations is normally distributed, what percentage of the observations differ from the mean by (a) more than 1.3σ, (b) less than 0.52σ?

14. The I.Q.s of 600 applicants to a certain college are approximately normally distributed with a mean of 115 and a standard deviation of 12. If the college requires an I.Q. of at least 95, how many of these students will be rejected on this basis regardless of their other qualifications?

15. The average rainfall, recorded to the nearest hundredth of an inch, in Roanoke, Virginia, for the month of March is 3.63 inches. Assuming a normal distribution with a standard deviation of 1.03 inches, find the probability that next March Roanoke receives

(a) Less than 0.72 inch of rain.

(b) More than 2 inches but not over 3 inches.

(c) More than 5.3 inches.

16. The average life of a certain type of small motor is 10 years, with a standard deviation of 2 years. The manufacturer replaces free all motors that fail while under guarantee. If he is willing to replace only 3% of the motors that fail, how long a guarantee should he offer? Assume that the lives of the motors follow a normal distribution.

17. Find the error in approximating $\sum_{x=1}^{4} b(x; 20, 0.1)$ by the normal-curve approximation.

18. A coin is tossed 400 times. Use the normal-curve approximation to find the probability of obtaining:

(a) Between 185 and 210 heads inclusive.

(b) Exactly 205 heads.

(c) Less than 176 or more than 227 heads.

19. A pair of dice is rolled 180 times. What is the probability that a total of 7 occurs:

(a) At least 25 times.

(b) Between 33 and 41 times inclusive.

(c) Exactly 30 times.

20. The probability that a patient recovers from a delicate heart operation is 0.9. What is the probability that between 84 and 95 inclusive of the next 100 patients having this operation survive?

21. A pheasant hunter claims that he brings down 75% of the birds he shoots at. If 35 of the next 80 pheasants shot at escape, what do we conclude about the hunter's claimed accuracy?

22. A drug manufacturer claims that a certain drug cures a blood disease on the average 80% of the time. To check the claim, government testers used the drug on a sample of 100 individuals and decide to accept the claim if 75 or more are cured.

(a) What is the probability that the claim will be rejected when the cure probability is in fact 0.8?

(b) What is the probability that the claim will be accepted by the government when the cure probability is as low as 0.7?

Compare with # 9 p 98

23. A survey of the residents in a United States city showed that 20% preferred a white telephone over any other color available. What is the probability that between 170 and 185 inclusive of the next 1000 telephones installed in this city will be white?

24. One sixth of the male freshmen entering a large state school are out-of-state students. If the students are assigned at random to the dormitories, 180 to a building, what is the probability that in a given dormitory at least one fifth of the students are from out of state?

25. A certain pharmaceutical company knows that, on the average, 5% of a certain type of pill has an ingredient that is below the minimum strength and thus unacceptable. What is the probability that fewer than 10 in a sample of 200 pills will be unacceptable?

SAMPLING THEORY CHAPTER 7

7.1 Populations and Samples

The outcome of a statistical experiment may be recorded either as a numerical value or as a descriptive representation. When a pair of dice is tossed and the total is the outcome of interest, we record a numerical value. However, if the students in a certain school are given blood tests and the type of blood is of interest, then a descriptive representation might be the most useful. A person's blood can be classified in 8 ways. It must be AB, A, B, or O, with a plus or minus sign, depending on the presence or absence of the Rh antigen.

The statistician works primarily with numerical observations. For the experiment involving the blood types, he will probably let numbers 1 to 8 represent each blood type and then record the appropriate number for each student. In the classification of blood types we can only have as many observations as there are students in the school. The experiment, therefore, results in a finite number of observations. In the die-tossing experiment we are interested in recording the total that occurs. Therefore, if we toss the dice indefinitely, we obtain an infinite set of values, each representing the result of a single toss of a pair of dice.

The totality of observations with which we are concerned, whether finite or infinite, constitutes what we call a *population*. In past years the word "population" referred to observations obtained from statistical studies involving people. Today the statistician uses the term to refer to observations relevant to anything of interest, whether it be groups of people, animals, or objects.

> DEFINITION *A* population *consists of the totality of the observations with which we are concerned.*

The number of observations in the population is defined to be the *size* of the population. If there are 600 students in the school that are classified according to blood type, we say we have a population of size 600. The die-tossing experiment generates a population whose size is infinite. The numbers on the cards in a deck, the heights of residents in a certain city, and the lengths of fish in a particular lake are examples of populations with finite size. In each case the total number of observations is a finite number. The observations obtained by measuring the atmospheric pressure every day from the past on into the future, or all measurements on the depth of a lake from any conceivable position, are examples of populations whose sizes are infinite. Some finite populations are so large that in theory we assume them to be infinite. This is true if you consider the population of lives of a certain type of storage battery being manufactured for mass distribution throughout the country.

Each observation in a population is a value of a random variable X having a probability distribution $f(x)$. If one is inspecting items coming off an assembly line for defects, then each observation in the population might be a value 0 or 1 of the binomial random variable X with probability distribution

$$b(x; 1, p) = p^x q^{1-x}, \qquad x = 0, 1,$$

where 0 indicates a nondefective item and 1 indicates a defective item. Of course it is assumed that p, the probability that any item is defective, remains constant from trial to trial. In the blood-type experiment the random variable X represents the type of blood by assuming a value from 1 to 8. Each student is given one of the values of the discrete random variable. The lives of the storage batteries are values assumed by a continuous random variable having perhaps a normal distribution. When we speak hereafter about a "binomial population," a "normal population," or in general, the "population $f(x)$," we shall mean a population whose observations are values of a random variable having a binomial distribution, a normal distribution, or the probability distribution $f(x)$. Hence the mean and variance of a random variable or probability distribution are also referred to as the mean and variance of the corresponding population.

The statistician is interested in arriving at conclusions concerning unknown population parameters. In a normal population, for example, the parameters μ and σ^2 may be unknown and are to be estimated from the information provided by a sample selected from the population. This takes us into the theory of sampling.

DEFINITION *A* sample *is a subset of the population.*

If our inferences are to be accurate, we must understand the relation of a sample to its population. Certainly the sample should be representative of the population. Such a sample is said to be *unbiased*. It should be a *random sample* in the sense that the observations are made independently and at random.

DEFINITION *A* random sample *of n observations is a sample that is chosen in such a way that each subset of n observations of the population has the same probability of being selected.*

We may wish to arrive at a conclusion concerning the proportion of people in the United States who prefer a certain brand of coffee. It would be impossible to question every American and compute the parameter representing the true proportion. Instead, a large random sample is selected and the proportion of this sample favoring the brand of coffee in question is calculated. This value is now used to make some inference concerning the true proportion.

A value computed from a sample is called a *statistic*. Since many random samples are possible from the same population, we would expect the statistic to vary somewhat from sample to sample. Hence a statistic is a *random variable*.

DEFINITION *A* statistic *is a random variable that depends only on the observed sample.*

A statistic is usually represented by ordinary Latin letters. The sample proportion in the above illustration is a statistic that is commonly represented by $\hat{P}$. The value of the random variable $\hat{P}$ for the given sample is denoted by $\hat{p}$. In order to use $\hat{p}$ to estimate, with some degree of accuracy, the true proportion of p of people in the United States who prefer the given brand of coffee, we must first know more about the probability distribution of the statistic $\hat{P}$.

7.2 Some Useful Statistics

In Chapter 4 we introduced the two parameters μ and σ^2, which measure the center and variability of a probability distribution. These are constant population parameters and are in no way affected or influenced by the observations of a random sample. We shall now define some important statistics whose values describe corresponding measures of a random sample. The most frequently used statistics for measuring the center of a set of data, arranged in order of magnitude, are the *mean, median*, and *mode*. The most important of these and the one we shall consider first is the mean. It is common practice to represent the sample mean by the statistic $\bar{X}$.

> **DEFINITION** *If $x_1, x_2, \ldots, x_n$, not necessarily all distinct, represents a random sample of size n, then the* sample mean *has the value*
>
> $$\bar{x} = \frac{\sum_{i=1}^{n} x_i}{n}.$$

Example 7.1 Find the mean of the random sample whose observations are 15, 22, and 20.

Solution. The observed value $\bar{x}$ of the statistic $\bar{X}$ is

$$\bar{x} = \frac{15 + 22 + 20}{3} = 19.$$

It is sometimes convenient to add (or subtract) a constant to all our observations and then compute the mean. How is this new mean related to the mean of the original set of observations? If we let $y_i = x_i + a$, then

$$\bar{y} = \frac{\sum_{i=1}^{n} y_i}{n} = \frac{\sum_{i=1}^{n} (x_i + a)}{n} = \bar{x} + a.$$

Therefore the addition (or subtraction) of a constant to all observations changes the mean by the same amount. To find the mean of the numbers $-5, -3, 1, 4,$ and 6, we might add 5 first to give the set of all positive values 0, 2, 6, 9, and 11 that have a mean of 5.6. Therefore the original numbers have a mean of $5.6 - 5 = 0.6$.

Now, suppose we let $y_i = ax_i$. It follows that

$$\bar{y} = \frac{\sum_{i=1}^{n} y_i}{n} = \frac{\sum_{i=1}^{n} ax_i}{n} = a\bar{x}.$$

Therefore, if all observations are multiplied or divided by a constant, the new observations will have a mean that is the same constant multiple of the original mean. The mean of the numbers 4, 6, 14 is equal to 8, and therefore, after dividing by 2, the mean of the set 2, 3, and 7 must be $\frac{8}{2} = 4$.

The second most useful statistic for measuring the center of a set of data is the median. We shall designate the median by the symbol $\tilde{X}$.

> **DEFINITION** *If $x_1, x_2, \ldots, x_n$ represents a random sample of size n, arranged in increasing order of magnitude, then the* sample median *is the middle value if n is odd, or the arithmetic mean of the two middle values if n is even.*

Example 7.2 Find the median for the random sample whose observations are 5, 8, 3, 9, 5, 6, and 8.

    Solution. Arranging the observations in order of magnitude, 3, 5, 5, 6, 8, 8, and 9, gives $\tilde{x} = 6$.

Example 7.3 Find the median for the random sample whose observations are 4, 10, 8, and 7.

    Solution. If we arrange the observations in order of magnitude, 4, 7, 8, and 10, the median is the arithmetic mean of the two middle values. Therefore, $\tilde{x} = (7 + 8)/2 = 7.5$.

The third and final statistic for measuring the center of a random sample that we shall discuss is the mode, designated by the statistic M.

> **DEFINITION** *If $x_1, x_2, \ldots, x_n$, not necessarily all different, represents a random sample of size n, then the* mode *is that value of the sample which occurs most often or with the greatest frequency. The mode may not exist, and when it does, it is not necessarily unique.*

Example 7.4 The mode for the random sample whose observations are 3, 4, 4, 6, 7, 7, 7, 8, 8, and 9 is $m = 7$.

Example 7.5 The observations 2, 5, 5, 5, 5, 6, 7, 7, 8, 8, 8, 8, and 9 have two modes, 5 and 8, since both 5 and 8 occur with the greatest frequency. The distribution of the sample is said to be *bimodal*.

When the mode could be either of two adjacent numbers arranged in order of magnitude, we take the arithmetic mean of the two numbers as the mode.

Therefore the modes of the observations 3, 5, 5, 5, 6, 6, 6, 7, 9, 9, and 9 are $(5 + 6)/2 = 5.5$ and 9.

In summary let us consider the relative merits of the mean, median, and mode. The mean is the most commonly used measure of a set of data in statistics. It is easy to calculate and employs all available information. The distributions of sample means are well known, and consequently the methods used in statistical inference are based on the sample mean. The only real disadvantage to the mean is that it may be affected adversely by extreme values. That is, if most contributions to a charity are less than $5, then a very large contribution, say $10,000, would produce an average donation that is considerably higher than the majority of gifts.

The median has the advantage of being easy to compute. It is not influenced by extreme values and would give a truer average in the case of the charitable contributions. In dealing with samples selected from populations, the sample means will not vary as much from sample to sample as will the medians. Therefore, if we are attempting to estimate the center of a population based on a sample value, the mean is more stable than the median. Hence a sample mean is likely to be closer to the population mean than the sample median would be to the population median.

The mode is the least used measure of the three. For small sets of data its value is almost useless if, in fact, it exists at all. Only in the case of a large mass of data does it have a significant meaning. Its only advantage is that it requires no calculation.

The three statistics defined above do not by themselves give an adequate description of the distribution of our data. We need to know how the observations spread out from the average. It is quite possible to have two sets of observations with the same mean or median that differ considerably in the variability of their measurements about the average.

Consider the following measurements, in ounces, for two samples of orange juice bottled by companies A and B:

Sample A	31	32	30	33	34
Sample B	34	32	28	29	37

Both samples have the same mean, 32. It is quite obvious that company A bottles orange juice with a more uniform content than company B. We say that the variability or the dispersion of the observations from the average is less for sample A than for sample B. Therefore, in buying orange juice, we would feel more confident that the bottle we select will be closer to the advertised average if we buy from company A.

The most important statistics for measuring the variability of a random sample are the *range* and the *variance*. The simplest of these to compute is the range.

DEFINITION *The* range *of a random sample* $x_1, x_2, \ldots, x_n$ *is the difference between the largest and the smallest observations in the set.*

Example 7.6 The range of the set of observations 10, 12, 12, 18, 19, 22, and 24 is $24 - 10 = 14$.

In the case of the companies bottling orange juice, the range for company A is 4 compared to a range of 9 for company B, indicating a greater spread in the values for company B.

The range is a poor measure of variability, particularly if the size of the sample is large. It considers only the extreme values and tells us nothing about the distribution of values in between. Consider, for example, the following two sets of data, both with a range of 12:

$$3, 4, 5, 6, 8, 9, 10, 12, 15,$$
$$3, 8, 8, 9, 9, 9, 10, 10, 15.$$

In the first set the mean and median are both 8, but the numbers vary over the entire interval from 3 to 15. In the second set the mean and median are both 9, but most of the values are closer to the average. Although the range fails to measure this variability between the upper and lower observations, it does have some useful applications. In industry the range for measurements on items coming off an assembly line might be specified in advance. As long as all measurements fall within the specified range, the process is said to be in control.

To overcome the disadvantage of the range, we shall consider a measure of variability, the *sample variance*, that considers the positions of each observation relative to the sample mean. The sample variance, denoted by the statistic S^2, assumes the value s^2 for a given random sample.

DEFINITION *If* $x_1, x_2, \ldots, x_n$ *represents a random sample of size n, then the* sample variance *has the value*

$$s^2 = \frac{\sum_{i=1}^{n} (x_i - \bar{x})^2}{n - 1}.$$

Note that s^2 is essentially defined to be the average of the squares of the deviations of the observations from their mean. The reason for using $n - 1$ as a divisor, rather than the more obvious choice n, will become apparent in Chapter 8.

Example 7.7 Find the variance of the random sample whose observations are 12, 15, 17, and 20.

Solution

$$\bar{x} = \frac{12 + 15 + 17 + 20}{4} = 16.$$

Therefore,

$$s^2 = \frac{\sum_{i=1}^{4} (x_i - 16)^2}{3}$$

$$= \frac{(12 - 16)^2 + (15 - 16)^2 + (17 - 16)^2 + (20 - 16)^2}{3}$$

$$= \frac{(-4)^2 + (-1)^2 + (1)^2 + (4)^2}{3}$$

$$= \frac{34}{3}.$$

If $\bar{x}$ is a decimal number that has been rounded off, we accumulate a large error using the sample-variance formula in the above form. To avoid this, let us derive the more useful computational formula, as given in the following theorem.

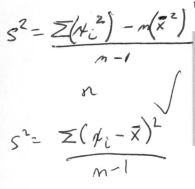

THEOREM 7.1 *If s^2 is the variance of a random sample of size n, we may write*

$$s^2 = \frac{n \sum_{i=1}^{n} x_i^2 - \left(\sum_{i=1}^{n} x_i \right)^2}{n(n-1)}.$$

Proof. By definition

$$s^2 = \frac{\sum_{i=1}^{n} (x_i - \bar{x})^2}{n - 1} = \frac{\sum_{i=1}^{n} (x_i^2 - 2\bar{x}x_i + \bar{x}^2)}{n - 1}.$$

Applying Theorems 4.1, 4.2, and 4.3, we have

$$s^2 = \frac{\sum_{i=1}^{n} x_i^2 - 2\bar{x} \sum_{i=1}^{n} x_i + n\bar{x}^2}{n - 1}.$$

Replacing $\bar{x}$ by $\sum_{i=1}^{n} x_i/n$, and multiplying numerator and denominator by n, we obtain

$$s^2 = \frac{n \sum_{i=1}^{n} x_i^2 - \left(\sum_{i=1}^{n} x_i \right)^2}{n(n-1)}.$$

The *sample standard deviation*, denoted by the statistic S, is defined to be the positive square root of the sample variance.

Example 7.8 Find the variance of the sample whose observations are 3, 4, 5, 6, 6, and 7.

Solution. In tabular form we write

x_i	x_i^2	
3	9	
4	16	
5	25	
6	36	
6	36	
7	49	
31	171	$n = 6$

Hence

$$s^2 = \frac{(6)(171) - (31)^2}{(6)(5)} = \frac{13}{6}.$$

7.3 Sampling Distributions

The field of inductive statistics is basically concerned with generalizations and predictions. For example, we might claim, based on the opinions of several people interviewed on the street, that in a forthcoming election 60% of the eligible voters in a city favor a certain candidate. In this case we are dealing with a random sample of opinions from a very large finite population. As a second illustration we might state that the average cost to build a residence in a certain city is between 25 and 30 thousand dollars, based on the estimates of 3 contractors selected at random from the 25 now building in this area. The population being sampled here is again finite but very small. Finally let us consider a soft-drink-dispensing machine in which the variance of the amounts of drink dispensed is being held to a known value. A company official computes the variance of 40 drinks and decides on the basis of this value whether the desired variability is being maintained. The 40 drinks represent a sample from the infinite population of possible drinks that will be dispensed by this machine.

In each of the above examples we have computed a statistic from a sample selected from the population, and from these statistics we made various statements concernining the values of parameters that may or may not be true. Generalizations from a statistic to a parameter can be made with confidence only if we understand the fluctuating behavior of our statistic when computed for different random samples from the same population. The distribution of the statistic in question will depend on the size of the population, the size of the samples, and the method of choosing the random samples. If the size of the population is large or infinite, the statistic has the same distribution whether we sample with or without replacement. On the other hand, sampling with replacement from a small finite population gives a slightly different distribution for the statistic than if we sample without replacement. Sampling with replacement from a finite population is equivalent to sampling from an infinite population, since there is no limit on the possible size of the sample selected.

> **DEFINITION** *The probability distribution of a statistic is called a* sampling distribution.

> **DEFINITION** *The standard deviation of the sampling distribution of a statistic is called the* standard error *of the statistic.*

The probability distribution of $\bar{X}$ is called the *sampling distribution of the mean*, and the standard error of the mean is the standard deviation of the sampling distribution of $\bar{X}$. Every sample of size n selected from a specified population provides a value s of the statistic S, the sample standard deviation. The standard error of the sample standard deviation is then the standard deviation of the statistic S.

In this chapter we shall study several of the more important sampling distributions of frequently used statistics. The applications of these sampling distributions to problems of statistical inference will be considered in Chapters 8 and 9.

7.4 Sampling Distributions of the Mean

The first important sampling distribution to be considered is that of the mean $\bar{X}$. To illustrate, we shall sample from a discrete uniform population consisting of the values 0, 1, 2, and 3. Clearly the four observations making up the population are values of a random variable X having the probability distribution

$$f(x) = \tfrac{1}{4}, \qquad x = 0, 1, 2, 3,$$

with mean

$$\mu = E(X) = \sum_{x=0}^{3} xf(x)$$

$$= \frac{0 + 1 + 2 + 3}{4} = \frac{3}{2}$$

and variance

$$\sigma^2 = E[(X - \mu)^2] = \sum_{x=0}^{3} (x - \mu)^2 f(x)$$

$$= \frac{(0 - \tfrac{3}{2})^2 + (1 - \tfrac{3}{2})^2 + (2 - \tfrac{3}{2})^2 + (3 - \tfrac{3}{2})^2}{4}$$

$$= \tfrac{5}{4}.$$

The frequency histogram of the population is shown in Figure 7.1.

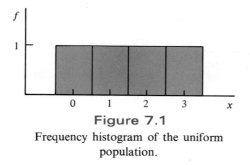

Figure 7.1

Frequency histogram of the uniform
population.

Suppose that we list all possible samples of size 2, *with replacement*, and then for each sample compute $\bar{x}$. The 16 possible samples and their means are given in Table 7.1. The statistic $\bar{X}$ clearly assumes values $\bar{x}$ fluctuating from

Table 7.1

Means of Random Samples with Replacement

No.	Sample	$\bar{x}$	No.	Sample	$\bar{x}$
1	0, 0	0	9	2, 0	1.0
2	0, 1	0.5	10	2, 1	1.5
3	0, 2	1.0	11	2, 2	2.0
4	0, 3	1.5	12	2, 3	2.5
5	1, 0	0.5	13	3, 0	1.5
6	1, 1	1.0	14	3, 1	2.0
7	1, 2	1.5	15	3, 2	2.5
8	1, 3	2.0	16	3, 3	3.0

Table 7.2

Sampling Distribution of $\bar{X}$
with Replacement

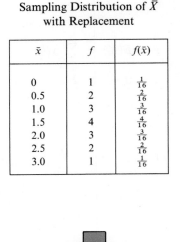

$\bar{x}$	f	$f(\bar{x})$
0	1	$\frac{1}{16}$
0.5	2	$\frac{2}{16}$
1.0	3	$\frac{3}{16}$
1.5	4	$\frac{4}{16}$
2.0	3	$\frac{3}{16}$
2.5	2	$\frac{2}{16}$
3.0	1	$\frac{1}{16}$

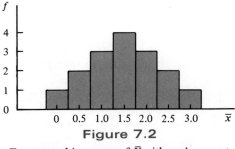

Figure 7.2

Frequency histogram of $\bar{X}$ with replacement.

0 to 3. The sampling distribution of $\bar{X}$ is given in Table 7.2 and the frequency histogram of the population of sample means of size 2 is shown in Figure 7.2.

It can be seen that the histogram of the sampling distribution of $\bar{X}$ may be approximated very closely by a normal curve with mean

$$\mu_{\bar{X}} = \sum \bar{x}f(\bar{x}) = \tfrac{3}{2} = \mu$$

and variance

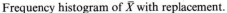

$$\sigma_{\bar{X}}^2 = \sum (\bar{x} - \tfrac{3}{2})^2 f(\bar{x}) = \tfrac{5}{8} = \frac{\frac{5}{4}}{2} = \frac{\sigma^2}{n}.$$

The mean and variance of $\bar{X}$ have been computed from the values in Table 7.2.

One could easily show that the sampling distribution of the means of 64 possible samples of size 3, selected with replacement, will give a closer approximation to a normal curve with $\mu_{\bar{X}} = \tfrac{3}{2}$ and $\sigma_{\bar{X}}^2 = \tfrac{5}{12}$. The mean of the variable $\bar{X}$ is always equal to the mean of the population from which the random samples are chosen and in no way depends on the size of the sample. The variance of $\bar{X}$, however, does depend on the sample size and is equal to the original population variance σ^2 divided by n. Consequently the larger the sample size, the smaller the standard error of $\bar{X}$, and the closer a particular $\bar{x}$

is likely to be to μ. Hence $\bar{x}$ could be used as an estimate of μ. The results of the above example illustrate the following well-known theorem.

> **THEOREM** 7.2 *If all possible random samples of size n are drawn with replacement from a finite population of size N with mean μ and standard deviation σ, then the sampling distribution of the mean $\bar{X}$ will be approximately normally distributed with mean $\mu_{\bar{X}} = \mu$ and standard deviation $\sigma_{\bar{X}} = \sigma/\sqrt{n}$. Hence*
>
> $$z = \frac{\bar{x} - \mu}{\sigma/\sqrt{n}}$$
>
> *is a value of a standard normal variable Z.*

Theorem 7.2 is valid for any finite population when $n \geq 30$. If $n < 30$, the results will be valid only if the population being sampled is not too different from a normal population. If the population is known to be bell-shaped, the sampling distribution of $\bar{X}$ will be approximately a normal distribution, regardless of the size of the sample.

Example 7.9 Given the population 1, 1, 1, 3, 4, 5, 6, 6, 6, and 7, find the probability that a random sample of size 36, selected with replacement, will yield a sample mean greater than 3.8 but less than 4.5 if the mean is measured to the nearest tenth.

Solution. The probability distribution of our population may be written as

x	1	3	4	5	6	7
$\Pr(X = x)$	0.3	0.1	0.1	0.1	0.3	0.1

Calculating the mean and variance by standard procedures, we find $\mu = 4$ and $\sigma^2 = 5$. The sampling distribution of $\bar{X}$ may be approximated by the normal distribution with mean $\mu_{\bar{X}} = \mu = 4$ and variance $\sigma_{\bar{X}}^2 = \sigma^2/n \doteq \frac{5}{36}$. Taking the square root, we find the standard deviation to be $\sigma_{\bar{X}} = 0.37$. The probability that $\bar{X}$ is greater than 3.8 and less than 4.5 is given by the area of the shaded region in Figure 7.3. The z values corresponding to $\bar{x}_1 = 3.85$ and $\bar{x}_2 = 4.45$ are

$$z_1 = \frac{3.85 - 4}{0.37} = -0.405,$$

$$z_2 = \frac{4.45 - 4}{0.37} = 1.216.$$

$E(x) = \mu = \sum_{i=1}^{m} x_i \, f(x_i)$

$Var(x) = \sigma^2 = E(x^2) - (E(x))^2$

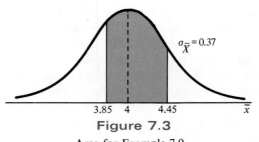

Figure 7.3

Area for Example 7.9.

Therefore,

$$\Pr(3.8 < \bar{X} < 4.5) \simeq \Pr(-0.405 < Z < 1.216)$$
$$= \Pr(Z < 1.216) - \Pr(Z < -0.405)$$
$$= 0.8880 - 0.3427$$
$$= 0.5453.$$

To verify Example 7.9, we could write the values of our population on tags and place them in a box from which we draw samples of size 36 with replacement. If we drew 100 samples each of size 36 and computed the sample means, we obtain what is known as an *experimental sampling distribution* of $\bar{X}$. From the experimental sampling distribution we should find that approx-

Table 7.3

Means of Random Samples Without Replacement

No.	Sample	$\bar{x}$	No.	Sample	$\bar{x}$
1	0, 1	0.5	7	1, 0	0.5
2	0, 2	1.0	8	2, 0	1.0
3	0, 3	1.5	9	3, 0	1.5
4	1, 2	1.5	10	2, 1	1.5
5	1, 3	2.0	11	3, 1	2.0
6	2, 3	2.5	12	3, 2	2.5

Table 7.4

Sampling Distribution of $\bar{X}$
Without Replacement

$\bar{x}$	f	$f(\bar{x})$
0.5	2	$\frac{1}{6}$
1.0	2	$\frac{1}{6}$
1.5	4	$\frac{1}{3}$
2.0	2	$\frac{1}{6}$
2.5	2	$\frac{1}{6}$

imately 55%, or 55 of our 100 sample means, fall within the interval from 3.85 to 4.45.

Suppose that we now draw all possible samples of size 2 from our uniform population, *without replacement*, and then for each compute $\bar{x}$. The 12 possible samples and their means are given in Table 7.3. The statistic $\bar{X}$ now assumes values that fluctuate from 0.5 to 2.5. The sampling distribution of $\bar{X}$ is given in Table 7.4 and the histogram of the population of sample means is shown in Figure 7.4.

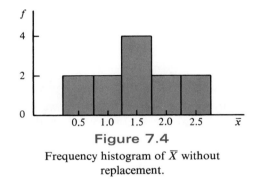

Figure 7.4

Frequency histogram of $\bar{X}$ without replacement.

We cannot expect the sampling distribution of $\bar{X}$ to approximate the normal distribution very closely when the samples are selected without replacement from a small finite population unless the population is bell-shaped. However, the mean and variance of $\bar{X}$ are

$$\mu_{\bar{X}} = \sum \bar{x} f(\bar{x}) = \tfrac{3}{2} = \mu,$$

$$\sigma_{\bar{X}}^2 = \sum (\bar{x} - \tfrac{3}{2})^2 f(\bar{x}) = \tfrac{5}{12} = \frac{\tfrac{5}{4}}{2}\left(\frac{4-2}{4-1}\right) = \frac{\sigma^2}{n}\left(\frac{N-n}{N-1}\right),$$

regardless of the size or form of the original population. When $n \geq 30$ and the population is at least twice the sample size, we may apply the following theorem.

THEOREM 7.3   *If all possible random samples of size n are drawn, without replacement, from a finite population of size N with mean μ and standard deviation σ, then the sampling distribution of the sample mean $\bar{X}$ will be approximately normally distributed with a mean and standard deviation given by*

$$\mu_{\bar{X}} = \mu,$$

$$\sigma_{\bar{X}} = \frac{\sigma}{\sqrt{n}}\sqrt{\frac{N-n}{N-1}}.$$

Example 7.10 Given the population 1, 1, 1, 3, 4, 5, 6, 6, 6, and 7, find the mean and standard deviation for the sampling distribution of means for samples of size 4 selected at random without replacement. Between what two values would you expect at least $\frac{3}{4}$ of the sample means to fall?

Solution. From Example 7.9 we know that $\mu = 4$ and $\sigma^2 = 5$. Using Theorem 7.3, the mean and standard deviation for the sampling distribution of means are

$$\mu_{\bar{X}} = 4,$$

$$\sigma_{\bar{X}} = \frac{\sqrt{5}}{\sqrt{4}} \sqrt{\frac{10 - 4}{10 - 1}} = 0.85.$$

Applying Chebyshev's theorem, we would expect at least $\frac{3}{4}$ of the sample means to fall in the interval $\mu_{\bar{X}} \pm 2\sigma_{\bar{X}} = 4 \pm (2)(0.85)$, or between 2.3 and 5.7.

For large N, relative to the sample size n, the factor $(N - n)/(N - 1)$ of the variance in Theorem 7.3 approaches 1, and $\sigma_{\bar{X}}^2$ approaches σ^2/n. Hence for large or infinite populations, whether discrete or continuous, we state the following theorem.

THEOREM 7.4 *If random samples of size n are drawn from a large or infinite population with mean μ and variance σ^2, then the sampling distribution of the sample mean $\bar{X}$ is approximately normally distributed with mean $\mu_{\bar{X}} = \mu$ and standard deviation $\sigma_{\bar{X}} = \sigma/\sqrt{n}$. Hence*

$$z = \frac{\bar{x} - \mu}{\sigma/\sqrt{n}}$$

is a value of a standard normal variable Z.

The normal approximation in Theorem 7.4 will be good if $n \geq 30$ regardless of the shape of the population. If $n < 30$, the approximation is good only if the population is not too different from a normal population. If the population is known to be normal, the sampling distribution of $\bar{X}$ will follow a normal distribution exactly, no matter how small the size of the samples.

Example 7.11 An electrical firm manufactures light bulbs that have a length of life that is approximately normally distributed, with mean equal to 800 hours and a standard deviation of 40 hours. Find the probability that a random sample of 16 bulbs will have an average life of less than 775 hours.

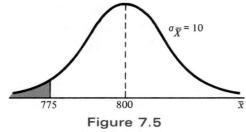

Figure 7.5
Area for Example 7.11.

Solution. The sampling distribution of $\bar{X}$ will be approximately normal, with $\mu_{\bar{x}} = 800$ and $\sigma_{\bar{x}} = 40/\sqrt{16} = 10$. The desired probability is given by the area of the shaded region in Figure 7.5. Corresponding to $\bar{x} = 775$, we find

$$z = \frac{775 - 800}{10} = -2.5,$$

and, therefore,

$$Pr(\bar{X} < 775) = Pr(Z < -2.5)$$
$$= 0.006.$$

7.5 Sampling Distribution of the Differences of Means

Suppose that we now have two populations, the first with mean μ_1 and variance σ_1^2, and the second with mean μ_2 and variance σ_2^2. Let the values of the variable $\bar{X}_1$ represent the means of random samples of size n_1 drawn from the first population, and the values of $\bar{X}_2$ represent the means of random samples of size n_2 drawn from the second population, independent of the samples from the first population. The distribution of the differences $\bar{x}_1 - \bar{x}_2$ between the two sets of independent sample means is called the *sampling distribution* of the statistic $\bar{X}_1 - \bar{X}_2$.

To illustrate, let the first population be 3, 4, and 5, with a mean

$$\mu_1 = \frac{3 + 4 + 5}{3} = 4$$

and a variance

$$\sigma_1^2 = \frac{(3 - 4)^2 + (4 - 4)^2 + (5 - 4)^2}{3} = \frac{2}{3}.$$

The second population consists of the two values 0 and 3, with a mean

Table 7.5

Means of Random Samples with Replacement
from Two Finite Populations

	From Population 1			From Population 2	
No.	*Sample*	$\bar{x}_1$	*No.*	*Sample*	$\bar{x}_2$
1	3, 3	3.0	1	0, 0, 0	0
2	3, 4	3.5	2	0, 0, 3	1
3	3, 5	4.0	3	0, 3, 0	1
4	4, 3	3.5	4	3, 0, 0	1
5	4, 4	4.0	5	0, 3, 3	2
6	4, 5	4.5	6	3, 0, 3	2
7	5, 3	4.0	7	3, 3, 0	2
8	5, 4	4.5	8	3, 3, 3	3
9	5, 5	5.0			

Table 7.6

Differences of Independent Means

	$\bar{x}_1$								
$\bar{x}_2$	3.0	3.5	4.0	3.5	4.0	4.5	4.0	4.5	5.0
0	3.0	3.5	4.0	3.5	4.0	4.5	4.0	4.5	5.0
1	2.0	2.5	3.0	2.5	3.0	3.5	3.0	3.5	4.0
1	2.0	2.5	3.0	2.5	3.0	3.5	3.0	3.5	4.0
1	2.0	2.5	3.0	2.5	3.0	3.5	3.0	3.5	4.0
2	1.0	1.5	2.0	1.5	2.0	2.5	2.0	2.5	3.0
2	1.0	1.5	2.0	1.5	2.0	2.5	2.0	2.5	3.0
2	1.0	1.5	2.0	1.5	2.0	2.5	2.0	2.5	3.0
3	0	0.5	1.0	0.5	1.0	1.5	1.0	1.5	2.0

$$\mu_2 = \frac{0 + 3}{2} = \frac{3}{2}$$

and variance

$$\sigma_2^2 = \frac{(0 - \frac{3}{2})^2 + (3 - \frac{3}{2})^2}{2} = \frac{9}{4}.$$

From the first population we draw all possible samples of size $n_1 = 2$ with replacement, and, for each sample, the mean $\bar{x}_1$ is computed. Similarly, for the second population, all possible samples of size $n_2 = 3$ are drawn with replacement, and, for each of these, we compute the mean $\bar{x}_2$. The two sets of possible samples and their means are given in Table 7.5. The 72 possible differences $\bar{x}_1 - \bar{x}_2$ are given in Table 7.6, and the frequency distribution of

Table 7.7
Sampling Distribution of
$\bar{X}_1 - \bar{X}_2$ with Replacement

$\bar{x}_1 - \bar{x}_2$	f
0	1
0.5	2
1.0	6
1.5	8
2.0	13
2.5	12
3.0	13
3.5	8
4.0	6
4.5	2
5.0	1

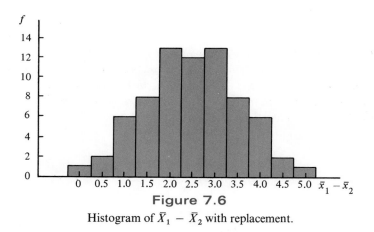

Figure 7.6
Histogram of $\bar{X}_1 - \bar{X}_2$ with replacement.

$\bar{X}_1 - \bar{X}_2$ is given in Table 7.7 with the corresponding histogram shown in Figure 7.6.

It is evident that the random variable $\bar{X}_1 - \bar{X}_2$ is approximately normally distributed. This approximation improves as n_1 and n_2 increase. Applying the corollary of Theorem 4.7 and then Theorem 7.2, we find the mean of the differences of independent sample means to be

$$\mu_{\bar{X}_1 - \bar{X}_2} = E(\bar{X}_1 - \bar{X}_2) = E(\bar{X}_1) - E(\bar{X}_2) = \mu_1 - \mu_2 = 4 - 1.5 = 2.5,$$

a result easily verified from the data in Table 7.7. Also, using Corollary 2 of Theorem 4.14 and then Theorem 7.2, we find the variance of the differences of independent means to be

$$\sigma^2_{\bar{X}_1 - \bar{X}_2} = \sigma^2_{\bar{X}_1} + \sigma^2_{\bar{X}_2} = \frac{\sigma^2_1}{n_1} + \frac{\sigma^2_2}{n_2}.$$

Thus, in our illustration,

$$\sigma^2_{\bar{X}_1 - \bar{X}_2} = \frac{\frac{2}{3}}{2} + \frac{\frac{9}{4}}{3} = \frac{13}{12},$$

a result easily checked by computing the variance for the data in Table 7.7.

The results obtained for the sampling distribution of $\bar{X}_1 - \bar{X}_2$ by sampling with replacement from a finite population are also valid for infinite populations, continuous or discrete, and for finite populations when sampling is without replacement provided the population sizes, N_1 and N_2, are large relative to the sample sizes, n_1 and n_2, respectively. However, if the populations are small and sampling is without replacement, then we must compute $\sigma_{\bar{X}_1}$ and $\sigma_{\bar{X}_2}$ by the formula for $\sigma_{\bar{X}}$ in Theorem 7.3.

In this text we shall concern ourselves with the sampling distribution of the differences of independent means only when the size of the populations from which the samples are selected is large.

THEOREM 7.5 *If independent samples of size n_1 and n_2 are drawn from two large or infinite populations, discrete or continuous, with means μ_1 and μ_2, and variances σ_1^2 and σ_2^2, respectively, then the sampling distribution of the differences of means, $\bar{X}_1 - \bar{X}_2$, is approximately normally distributed with mean and standard deviation given by*

$$\mu_{\bar{X}_1 - \bar{X}_2} = \mu_1 - \mu_2,$$

$$\sigma_{\bar{X}_1 - \bar{X}_2} = \sqrt{\frac{\sigma_1^2}{n_1} + \frac{\sigma_2^2}{n_2}}.$$

Hence

$$z = \frac{(\bar{x}_1 - \bar{x}_2) - (\mu_1 - \mu_2)}{\sqrt{(\sigma_1^2/n_1) + (\sigma_2^2/n_2)}}$$

is a value of a standard normal variable Z.

If both n_1 and n_2 are greater than or equal to 30, the normal approximation for the distribution of $\bar{X}_1 - \bar{X}_2$ is very good.

Example 7.12 The television picture tubes of manufacturer *A* have a mean lifetime of 6.5 years and a standard deviation of 0.9 year, while those of manufacturer *B* have a mean lifetime of 6.0 years and a standard deviation of 0.8 year. What is the probability that a random sample of 36 tubes from manufacturer *A* will have a mean lifetime that is at least 1 year more than the mean lifetime of a sample of 49 tubes from manufacturer *B*?

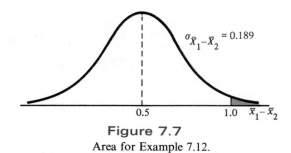

Figure 7.7
Area for Example 7.12.

Solution. We are given the following information:

Population 1	Population 2
$\mu_1 = 6.5$	$\mu_2 = 6.0$
$\sigma_1 = 0.9$	$\sigma_2 = 0.8$
$n_1 = 36$	$n_2 = 49$

Using Theorem 7.5, the sampling distribution of $\bar{X}_1 - \bar{X}_2$ will have a mean and standard deviation given by

$$\mu_{\bar{X}_1 - \bar{X}_2} = 6.5 - 6.0 = 0.5,$$

$$\sigma_{\bar{X}_1 - \bar{X}_2} = \sqrt{\frac{0.81}{36} + \frac{0.64}{49}} = 0.189.$$

The probability that the mean of 36 tubes from manufacturer A will be at least 1 year longer than the mean of 49 tubes from manufacturer B is given by the area of the shaded region in Figure 7.7. Corresponding to the value $\bar{x}_1 - \bar{x}_2 = 1.0$, we find

$$z = \frac{1.0 - 0.5}{0.189} = 2.646,$$

and hence

$$\Pr(\bar{X}_1 - \bar{X}_2 \geq 1.0) = \Pr(Z > 2.646)$$
$$= 1 - \Pr(Z < 2.646)$$
$$= 1 - 0.9959$$
$$= 0.0041.$$

The distribution of $\bar{X}_1 - \bar{X}_2$ in Theorem 7.5 suggests a more general result, which we state without proof in the following theorem.

THEOREM 7.6 *If the random variables X and Y are independent and normally distributed with means μ_X and μ_Y and variances σ_X^2 and σ_Y^2, respectively, then the distribution of the difference $X - Y$ is normally distributed with mean*

$$\mu_{X-Y} = \mu_X - \mu_Y$$

and variance

$$\sigma_{X-Y}^2 = \sigma_X^2 + \sigma_Y^2.$$

7.6 *t* Distribution

Most of the time we are not fortunate enough to know the variance of the population from which we select our random samples. For samples of size $n \geq 30$, a good estimate of σ^2 is provided by calculating s^2. What then happens to the z values of Theorem 7.4 if we replace σ^2 by s^2? As long as s^2 is a good estimate of σ^2 and does not vary much from sample to sample, which is usually the case for $n \geq 30$, the values $(\bar{x} - \mu)/(s/\sqrt{n})$ are still approximately distributed as a standard normal variable Z, and Theorem 7.4 is valid.

If the sample size is small ($n < 30$), the values of s^2 fluctuate considerably from sample to sample and the distribution of the values $(\bar{x} - \mu)/(s/\sqrt{n})$ is no longer a standard normal distribution. We are now dealing with the distribution of a statistic that we shall call T, whose values are given by

$$t = \frac{\bar{x} - \mu}{s/\sqrt{n}}.$$

In 1908 W. S. Gosset published a paper in which he derived the equation of the probability distribution of T. At the time, Gosset was employed by an Irish brewery that disallowed publication of research by members of its staff. To circumvent this restriction, he published his work secretly under the name "Student." Consequently the distribution of T is usually called the *Student-t distribution*, or simply the *t distribution*. In deriving the equation of this distribution, Gosset assumed the samples were selected from a normal population. Although this would seem to be a very restrictive assumption, it can be shown that nonnormal populations possessing bell-shaped distributions will still provide values of T that approximate the t distribution very closely.

The actual mathematical formula is omitted here, since the areas under the curve have been tabulated in sufficient detail to meet the requirements of most problems. However, to evaluate probabilities associated with the t distribution we need to understand some of the characteristics of the t curve.

The distribution of T is similar to the distribution of Z, in that they both are symmetrical about a mean of zero. Both distributions are bell-shaped but the t distribution is more variable, owing to the fact that the t values depend on the fluctuations of two quantities, $\bar{x}$ and s^2, whereas the z values depend only on the changes of $\bar{x}$ from sample to sample. The distribution of T differs from that of Z in that the variance depends on the sample size n and is always greater than 1. Only when the sample size $n \to \infty$ will the two distributions become the same.

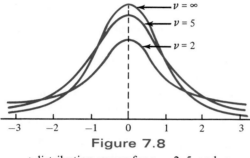

Figure 7.8

t distribution curves for $v = 2, 5,$ and ∞.

The divisor, $n - 1$, that appears in the formula for s^2 is called the number of *degrees of freedom* associated with s^2. If $\bar{x}$ and s^2 are computed from samples of size n, the corresponding values of t are said to belong to a t distribution with v degrees of freedom, where $v = n - 1$. Thus we have a different t curve or t distribution for each possible sample size, a curve that becomes more and more like the standard normal curve as $n \to \infty$. In Figure 7.8 we show the relationship between a standard normal distribution ($v = \infty$) and t distributions with 2 and 5 degrees of freedom.

The curve in Figure 7.8 for $v = 5$ represents the distribution of all t values computed from repeated random samples of size 6 taken from a normal population. Similarly the curve for $v = 2$ represents the distribution of all t values computed from samples of size 3.

THEOREM 7.7 *If $\bar{x}$ and s^2 are the mean and variance, respectively, of a random sample of size n taken from a normal population having the mean μ and unknown variance σ^2, then*

$$t = \frac{\bar{x} - \mu}{s/\sqrt{n}}$$

is a value of a random variable T having the t distribution with $v = n - 1$ degrees of freedom.

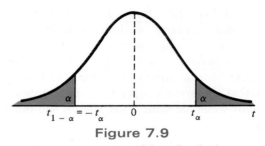

Figure 7.9

Symmetry property of the t distribution.

The probability that a random sample produces a t value falling between any two specified values is equal to the area under the curve of the t distribution between the two ordinates corresponding to the specified values. It would be a tedious task to attempt to set up separate tables giving the areas between every conceivable pair of ordinates for all values of $n < 30$. Table A.5 gives only those t values above which we find a specified area α, where α is 0.1, 0.05, 0.025, 0.01, or 0.005. This table is set up differently from the table of normal-curve areas in that the areas are now the column headings and the entries are the t values. The left column gives the degrees of freedom. It is customary to let t_α represent the t value above which we find an area equal to α. Hence the t value with 10 degrees of freedom leaving an area of 0.025 to the right is $t_{0.025} = 2.228$. Since the t distribution is symmetric about a mean of zero, we have $t_{1-\alpha} = -t_\alpha$; that is, the t value leaving an area of $1 - \alpha$ to the right and therefore an area of α to the left is equal to the negative t value that leaves an area of α in the right tail of the distribution (see Figure 7.9). For a t distribution with 10 degrees of freedom we have $t_{0.975} = -t_{0.025} = -2.228$. This means that the t value of a random sample of size 11, selected from a normal population, will fall between -2.228 and 2.228, with probability equal to 0.95.

Exactly 95% of a t distribution with $n - 1$ degrees of freedom lies between $-t_{0.025}$ and $t_{0.025}$. Therefore a t value falling below $-t_{0.025}$ or above $t_{0.025}$ would tend to make one believe that either a rare event has taken place or our assumption about μ is in error. The importance of μ will determine the length of the interval for an acceptable t value. In other words, if you do not mind having the true mean slightly different than what you claim it to be, you might choose a wide interval from $-t_{0.01}$ to $t_{0.01}$ in which the t value should fall. A t value falling at either end of the interval, but within the interval, would lead us to believe that our assumed value for μ is correct, although it is very probable that some other value close to μ is the true value. If μ is to be known with a high degree of accuracy, a short interval such as $-t_{0.05}$ to $t_{0.05}$ should be used. In this case, a t value falling outside the interval would lead you to believe that the assumed value of μ is in error when it is entirely possible that it is correct. The problems connected with the establishment of proper intervals in testing hypotheses concerning the parameter μ will be treated in Chapter 9.

Example 7.13 A manufacturer of light bulbs claims that his bulbs will burn on the average 500 hours. To maintain this average, he tests 25 bulbs each month. If the computed t value falls between $-t_{0.05}$ and $t_{0.05}$, he is satisfied with his claim. What conclusion should he draw from a sample that has a mean $\bar{x} = 518$ hours and a standard deviation $s = 40$ hours? Assume the distribution of burning times to be approximately normal.

Solution. From Table A.5 we find $t_{0.05} = 1.711$ for 24 degrees of freedom. Therefore the manufacturer is satisfied with his claim if a sample of 25 bulbs yields a t value between -1.711 and 1.711. If $\mu = 500$, then

$$t = \frac{518 - 500}{40/\sqrt{25}} = 2.25,$$

a value well above 1.711. The probability of obtaining a t value, with $v = 24$, equal to or greater than 2.25 is approximately 0.02. If $\mu > 500$, the value of t computed from the sample would be more reasonable. Hence the manufacturer is likely to conclude that his bulbs are a better product than he thought.

7.7 Chi-Square Distribution

If samples of size n are drawn repeatedly from a normal population with variance σ^2, and the sample variance s^2 is computed for each sample, we obtain the values of a statistic S^2. The sampling distribution of S^2 has little practical application in statistics. Instead we shall consider the distribution of a random variable X^2, called *chi-square*, whose values are calculated from each sample by the formula

$$\chi^2 = \frac{(n - 1)s^2}{\sigma^2}.$$

The distribution of X^2 is referred to as the *chi-square distribution*, with $v = n - 1$ degrees of freedom. As before, v is equal to the divisor in the calculation of s^2.

Obviously the χ^2 values cannot be negative, and therefore the curve of the chi-square distribution cannot be symmetric about $\chi^2 = 0$. The mathematical equation for the curve is rather complicated and fortunately may be omitted. One could easily obtain an experimental sampling distribution of X^2 by selecting several random samples of size n from a normal population and computing the χ^2 value for each sample. The χ^2 curve could then be approximated by drawing a smooth curve over the histogram of the χ^2 values. The

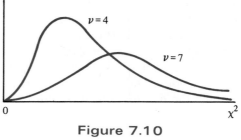

Figure 7.10

Chi-square curves for $v = 4$ and $v = 7$.

curve will have the appearance of those illustrated in Figure 7.10, provided that samples of size 5 or 8 are selected.

The curve in Figure 7.10 for $v = 4$ represents the distribution of χ^2 values computed from all possible samples of size 5 from a normal population having the variance σ^2. Similarly the curve for $v = 7$ represents the distribution of χ^2 values computed from all possible samples of size 8.

> **THEOREM 7.8** *If s^2 is the variance of a random sample of size n taken from a normal population having the variance σ^2, then*
>
> $$\chi^2 = \frac{(n-1)s^2}{\sigma^2}$$
>
> *is a value of a random variable X^2 having the chi-square distribution with $v = n - 1$ degrees of freedom.*

The probability that a random sample produces a χ^2 value greater than some specified value is equal to the area under the curve to the right of this value. It is customary to let χ_α^2 represent the χ^2 value above which we find an area of α. This is illustrated by the shaded region in Figure 7.11.

Table A.6 gives values of χ_α^2 for various values of α and v. The areas, α, are the column headings; the degrees of freedom, v, are given in the left column; and the table entries are the χ^2 values. Hence the χ^2 value with 7 degrees of freedom, leaving an area of 0.05 to the right, is $\chi_{0.05}^2 = 14.067$. Owing to lack of symmetry, we must also use the tables to find $\chi_{0.95}^2 = 2.167$ for $v = 7$.

Exactly 95% of a chi-square distribution with $n - 1$ degrees of freedom lies between $\chi_{0.975}^2$ and $\chi_{0.025}^2$. A χ^2 value falling to the right of $\chi_{0.025}^2$ is not likely to occur unless our assumed value of σ^2 is too small. Likewise a χ^2 value falling to the left of $\chi_{0.975}^2$ is unlikely unless our assumed value of σ^2 is too large. In other words it is possible to have a χ^2 value to the left of $\chi_{0.975}^2$ or to the right of $\chi_{0.025}^2$ when σ^2 is correct, but if this should occur it is more probable that the assumed value of σ^2 is in error.

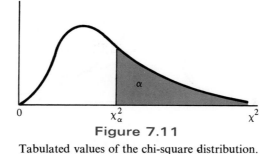

Figure 7.11

Tabulated values of the chi-square distribution.

Example 7.14 A manufacturer of car batteries guarantees that his batteries will last, on the average, 3 years with a standard deviation of 1 year. If 5 of these batteries have lifetimes of 1.9, 2.4, 3.0, 3.5, and 4.2 years, is the manufacturer still convinced that his batteries have a standard deviation of 1 year?

Solution. We first find the sample variance:

$$s^2 = \frac{(5)(48.26) - (15)^2}{(5)(4)} = 0.815.$$

Then

$$\chi^2 = \frac{(4)(0.815)}{1} = 3.26$$

is a value from a chi-square distribution with 4 degrees of freedom. Since 95% of the χ^2 values with 4 degrees of freedom fall between 0.484 and 11.143, the computed value with $\sigma^2 = 1$ is reasonable, and therefore the manufacturer has no reason to suspect that the standard deviation is different than 1 year.

7.8 *F* Distribution

One of the most important distributions in applied statistics is the *F distribution*. Theoretically we might define the *F* distribution to be the ratio of two independent chi-square distributions, each divided by their degrees of freedom. Hence if *f* is a value of the random variable *F*, we have

$$f = \frac{\chi_1^2/v_1}{\chi_2^2/v_2} = \frac{s_1^2/\sigma_1^2}{s_2^2/\sigma_2^2} = \frac{\sigma_2^2 s_1^2}{\sigma_1^2 s_2^2},$$

where χ_1^2 is a value of a chi-square distribution with $v_1 = n_1 - 1$ degrees of

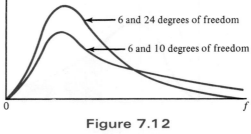

Figure 7.12
Typical F distributions.

freedom and χ_2^2 is a value of a chi-square distribution with $v_2 = n_2 - 1$ degrees of freedom. We say that f is a value of the F distribution with v_1 and v_2 degrees of freedom.

To obtain an f value, first select a random sample of size n_1 from a normal population having a variance σ_1^2 and compute s_1^2/σ_1^2. An independent sample of size n_2 is then selected from a second normal population with variance σ_2^2 and s_2^2/σ_2^2 is computed. The ratio of the two quantities s_1^2/σ_1^2 and s_2^2/σ_2^2 produces an f value. The distribution of all possible f values where s_1^2/σ_1^2 is the numerator and s_2^2/σ_2^2 is the denominator is called the F distribution, with v_1 and v_2 degrees of freedom. If we consider all possible ratios where s_2^2/σ_2^2 is the numerator and s_1^2/σ_1^2 is the denominator, then we have a distribution of values that also possess an F distribution but with v_2 and v_1 degrees of freedom. The number of degrees of freedom associated with the sample variance in the numerator is always stated first, followed by the number of degrees of freedom associated with the sample variance in the denominator. Thus the curve of the F distribution depends not only on the two parameters v_1 and v_2 but also on the order in which we state them. Once these two values are given, we can identify the curve. Typical f curves are shown in Figure 7.12.

The foregoing discussion may be summarized in the following theorem.

> **THEOREM 7.9** *If s_1^2 and s_2^2 are the variances of independent random samples of size n_1 and n_2 taken from normal populations with variances σ_1^2 and σ_2^2, respectively, then*
>
> $$f = \frac{s_1^2/\sigma_1^2}{s_2^2/\sigma_2^2} = \frac{\sigma_2^2 s_1^2}{\sigma_1^2 s_2^2}$$
>
> *is a value of a random variable F having the F distribution with $v_1 = n_1 - 1$ and $v_2 = n_2 - 1$ degrees of freedom.*

Let f_α be the f value above which we find an area equal to α. This is illustrated by the shaded region in Figure 7.13. Table A.7 gives values of f_α only for $\alpha = 0.05$ and $\alpha = 0.01$ for various combinations of the degrees of freedom

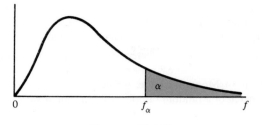

Figure 7.13

Tabulated values of the F distribution.

v_1 and v_2. Hence the f value with 6 and 10 degrees of freedom, leaving an area of 0.05 to the right, is $f_{0.05} = 3.22$. By means of the following theorem, Table A.7 can be used to find values of $f_{0.95}$ and $f_{0.99}$.

> **THEOREM 7.10** *Writing $f_\alpha(v_1, v_2)$ for f_α with v_1 and v_2 degrees of freedom, then*
>
> $$f_{1-\alpha}(v_1, v_2) = \frac{1}{f_\alpha(v_2, v_1)}.$$

Thus the f value with 6 and 10 degrees of freedom, leaving an area of 0.95 to the right, is

$$f_{0.95}(6, 10) = \frac{1}{f_{0.05}(10, 6)} = \frac{1}{4.06} = 0.246.$$

The F distribution is applied primarily in Chapter 11 in the analysis of variance, where we wish to test the equality of several means simultaneously. In Chapters 8 and 9 we shall use the F distribution to make inferences concerning the variance of two normal populations.

EXERCISES

1. Define suitable populations from which the following samples are selected:
   (a) One thousand homes are called by telephone in the city of Richmond and asked to name the television program that they are now watching.
   (b) A coin is flipped 50 times and 32 heads are recorded.
   (c) Two hundred pairs of a new type of combat boot were tested for durability in Vietnam and, on the average, lasted 2 months.
   (d) On the practice tee a golfer hits the ball 210, 260, 244, and 225 yards with his driver.

2. In a random sample of 18 students at Roanoke College the following number of days absences were recorded for the previous semester: 1, 3, 4, 0, 4, 2, 3, 1, 2, 3, 0, 4, 1, 1, 1, 5, 1, and 0. Find the mean, median, and mode.

3. The number of trout caught by 8 fishermen on the first day of the trout season are 7, 4, 6, 7, 4, 4, 8, and 7. If these 8 values represent the catch of a random sample of fishermen at Smith Mountain Lake, define a suitable population. If the values represent the catch of a random sample of fisherman at various lakes and streams in Montgomery County, define a suitable population. Find the mean, median, and mode for the data.

4. The number of goals scored by a random sample of 16 hockey players from the National Hockey League for a given season are 5, 3, 21, 10, 2, 7, 0, 30, 19, 6, 7, 4, 10, 5, 7, and 24. Find the mean, median, and mode.

5. Find the mean, median, and mode for the sample 18, 10, 11, 98, 22, 15, 11, 25, and 17. Which value appears to be the best measure of the center of our data? Give reasons for your preference.

6. Calculate the range and variance for the data of Exercise 2.

7. Calculate the range and variance for the data of Exercise 4.

8. The grade-point averages of 15 college seniors selected at random from the graduating class are as follows:

2.3	3.4	2.9
2.6	2.1	2.4
3.1	2.7	2.6
1.9	2.0	3.6
2.1	1.8	2.1

Calculate the standard deviation.

9. Show that the sample variance is unchanged if a constant is added to or subtracted from each value in the sample.

10. If each observation in a sample is multiplied by k, show that the sample variance becomes k^2 times its original value.

11. (a) Calculate the variance of the sample 3, 5, 8, 7, 5, and 7.
   (b) Without calculating, state the variance of the sample 6, 10, 16, 14, 10, and 14.
   (c) Without calculating, state the variance of the sample 25, 27, 30, 29, 27, and 29.

12. The *mean deviation* of a sample of n observations is defined to be $\sum_{i=1}^{n} |x_i - \bar{x}|/n$. Find the mean deviation of the sample 2, 3, 5, 7, and 8.

13. A finite population consists of the numbers 2, 4, and 6.
   (a) Construct a frequency histogram for the sampling distribution of $\bar{X}$ when samples of size 4 are drawn with replacement.
   (b) Verify that $\mu_{\bar{X}} = \mu$ and $\sigma_{\bar{X}}^2 = \sigma^2/n$.
   (c) Between what two values would you expect the middle 68% of the sample means to fall?

14. If in Exercise 13, a sample of size 54 is drawn with replacement, what is the probability that the sample mean will be greater than 4.1 but less than 4.4? Assume the means to be measured to the nearest tenth.

15. A population consists of the numbers 1, 1, 1, 2, 2, 3, and 4.
    (a) List all possible samples of size 2 that can be drawn from this population without replacement.
    (b) Verify that $\mu_{\bar{X}} = \mu$ and $\sigma_{\bar{X}} = (\sigma/\sqrt{n})\sqrt{(N - n)/(N - 1)}$.
    (c) Between what two values would you expect at least $\frac{8}{9}$ of the sample means to fall?

16. If all possible samples of size 16 are drawn from a normal population with mean equal to 50 and standard deviation equal to 5, what is the probability that a sample mean $\bar{X}$ will fall in the interval from $\mu_{\bar{X}} - 1.9\sigma_{\bar{X}}$ to $\mu_{\bar{X}} - 0.4\sigma_{\bar{X}}$? Assume that the sample means can be measured to any degree of accuracy.

17. If the size of a sample is 36 and the standard error of the mean is 2, what must the size of the sample become if the standard error is to be reduced to 1.2?

18. A soft-drink machine is regulated so that the amount of drink dispensed is approximately normally distributed with a mean of 7 ounces per cup and a standard deviation equal to 0.5 ounce. Periodically the machine is checked by taking a sample of 9 drinks and computing the average content. If the mean, $\bar{X}$, of the 9 drinks falls within the interval $\mu_{\bar{X}} \pm 2\sigma_{\bar{X}}$, the machine is thought to be operating satisfactorily; otherwise, adjustments must be made. What action should one take if a sample of 9 drinks has a mean content of 7.4 ounces?

19. The heights of 1000 students are approximately normally distributed with a mean of 68.5 inches and a standard deviation of 2.7 inches. If 200 random samples of size 25 are drawn from this population and the means recorded to the nearest tenth of an inch, determine:
    (a) The expected mean and standard deviation of the sampling distribution of the mean.
    (b) The number of sample means that fall between 67.9 and 69.2 inclusive.
    (c) The number of sample means falling below 67.0.

20. Let $\bar{X}_1$ represent the mean of a sample of size $n_1 = 2$, with replacement, from the finite population 2, 3, and 7. Similarly let $\bar{X}_2$ represent the mean of a sample of size $n_2 = 2$, with replacement, from the population 1, 1, and 3.
    (a) Construct a frequency histogram for the sampling distribution of $\bar{X}_1 - \bar{X}_2$.
    (b) Find $\mu_{\bar{X}_1 - \bar{X}_2}$ and $\sigma^2_{\bar{X}_1 - \bar{X}_2}$.

21. A random sample of size 25 is taken from a normal population having a mean of 80 and a standard deviation of 5. A second random sample of size 36 is taken from a different normal population having a mean of 75 and a standard deviation of 3. Find the probability that the sample mean computed from the 25 measurements will exceed the sample mean computed from the 36 measurements by at least 3.4 but less than 5.9. Assume the means to be measured to the nearest tenth.

22. The mean score for freshmen on an aptitude test, at a certain college, is 540, with a standard deviation of 50. What is the probability that two groups of students selected at random, consisting of 32 and 50 students, respectively, will differ in their mean scores by (a) more than 20 points, (b) an amount between 5 and 10 points? Assume the means to be measured to any degree of accuracy.

23. (a) Find $t_{0.025}$ when $v = 17$.
    (b) Find $t_{0.99}$ when $v = 10$.
    (c) Find t_α such that $\Pr(-t_\alpha < T < t_\alpha) = 0.90$ when $v = 23$.

24. A normal population with unknown variance has a mean of 20. Is one likely to obtain a random sample of size 9 from this population with a mean of 24 and a standard deviation of 4.1? If not, what conclusion would you draw?

25. A cigarette manufacturer claims that his cigarettes have an average nicotine content of 18.3 milligrams. If a random sample of 8 cigarettes of this type have nicotine contents of 20, 17, 21, 19, 22, 21, 20, and 16 milligrams, would you agree with the manufacturer's claim?

26. For a chi-square distribution find:
    (a) $\chi^2_{0.01}$ with $v = 18$.
    (b) $\chi^2_{0.975}$ with $v = 29$.
    (c) χ^2_α such that $Pr(X^2 < \chi^2_\alpha) = 0.99$ with $v = 4$.

27. Find the probability that a random sample of 25 observations, from a normal population with variance $\sigma^2 = 6$, will have a variance s^2 (a) greater than 9.1, (b) between 3.462 and 10.745. Assume the sample variances to be continuous measurements.

28. A placement test has been given for the past 5 years to college freshmen with a mean $\mu = 74$ and a variance $\sigma^2 = 8$. Would a school consider these values valid today if 20 students obtained a mean $\bar{x} = 72$ and a variance $s^2 = 16$ on this test?

29. For an F distribution find:
    (a) $f_{0.05}$ with $v_1 = 7$ and $v_2 = 15$.
    (b) $f_{0.05}$ with $v_1 = 15$ and $v_2 = 7$.
    (c) $f_{0.01}$ with $v_1 = 24$ and $v_2 = 19$.
    (d) $f_{0.95}$ with $v_1 = 19$ and $v_2 = 24$.
    (e) $f_{0.99}$ with $v_1 = 28$ and $v_2 = 12$.

30. If S_1^2 and S_2^2 represent the variances of independent random samples of size $n_1 = 25$ and $n_2 = 31$, taken from normal populations with variances $\sigma_1^2 = 10$ and $\sigma_2^2 = 15$, respectively, find the $Pr(S_1^2/S_2^2 > 1.26)$.

31. If S_1^2 and S_2^2 represent the variances of independent random samples of size $n_1 = 8$ and $n_2 = 12$, taken from normal populations with equal variances, find the $Pr(S_1^2/S_2^2 < 4.89)$.

ESTIMATION THEORY CHAPTER 8

8.1 Introduction

The theory of *statistical inference*, which has more recently become known as *decision theory*, may be defined to be those methods by which one makes inferences or generalizations about a population based on information obtained from samples selected from the population. In this chapter we shall consider inferences about unknown population parameters such as the mean, proportion, and the standard deviation by computing statistics from random samples and applying the theory of sampling distributions from Chapter 7.

Decision theory may be divided into two major areas: *estimation* and *tests of hypotheses*. We shall treat these two areas separately, dealing with the theory of estimation in this chapter and the theory of hypothesis testing in Chapter 9. To distinguish clearly between the two areas, consider the following examples. A candidate for public office may wish to estimate the true proportion of voters favoring him by obtaining the opinions from a random sample of 100 eligible voters. The fraction of voters in the sample favoring the candidate could be used as an estimate of the true proportion of the population of voters. A knowledge of the sampling distribution of a proportion enables one to establish the degree of accuracy of our estimate. This problem falls in the area of estimation.

Now consider the case in which a housewife is interested in finding out whether brand *A* floor wax is more scuff-resistant than brand *B* floor wax. She might hypothesize that brand *A* is better than brand *B* and, after proper testing, accept or reject this hypothesis. In this example we do not attempt to estimate a parameter, but instead we try to arrive at a correct decision about a prestated hypothesis. Once again we are dependent upon sampling theory to provide us with some measure of accuracy for our decision.

8.2 Classical Methods of Estimation

An estimate of a population parameter may be given as a *point estimate* or as an *interval estimate*. A point estimate of some population parameter θ is a single value $\hat{\theta}$ of a statistic $\hat{\Theta}$. For example the value $\bar{x}$ of the statistic $\bar{X}$, computed from a sample of size n, is a point estimate of the population parameter μ. Similarly $\hat{p} = x/n$ is a point estimate of the true proportion p for a binomial experiment.

An interval estimate of the parameter θ is an interval of the form $a < \theta < b$, where a and b depend on the point estimate $\hat{\theta}$ for the particular sample chosen and also on the sampling distribution of the statistic $\hat{\Theta}$. Thus a random sample of SAT verbal scores for students of the entering freshman class might produce an interval from 530 to 550 within which we expect to find the true average of all SAT verbal scores for the freshman class. The values of the end points, 530 and 550, will depend on the computed sample mean $\bar{x}$ and the sampling distribution of $\bar{X}$. As the sample size increases, we know that $\sigma_{\bar{X}}^2 = \sigma^2/n$ decreases, and consequently our estimate is likely to be closer to the parameter μ, resulting in a shorter interval. Thus the interval estimate indicates, by its length, the accuracy of the point estimate.

The statistic that one uses to obtain a point estimate is called an *estimator* or a *decision function*. Hence the decision function S, which is a function of the random sample, is an estimator of σ and the estimate s is the "*action*" taken. Different samples will generally lead to different actions or estimates.

DEFINITION *The set of all possible actions that can be taken in an estimation problem is called the* action *space or* decision *space.*

An estimator is not expected to estimate the population parameter without error. We do not expect $\bar{X}$ to estimate μ exactly, but we certainly hope that it is not too far off. For a particular sample it is possible to obtain a closer estimate of μ by using the median $\tilde{X}$ as an estimator. Consider, for instance, a sample consisting of the values 2, 5, and 11 from a population whose mean is 4 but supposedly unknown. We would estimate μ to be $\bar{x} = 6$, using the sample mean as our estimate, or $\tilde{x} = 5$, using the median as our estimate. In this case the estimator $\tilde{X}$ produces an estimate closer to the true parameter than that of the estimator $\bar{X}$. On the other hand, if our random sample contains the values, 2, 6, and 7, then $\bar{x} = 5$ and $\tilde{x} = 6$, so that $\bar{X}$ is now the better estimator. Not knowing the true value of μ, we must decide in advance whether to use $\bar{X}$ or $\tilde{X}$ as our estimator.

What are the desirable properties of a "good" decision function that would influence us to choose one estimator rather than another? Let $\hat{\Theta}$ be an estimator whose value $\hat{\theta}$ is a point estimate of some unknown population parameter θ. Certainly we would like the sampling distribution of $\hat{\Theta}$ to have a mean equal to the parameter estimated. In our definition of the sample variance in Chapter 7, it was necessary to divide by $n - 1$, rather than n, if we were to have $E(S^2) = \sigma^2$. An estimator possessing this property is said to be *unbiased*.

> DEFINITION *A statistic* $\hat{\Theta}$ *is said to be an* unbiased estimator *of the parameter* θ *if* $\mu_{\hat{\Theta}} = E(\hat{\Theta}) = \theta$.

If $\hat{\Theta}_1$ and $\hat{\Theta}_2$ are two unbiased estimators of the same population parameter θ, we would choose the estimator whose sampling distribution has the smallest variance. Hence, if $\sigma^2_{\hat{\Theta}_1} < \sigma^2_{\hat{\Theta}_2}$, we say that $\hat{\Theta}_1$ is a *more efficient* estimator than $\hat{\Theta}_2$.

> DEFINITION *If we consider all possible unbiased estimators of some parameter* θ, *the one with the smallest variance is called the* most efficient estimator *of* θ.

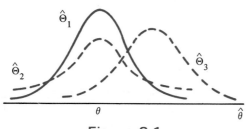

Figure 8.1

Sampling distributions of different estimators
of θ.

In Figure 8.1 we illustrate the sampling distributions of three different estimators $\hat{\Theta}_1$, $\hat{\Theta}_2$, and $\hat{\Theta}_3$, all estimating θ. It is clear that only $\hat{\Theta}_1$ and $\hat{\Theta}_2$ are unbiased, since their distributions are centered at θ. The estimator $\hat{\Theta}_1$ has a smaller variance than $\hat{\Theta}_2$ and is therefore more efficient. Hence our choice for an estimator of θ, among the three considered, would be $\hat{\Theta}_1$.

For normal populations one can show that both $\bar{X}$ and $\tilde{X}$ are unbiased estimators of the population mean μ, but the variance of $\bar{X}$ is smaller than the variance of $\tilde{X}$. Thus both estimates $\bar{x}$ and $\tilde{x}$ will, on the average, equal the population mean μ, but $\tilde{x}$ is likely to be closer to μ for a given sample.

Different samples yield different values of $\hat{\theta}$ and therefore produce different interval estimates of the population parameter θ. Some of these intervals will contain θ and others will not. The sampling distribution of $\hat{\Theta}$ will enable us to find a and b for all possible samples such that any specified fraction of these intervals will contain θ. Therefore, if a and b are computed so that 0.95 of all possible intervals, in repeated sampling, would contain θ, then we have a probability equal to 0.95 of selecting one of the samples that will produce an interval containing θ. This interval, computed from the selected random sample, is called a 95% *confidence interval*. In other words we are 95% confident that our computed interval does in fact contain the population parameter θ. Generally speaking, the distribution of $\hat{\Theta}$ enables us to compute the end points, a and b, so that any specified fraction $1 - \alpha$, $0 < \alpha < 1$, of the intervals computed from all possible samples contain the parameter θ. The interval computed from a particular sample is then called a $(1 - \alpha)100\%$ *confidence interval*. The fraction $1 - \alpha$ is called the *confidence coefficient*, and the end points, a and b, are called the *confidence limits* or *fiducial limits*.

The longer the confidence interval, the more confident we can be that the given interval contains the unknown parameter. Of course it is better to be 95% confident that the average life of a certain television tube is between 6 and 7 years than to be 99% confident that it is between 3 and 10 years. Ideally, we prefer a short interval with a high degree of confidence.

In the next several sections of this chapter we shall present the most efficient estimators of frequently used population parameters. Using these estimators and their sampling distributions, we shall construct the corresponding confidence intervals.

8.3 Estimating the Mean

A point estimator of the population mean μ is given by the statistic $\bar{X}$. The sampling distribution of $\bar{X}$ is centered at μ, and in most applications the variance is smaller than that of any other estimator. Thus the sample mean $\bar{x}$ will be used as a point estimate for the population mean μ. Recall that $\sigma_{\bar{X}}^2 = \sigma^2/n$, so that a large sample will yield a value of $\bar{X}$ that comes from a sampling distribution with a small variance. Hence $\bar{x}$ is likely to be a very accurate estimate of μ when n is large.

Let us now consider the interval estimate of μ. If our sample is selected from a normal population or, failing this, if n is sufficiently large, we can establish a confidence interval for μ by considering the sampling distribution of $\bar{X}$. According to Theorems 7.2 and 7.4 we can expect the sampling distribution of $\bar{X}$ to be approximately normally distributed with mean $\mu_{\bar{X}} = \mu$ and standard deviation $\sigma_{\bar{X}} = \sigma/\sqrt{n}$. Therefore we can assert with a probability of 0.95 that the standard normal variable Z will fall between -1.96 and 1.96, where

$$Z = \frac{\bar{X} - \mu}{\sigma/\sqrt{n}}.$$

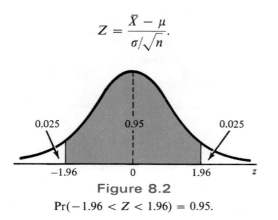

| 0.025 | 0.95 | 0.025 |

| −1.96 | 0 | 1.96 | z |

Figure 8.2

$\Pr(-1.96 < Z < 1.96) = 0.95.$

Referring to Figure 8.2, we write

$$\Pr(-1.96 < Z < 1.96) = 0.95.$$

Substituting for Z, we can state equivalently that

$$\Pr\left(-1.96 < \frac{\bar{X} - \mu}{\sigma/\sqrt{n}} < 1.96\right) = 0.95.$$

Multiplying each term in the inequality by $\sigma/\sqrt{n}$, we have

$$\Pr\left(\frac{-1.96\sigma}{\sqrt{n}} < \bar{X} - \mu < \frac{1.96\sigma}{\sqrt{n}}\right) = 0.95.$$

Subtract $\bar{X}$ from each term and multiply by -1 (reversing the sense of the inequalities) to obtain

$$\Pr\left(\bar{X} - \frac{1.96\sigma}{\sqrt{n}} < \mu < \bar{X} + \frac{1.96\sigma}{\sqrt{n}}\right) = 0.95.$$

We are now stating with a probability of 0.95 that the *random interval* $\bar{X} \pm 1.96\sigma/\sqrt{n}$ contains the parameter μ. From a particular sample of size n the mean $\bar{x}$ is computed and the 95% confidence interval for μ is

$$\bar{x} - \frac{1.96\sigma}{\sqrt{n}} < \mu < \bar{x} + \frac{1.96\sigma}{\sqrt{n}},$$

where 1.96 is the critical value, corresponding to a confidence coefficient of 0.95, and the end points $\bar{x} - 1.96\sigma/\sqrt{n}$ and $\bar{x} + 1.96\sigma/\sqrt{n}$ are the confidence limits. To obtain some other degree of confidence, one must choose a different confidence coefficient, which will in turn change the length of the interval.

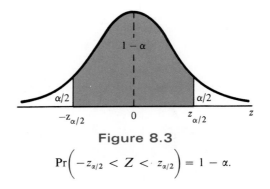

Figure 8.3

$$\Pr\left(-z_{\alpha/2} < Z < z_{\alpha/2}\right) = 1 - \alpha.$$

Let us now generalize the foregoing discussion and find a $(1 - \alpha)100\%$ confidence interval where $0 < \alpha < 1$. Writing $z_{\alpha/2}$ for the z value above which we find an area of $\alpha/2$, we can see from Figure 8.3 that

$$\Pr(-z_{\alpha/2} < Z < z_{\alpha/2}) = 1 - \alpha.$$

Hence

$$\Pr\left(-z_{\alpha/2} < \frac{\bar{X} - \mu}{\sigma/\sqrt{n}} < z_{\alpha/2}\right) = 1 - \alpha.$$

Multiplying each term in the inequality by $\sigma/\sqrt{n}$, and then subtracting $\bar{X}$ from each term and multiplying by -1, we have

$$\Pr\left(\bar{X} - z_{\alpha/2}\frac{\sigma}{\sqrt{n}} < \mu < \bar{X} + z_{\alpha/2}\frac{\sigma}{\sqrt{n}}\right) = 1 - \alpha.$$

A random sample of size n is selected from a population whose variance σ^2 is known and the mean $\bar{x}$ is computed to give the $(1 - \alpha)100\%$ confidence interval:

$$\bar{x} - z_{\alpha/2}\frac{\sigma}{\sqrt{n}} < \mu < \bar{x} + z_{\alpha/2}\frac{\sigma}{\sqrt{n}}.$$

CONFIDENCE INTERVAL FOR μ; σ KNOWN *A $(1 - \alpha)100\%$ confidence interval for μ is*

$$\bar{x} - z_{\alpha/2}\frac{\sigma}{\sqrt{n}} < \mu < \bar{x} + z_{\alpha/2}\frac{\sigma}{\sqrt{n}},$$

where $\bar{x}$ is the mean of a sample of size n from a population with known variance σ^2 and $z_{\alpha/2}$ is the value of the standard normal distribution leaving an area of $\alpha/2$ to the right.

For small samples selected from nonnormal populations, we cannot expect our degree of confidence to be accurate. However, for samples of size $n \geq 30$, regardless of the shape of most populations, sampling theory guarantees good results.

To compute a $(1 - \alpha)100\%$ confidence interval for μ, we have assumed that σ is known. Since this is generally not the case, we shall replace σ by the sample standard deviation s, provided $n \geq 30$.

Example 8.1 The mean and standard deviation for the quality grade-point averages of a random sample of 36 college seniors are calculated to be 2.6 and 0.3, respectively. Find the 95% and 99% confidence intervals for the mean of the entire senior class.

Solution. The point estimate of μ is $\bar{x} = 2.6$. Since the sample size is large, the standard deviation σ can be approximated by $s = 0.3$. The z value, leaving an area of 0.025 to the right and therefore an area of 0.975 to the left, is $z_{0.025} = 1.96$ (Table A.4). Hence the 95% confidence interval is

$$2.6 - (1.96)(0.3/\sqrt{36}) < \mu < 2.6 + (1.96)(0.3/\sqrt{36}),$$

which reduces to

$$2.50 < \mu < 2.70.$$

To find a 99% confidence interval, we find the z value leaving an area of 0.005 to the right and 0.995 to the left. Therefore, using Table A.4 again, $z_{0.005} = 2.575$, and the 99% confidence interval is

$$2.6 - (2.575)(0.3/\sqrt{36}) < \mu < 2.6 + (2.575)(0.3/\sqrt{36}),$$

or simply

$$2.47 < \mu < 2.73.$$

We now see that a longer interval is required to estimate μ with a higher degree of accuracy.

The $(1 - \alpha)100\%$ confidence interval provides an estimate of the accuracy of our point estimate. If μ is actually the center value of the interval, then $\bar{x}$ estimates μ without error. Most of the time, however, $\bar{x}$ will not be exactly equal to μ and the point estimate is in error. The size of this error will be the

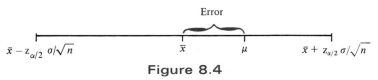

Figure 8.4

Error in estimating μ by $\bar{x}$.

absolute value of the difference between μ and $\bar{x}$, and we can be $(1 - \alpha)100\%$ confident that this difference will be less than $z_{\alpha/2}\sigma/\sqrt{n}$. We can readily see this if we draw a diagram of the confidence interval as in Figure 8.4.

> **THEOREM 8.1** *If $\bar{x}$ is used as an estimate of μ, we can be $(1 - \alpha)100\%$ confident that the error will be less than $z_{\alpha/2}\sigma/\sqrt{n}$.*

In Example 8.1 we are 95% confident that the sample mean $\bar{x} = 2.6$ differs from the true mean μ by an amount less than 0.1 and 99% confident that the difference is less than 0.13.

Frequently we wish to know how large a sample is necessary to ensure that the error in estimating μ will be less than a specified amount e. By Theorem 8.1 this means we must choose n such that $z_{\alpha/2}\sigma/\sqrt{n} = e$.

> **THEOREM 8.2** *If $\bar{x}$ is used as an estimate of μ, we can be $(1 - \alpha)100\%$ confident that the error will be less than a specified amount e when the sample size is*
>
> $$n = \left(\frac{z_{\alpha/2}\sigma}{e}\right)^2.$$

Strictly speaking, the formula in Theorem 8.2 is applicable only if we know the variance of the population from which we are to select our sample. Lacking this information, a preliminary sample of size $n \geq 30$ could be taken to provide an estimate of σ, then, using Theorem 8.2, we could determine approximately how many observations are needed to provide the desired degree of accuracy.

Example 8.2 How large a sample is required in Example 8.1 if we want to be 95% confident that our estimate of μ is off by less than 0.05?

Solution. The sample standard deviation $s = 0.3$ obtained from the preliminary sample of size 36 will be used for σ. Then, by Theorem 8.2,

$$n = \left[\frac{(1.96)(0.3)}{0.05}\right]^2 = 138.3.$$

Therefore we can be 95% confident that a random sample of size 139 will provide an estimate $\bar{x}$ differing from μ by an amount less than 0.05.

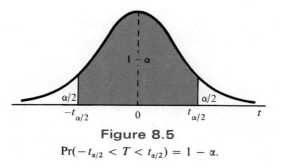

Figure 8.5

$$\Pr(-t_{\alpha/2} < T < t_{\alpha/2}) = 1 - \alpha.$$

Frequently we are attempting to estimate the mean of a population when the variance is unknown and it is impossible to obtain a sample of size $n \geq 30$. Cost can often be a factor that limits our sample size. As long as our population is approximately bell-shaped, confidence intervals can be computed when σ^2 is unknown and the sample size is small by using the sampling distribution of T, where

$$T = \frac{\bar{X} - \mu}{S/\sqrt{n}}.$$

The procedure is the same as for large samples except that we use the t distribution in place of the standard normal.

Referring to Figure 8.5, we can assert that

$$\Pr(-t_{\alpha/2} < T < t_{\alpha/2}) = 1 - \alpha,$$

where $t_{\alpha/2}$ is the t value with $n - 1$ degrees of freedom, above which we find an area of $\alpha/2$. Owing to symmetry, an equal area of $\alpha/2$ will fall to the left of $-t_{\alpha/2}$. Substituting for T, we write

$$\Pr\left(-t_{\alpha/2} < \frac{\bar{X} - \mu}{S/\sqrt{n}} < t_{\alpha/2}\right) = 1 - \alpha.$$

Multiplying each term in the inequality by $S/\sqrt{n}$, and then subtracting $\bar{X}$ from each term and multiplying by -1, we obtain

$$\Pr\left(\bar{X} - t_{\alpha/2}\frac{S}{\sqrt{n}} < \mu < \bar{X} + t_{\alpha/2}\frac{S}{\sqrt{n}}\right) = 1 - \alpha.$$

For our particular random sample of size n, the mean $\bar{x}$ and standard deviation s are computed and the $(1 - \alpha)100\%$ confidence interval is given by

$$\bar{x} - t_{\alpha/2}\frac{s}{\sqrt{n}} < \mu < \bar{x} + t_{\alpha/2}\frac{s}{\sqrt{n}}.$$

CONFIDENCE INTERVAL FOR μ; σ UNKNOWN AND $n < 30$ A $(1 - \alpha)100\%$ *confidence interval for μ is*

$$\bar{x} - t_{\alpha/2} \frac{s}{\sqrt{n}} < \mu < \bar{x} + t_{\alpha/2} \frac{s}{\sqrt{n}},$$

where $\bar{x}$ and s are the mean and standard deviation, respectively, of a sample of size $n \le 30$ from an approximate normal population and $t_{\alpha/2}$ is the value of the t distribution, with $v = n - 1$ degrees of freedom, leaving an area of $\alpha/2$ to the right.

Example 8.3 The weights of 7 similar boxes of cereal are 9.8, 10.2, 10.4, 9.8, 10.0, 10.2, and 9.6 ounces. Find a 95% confidence interval for the mean of all such boxes of cereal, assuming an approximate normal distribution.

Solution. The sample mean and standard deviation for the given data are

$$\bar{x} = 10.0 \quad \text{and} \quad s = 0.283.$$

Using Table A.5, we find $t_{0.025} = 2.447$ for $v = 6$ degrees of freedom. Hence the 95% confidence interval for μ is

$$10.0 - (2.447)(0.283/\sqrt{7}) < \mu < 10.0 + (2.447)(0.283/\sqrt{7}),$$

which reduces to

$$9.74 < \mu < 10.26.$$

8.4 Estimating the Difference Between Two Means

If we have two populations with means μ_1 and μ_2 and variances σ_1^2 and σ_2^2, respectively, a point estimator of the difference between μ_1 and μ_2 is given by the statistic $\bar{X}_1 - \bar{X}_2$. Therefore, to obtain a point estimate of $\mu_1 - \mu_2$, we shall select two independent random samples, one from each population, of size n_1 and n_2, and compute the difference, $\bar{x}_1 - \bar{x}_2$, of the sample means.

If our independent samples are selected from normal populations, or failing this, if n_1 and n_2 are both greater than 30, we can establish a confidence interval for $\mu_1 - \mu_2$ by considering the sampling distribution of $\bar{X}_1 - \bar{X}_2$.

According to Theorem 7.5 we can expect the sampling distribution of $\bar{X}_1 - \bar{X}_2$ to be approximately normally distributed with mean $\mu_{\bar{X}_1 - \bar{X}_2} = \mu_1 - \mu_2$ and standard deviation $\sigma_{\bar{X}_1 - \bar{X}_2} = \sqrt{(\sigma_1^2/n_1) + (\sigma_2^2/n_2)}$. Therefore we can assert with a probability of $1 - \alpha$ that the standard normal variable

$$Z = \frac{(\bar{X}_1 - \bar{X}_2) - (\mu_1 - \mu_2)}{\sqrt{(\sigma_1^2/n_1) + (\sigma_2^2/n_2)}}$$

will fall between $-z_{\alpha/2}$ and $z_{\alpha/2}$. Referring once again to Figure 8.3, we write

$$\Pr(-z_{\alpha/2} < Z < z_{\alpha/2}) = 1 - \alpha.$$

Substituting for Z, we state equivalently that

$$\Pr\left[-z_{\alpha/2} < \frac{(\bar{X}_1 - \bar{X}_2) - (\mu_1 - \mu_2)}{\sqrt{(\sigma_1^2/n_1) + (\sigma_2^2/n_2)}} < z_{\alpha/2}\right] = 1 - \alpha.$$

Multiplying each term of the inequality by $\sqrt{(\sigma_1^2/n_1) + (\sigma_2^2/n_2)}$, and then subtracting $\bar{X}_1 - \bar{X}_2$ from each term and multiplying by -1, we have

$$\Pr\left[(\bar{X}_1 - \bar{X}_2) - z_{\alpha/2}\sqrt{\frac{\sigma_1^2}{n_1} + \frac{\sigma_2^2}{n_2}} < \mu_1 - \mu_2 \right.$$

$$\left. < (\bar{X}_1 - \bar{X}_2) + z_{\alpha/2}\sqrt{\frac{\sigma_1^2}{n_1} + \frac{\sigma_2^2}{n_2}}\right]$$

$$= 1 - \alpha.$$

For any two independent random samples of size n_1 and n_2 selected from two populations whose variances σ_1^2 and σ_2^2 are known, the difference of the sample means, $\bar{x}_1 - \bar{x}_2$, is computed and the $(1 - \alpha)100\%$ confidence interval is given by

$$(\bar{x}_1 - \bar{x}_2) - z_{\alpha/2}\sqrt{\frac{\sigma_1^2}{n_1} + \frac{\sigma_2^2}{n_2}} < \mu_1 - \mu_2 < (\bar{x}_1 - \bar{x}_2) + z_{\alpha/2}\sqrt{\frac{\sigma_1^2}{n_1} + \frac{\sigma_2^2}{n_2}}.$$

CONFIDENCE INTERVAL FOR $\mu_1 - \mu_2$; σ_1^2 AND σ_2^2 KNOWN *A* *(1 − α)100% confidence interval for* $\mu_1 - \mu_2$ *is*

$$(\bar{x}_1 - \bar{x}_2) - z_{\alpha/2}\sqrt{\frac{\sigma_1^2}{n_1} + \frac{\sigma_2^2}{n_2}} < \mu_1 - \mu_2 < (\bar{x}_1 - \bar{x}_2) + z_{\alpha/2}\sqrt{\frac{\sigma_1^2}{n_1} + \frac{\sigma_2^2}{n_2}},$$

where $\bar{x}_1$ *and* $\bar{x}_2$ *are means of independent random samples of size* n_1 *and* n_2 *from populations with known variances* σ_1^2 *and* σ_2^2, *respectively, and* $z_{\alpha/2}$ *is the value of the standard normal curve leaving an area of* $\alpha/2$ *to the right.*

The degree of confidence is exact when samples are selected from normal populations. For nonnormal populations we obtain an approximate confidence interval that is very good when both n_1 and n_2 exceed 30. As before, if σ_1^2 and σ_2^2 are unknown and our samples are sufficiently large, we may replace σ_1^2 by s_1^2 and σ_2^2 by s_2^2 without appreciably affecting the confidence interval.

Example 8.4 A standardized chemistry test was given to 50 girls and 75 boys. The girls made an average grade of 76 with a standard deviation of 6, while the boys made an average grade of 82 with a standard deviation of 8. Find a 96% confidence interval for the difference $\mu_1 - \mu_2$, where μ_1 is the mean score of all boys and μ_2 is the mean score of all girls who might take this test.

Solution. The point estimate of $\mu_1 - \mu_2$ is $\bar{x}_1 - \bar{x}_2 = 82 - 76 = 6$. Since n_1 and n_2 are both large, we can substitute $s_1 = 8$ for σ_1 and $s_2 = 6$ for σ_2. Using $\alpha = 0.04$, we find $z_{0.02} = 2.054$ from Table A.4. Hence substitution in the formula

$$(\bar{x}_1 - \bar{x}_2) - z_{\alpha/2}\sqrt{\frac{\sigma_1^2}{n_1} + \frac{\sigma_2^2}{n_2}} < \mu_1 - \mu_2$$

$$< (\bar{x}_1 - \bar{x}_2) + z_{\alpha/2}\sqrt{\frac{\sigma_1^2}{n_1} + \frac{\sigma_2^2}{n_2}}$$

yields the 96% confidence interval

$$6 - 2.054\sqrt{\frac{64}{75} + \frac{36}{50}} < \mu_1 - \mu_2 < 6 + 2.054\sqrt{\frac{64}{75} + \frac{36}{50}}$$

or

$$3.42 < \mu_1 - \mu_2 < 8.58.$$

The above procedure for estimating the difference between two means is applicable if σ_1^2 and σ_2^2 are known or can be estimated from large samples. If the sample sizes are small, we must again resort to the t distribution to provide confidence intervals that are valid when the populations are approximately normally distributed.

Let us now assume that σ_1^2 and σ_2^2 are unknown and n_1 and n_2 are small (<30). If $\sigma_1^2 = \sigma_2^2 = \sigma^2$, we obtain a standard normal variable in the form

$$Z = \frac{(\bar{X}_1 - \bar{X}_2) - (\mu_1 - \mu_2)}{\sqrt{\sigma^2[(1/n_1) + (1/n_2)]}},$$

where σ^2 is to be estimated by combining or pooling the sample variances. Denoting the pooled estimator by S_p^2, we write

$$S_p^2 = \frac{(n_1 - 1)S_1^2 + (n_2 - 1)S_2^2}{n_1 + n_2 - 2}.$$

Substituting S_p^2 for σ^2 and noting that the divisor in S_p^2 is $n_1 + n_2 - 2$, we obtain the statistic

$$T = \frac{(\bar{X}_1 - \bar{X}_2) - (\mu_1 - \mu_2)}{S_p\sqrt{(1/n_1) + (1/n_2)}},$$

which has a t distribution with $v = n_1 + n_2 - 2$ degrees of freedom.

We can see by Figure 8.5 that

$$\Pr(-t_{\alpha/2} < T < t_{\alpha/2}) = 1 - \alpha,$$

where $t_{\alpha/2}$ is the t value with $n_1 + n_2 - 2$ degrees of freedom, above which we find an area of $\alpha/2$.

Substituting for T in the inequality, we write

$$\Pr\left[-t_{\alpha/2} < \frac{(\bar{X}_1 - \bar{X}_2) - (\mu_1 - \mu_2)}{S_p\sqrt{(1/n_1) + (1/n_2)}} < t_{\alpha/2}\right] = 1 - \alpha.$$

Multiplying each term of the inequality by $S_p\sqrt{(1/n_1) + (1/n_2)}$, and then subtracting $\bar{X}_1 - \bar{X}_2$ from each term and multiplying by -1, we obtain

$$\Pr\left[(\bar{X}_1 - \bar{X}_2) - t_{\alpha/2}S_p\sqrt{\frac{1}{n_1} + \frac{1}{n_2}} < \mu_1 - \mu_2\right.$$
$$\left. < (\bar{X}_1 - \bar{X}_2) + t_{\alpha/2}S_p\sqrt{\frac{1}{n_1} + \frac{1}{n_2}}\right]$$
$$= 1 - \alpha.$$

For any two independent random samples of size n_1 and n_2 selected from two normal populations, the difference of the sample means, $\bar{x}_1 - \bar{x}_2$, and the pooled standard deviation, s_p, are computed and the $(1 - \alpha)100\%$ confidence interval is given by

$$(\bar{x}_1 - \bar{x}_2) - t_{\alpha/2}s_p\sqrt{\frac{1}{n_1} + \frac{1}{n_2}} < \mu_1 - \mu_2 < (\bar{x}_1 - \bar{x}_2) + t_{\alpha/2}s_p\sqrt{\frac{1}{n_1} + \frac{1}{n_2}}.$$

SMALL SAMPLE CONFIDENCE INTERVAL FOR $\mu_1 - \mu_2$; $\sigma_1^2 = \sigma_2^2$, BUT UNKNOWN A $(1 - \alpha)100\%$ *confidence interval for* $\mu_1 - \mu_2$ *is*

$$(\bar{x}_1 - \bar{x}_2) - t_{\alpha/2}s_p \sqrt{\frac{1}{n_1} + \frac{1}{n_2}} < \mu_1 - \mu_2 < (\bar{x}_1 - \bar{x}_2) + t_{\alpha/2}s_p \sqrt{\frac{1}{n_1} + \frac{1}{n_2}},$$

where $\bar{x}_1$ *and* $\bar{x}_2$ *are the means of small independent samples of size* n_1 *and* n_2, *respectively, from approximate normal distributions,* s_p *the pooled standard deviation, and* $t_{\alpha/2}$ *the value of the t distribution with* $v = n_1 + n_2 - 2$ *degrees of freedom, leaving an area of* $\alpha/2$ *to the right.*

Example 8.5 A course in mathematics is taught to 12 students by the conventional classroom procedure. A second group of 10 students was given the same course by means of programed materials. At the end of the semester the same examination was given to each group. The 12 students meeting in the classroom made an average grade of 85 with a standard deviation of 4, while the 10 students using programed materials made an average of 81 with a standard deviation of 5. Find a 90% confidence interval for the difference between the population means, assuming the populations are approximately normally distributed with equal variances.

Solution. Let μ_1 and μ_2 represent the average grades of all students who might take this course by the classroom and programed presentations, respectively. We wish to find a 90% confidence interval for $\mu_1 - \mu_2$. Our point estimate of $\mu_1 - \mu_2$ is $\bar{x}_1 - \bar{x}_2 = 85 - 81 = 4$. The pooled estimate, s_p^2, of the common variance, σ^2, is

$$s_p^2 = \frac{(n_1 - 1)s_1^2 + (n_2 - 1)s_2^2}{n_1 + n_2 - 2}$$

$$= \frac{(11)(16) + (9)(25)}{12 + 10 - 2} = 20.05.$$

Taking the square root, $s_p = 4.478$. Using $\alpha = 0.1$, we find in Table A.5 that $t_{0.05} = 1.725$ for $v = n_1 + n_2 - 2 = 20$ degrees of freedom. Therefore, substituting in the formula

$$(\bar{x}_1 - \bar{x}_2) - t_{\alpha/2}s_p \sqrt{\frac{1}{n_1} + \frac{1}{n_2}} < \mu_1 - \mu_2$$

$$< (\bar{x}_1 - \bar{x}_2) + t_{\alpha/2}s_p \sqrt{\frac{1}{n_1} + \frac{1}{n_2}},$$

we obtain the 90% confidence interval

$$4 - (1.725)(4.478)\sqrt{\frac{1}{12} + \frac{1}{10}} < \mu_1 - \mu_2$$

$$< 4 + (1.725)(4.478)\sqrt{\frac{1}{12} + \frac{1}{10}},$$

which simplifies to

$$0.69 < \mu_1 - \mu_2 < 7.31.$$

Hence we are 90% confident that the interval from 0.69 to 7.31 contains the true difference of the average grades for the two methods of instruction. The fact that both confidence limits are positive indicates that the classroom method of learning this particular mathematics course is superior to the method using programed materials.

The procedure for constructing confidence intervals for $\mu_1 - \mu_2$ from small samples assumes the populations to be normal and the population variances to be equal. Slight departures from either of these assumptions do not seriously alter the degree of confidence for our interval. A procedure will be presented in Chapter 9 for testing the equality of two unknown population variances based on the information provided by the sample variances. If the population variances are considerably different, we still obtain good results when the populations are normal, provided $n_1 = n_2$. Therefore, in a planned experiment, one should make every effort to equalize the size of the samples.

Let us now consider the problem of finding an interval estimate of $\mu_1 - \mu_2$ for small samples when the unknown population variances are not likely to be equal, and it is impossible to select samples of equal size. The statistic most often used in this case is

$$T' = \frac{(\bar{X}_1 - \bar{X}_2) - (\mu_1 - \mu_2)}{\sqrt{(S_1^2/n_1) + (S_2^2/n_2)}},$$

which has approximately a t distribution with v degrees of freedom, where

$$v = \frac{(s_1^2/n_1 + s_2^2/n_2)^2}{[(s_1^2/n_1)^2/(n_1 - 1)] + [(s_2^2/n_2)^2/(n_2 - 1)]}.$$

Since v is seldom an integer, we round it off to the nearest whole number. Using the statistic T', we write

$$\Pr(-t_{\alpha/2} < T' < t_{\alpha/2}) \simeq 1 - \alpha,$$

where $t_{\alpha/2}$ is the value of the t distribution with v degrees of freedom, above which we find an area of $\alpha/2$. Substituting for T' in the inequality, and following the exact steps as before, we state the final result.

SMALL SAMPLE CONFIDENCE INTERVAL FOR $\mu_1 - \mu_2$; $\sigma_1^2 \neq \sigma_2^2$ AND UNKNOWN *An approximate* $(1 - \alpha)100\%$ *confidence interval for* $\mu_1 - \mu_2$ *is*

$$(\bar{x}_1 - \bar{x}_2) - t_{\alpha/2}\sqrt{\frac{s_1^2}{n_1} + \frac{s_2^2}{n_2}} < \mu_1 - \mu_2 < (\bar{x}_1 - \bar{x}_2) + t_{\alpha/2}\sqrt{\frac{s_1^2}{n_1} + \frac{s_2^2}{n_2}},$$

where $\bar{x}_1$ and s_1^2 and $\bar{x}_2$ and s_2^2 are the means and variances of small independent samples of size n_1 and n_2, respectively, from approximate normal distributions, and $t_{\alpha/2}$ is the value of the t distribution with

$$\nu = \frac{(s_1^2/n_1 + s_2^2/n_2)^2}{[(s_1^2/n_1)^2/(n_1 - 1)] + [(s_2^2/n_2)^2/(n_2 - 1)]}$$

degrees of freedom, leaving an area of $\alpha/2$ to the right.

Example 8.6 Records for the past 15 years have shown the average rainfall, in a certain region of the country for the month of May, to be 1.94 inches with a standard deviation of 0.45 inch. A second region of the country has had an average rainfall in May of 1.04 inches with a standard deviation of 0.26 inch during the past 10 years. Find a 95% confidence interval for the difference of the true average rainfalls in these two regions, assuming the observations came from normal populations with different variances.

Solution. For the first region we have $\bar{x}_1 = 1.94$, $s_1 = 0.45$, and $n_1 = 15$. For the second region $\bar{x}_2 = 1.04$, $s_2 = 0.26$, and $n_2 = 10$. We wish to find a 95% confidence interval for $\mu_1 - \mu_2$. Since the population variances are assumed to be unequal and our sample sizes are not the same, we can only find an approximate 95% confidence interval based on the t distribution with ν degrees of freedom, where

$$\nu = \frac{(s_1^2/n_1 + s_2^2/n_2)^2}{[(s_1^2/n_1)^2/(n_1 - 1)] + [(s_2^2/n_2)^2/(n_2 - 1)]}$$

$$= \frac{(0.2025/15 + 0.0676/10)^2}{[(0.2025/15)^2/14] + [(0.0676/10)^2/9]}$$

$$= 22.7 \simeq 23.$$

Our point estimate of $\mu_1 - \mu_2$ is $\bar{x}_1 - \bar{x}_2 = 1.94 - 1.04$

$1 - .95 = ?$

$= 0.90.$ Using $\alpha \overset{?}{=} 0.05,$ we find in Table A.5 that $t_{0.025} = 2.069$ for $v = 23$ degrees of freedom. Therefore, substituting in the formula

$$(\bar{x}_1 - \bar{x}_2) - t_{\alpha/2}\sqrt{\frac{s_1^2}{n_1} + \frac{s_2^2}{n_2}} < \mu_1 - \mu_2$$

$$< (\bar{x}_1 - \bar{x}_2) + t_{\alpha/2}\sqrt{\frac{s_1^2}{n_1} - \frac{s_2^2}{n_2}},$$

we obtain the approximate 95% confidence interval

$$0.90 - 2.069\sqrt{\frac{0.2025}{15} + \frac{0.0676}{10}} < \mu_1 - \mu_2$$

$$< 0.90 + 2.069\sqrt{\frac{0.2025}{15} + \frac{0.0676}{10}},$$

which simplifies to

$$0.61 < \mu_1 - \mu_2 < 1.19.$$

Hence we are 95% confident that the interval from 0.61 to 1.19 contains the true difference of the average rainfall for the two regions.

We conclude this section by considering estimation procedures for the difference of two means when the samples are not independent and the variances of the two populations are not necessarily equal. This will be true if the observations in the two samples occur in pairs so that the two observations are related. For instance, if we run a test on a new diet using 15 individuals, the weights before and after completion of the test form our two samples. Observations in the two samples made on the same individual are related and hence form a pair. To determine if the diet is effective, we must consider the differences, d_i, of paired observations. These differences are the values of a random sample $d_1, d_2, \ldots, d_n$ from a population which we shall assume to be normal with mean μ_D and unknown variance σ_D^2. We estimate σ_D^2 by s_d^2, the variance of the differences constituting the sample. Therefore s_d^2 is a value of the statistic S_d^2 that fluctuates from sample to sample. The point estimate of $\mu_1 - \mu_2 = \mu_D$ is given by $\bar{d}$.

A $(1 - \alpha)100\%$ confidence interval for μ_D can be established by writing

$$\Pr(-t_{\alpha/2} < T < t_{\alpha/2}) = 1 - \alpha,$$

where

$$T = \frac{\bar{D} - \mu_D}{S_d/\sqrt{n}},$$

and $t_{\alpha/2}$, as before, is a value of the t distribution with $n - 1$ degrees of freedom.

It is now a routine procedure to replace T, by its definition, in the above inequality and carry out the mathematical steps that lead to the $(1 - \alpha)100\%$ confidence interval:

$$\bar{d} - t_{\alpha/2} \frac{s_d}{\sqrt{n}} < \mu_D < \bar{d} + t_{\alpha/2} \frac{s_d}{\sqrt{n}}.$$

CONFIDENCE INTERVAL FOR $\mu_1 - \mu_2 = \mu_D$ **FOR PAIRED OBSERVA-TIONS** A $(1 - \alpha)100\%$ *confidence interval for* μ_D *is*

$$\bar{d} - t_{\alpha/2} \frac{s_d}{\sqrt{n}} < \mu_D < \bar{d} + t_{\alpha/2} \frac{s_d}{\sqrt{n}},$$

where $\bar{d}$ and s_d are the mean and standard deviation of the differences of n pairs of measurements and $t_{\alpha/2}$ is the value of the t distribution with $v = n - 1$ degrees of freedom, leaving an area of $\alpha/2$ to the right.

Example 8.7 Twenty college freshmen were divided into 10 pairs, each member of the pair having approximately the same I.Q. One of each pair was selected at random and assigned to a mathematics section using programed materials only. The other member of each pair was assigned to a section in which the professor lectured. At the end of the semester each group was given the same examination and the following results were recorded.

Pair	Programed Materials	Lectures	d
1	76	81	−5
2	60	52	8
3	85	87	−2
4	58	70	−12
5	91	86	5
6	75	77	−2
7	82	90	−8
8	64	63	1
9	79	85	−6
10	88	83	5

Find a 98% confidence interval for the true difference in the two learning procedures.

Solution. We wish to find a 98% confidence interval for $\mu_1 - \mu_2$, where μ_1 and μ_2 represent the average grades of all students by the programed and lecture method of presentation, respectively. Since the observations are paired, $\mu_1 - \mu_2 = \mu_D$. The point estimate of μ_D is given by $\bar{d} = -1.6$. The variance, s_d^2, of the sample differences is

$$\bar{d} = \frac{\sum d_i}{m}$$

$$s_d^2 = \frac{n \sum d_i^2 - (\sum d_i)^2}{n(n-1)} = \frac{(10)(392) - (-16)^2}{(10)(9)} = 40.7.$$

By taking the square root, $s_d = 6.38$. Using $\alpha = 0.02$, we find in Table A.5 that $t_{0.01} = 2.821$ for $v = n - 1 = 9$ degrees of freedom. Therefore, substituting in the formula

$$\bar{d} - t_{\alpha/2} \frac{s_d}{\sqrt{n}} < \mu_D < \bar{d} + t_{\alpha/2} \frac{s_d}{\sqrt{n}},$$

we obtain the 98% confidence interval

$$-1.6 - (2.821)(6.38/\sqrt{10}) < \mu_D$$
$$< -1.6 + (2.821)(6.38/\sqrt{10}),$$

or simply

$$-7.29 < \mu_D < 4.09.$$

Hence we are 98% confident that the interval from -7.29 to 4.09 contains the true difference of the average grades for the two methods of instruction. Since this interval allows for the possibility of μ_D being equal to zero, we are unable to state that one method of instruction is better than the other, even though this particular sample of differences shows the lecture procedure to be superior.

8.5 Estimating a Proportion

A point estimator of the proportion p in a binomial experiment is given by the statistic $\hat{P} = X/n$. Therefore the sample proportion $\hat{p} = x/n$ will be used as the point estimate for the parameter p.

If the unknown proportion p is not expected to be too close to zero or 1, we can establish a confidence interval for p by considering the sampling distribution of $\hat{P}$, which of course is the same as that of the random variable X except for a change in scale. Hence, by Theorem 6.1, for n sufficiently large, the distribution of $\hat{P}$ is approximately normally distributed with mean

$$\mu_{\hat{P}} = E(\hat{P}) = E\left(\frac{X}{n}\right) = \frac{np}{n} = p$$

and variance

$$\sigma_{\hat{P}}^2 = \sigma_{\bar{X}/n}^2 = \frac{\sigma_X^2}{n^2} = \frac{npq}{n^2} = \boxed{\frac{pq}{n}}.$$

Therefore we can assert that

$$\Pr(-z_{\alpha/2} < Z < z_{\alpha/2}) = 1 - \alpha,$$

where

$$Z = \frac{\hat{P} - p}{\sqrt{pq/n}}$$

and $z_{\alpha/2}$ is the value of the standard normal curve above which we find an area of $\alpha/2$. Substituting for Z, we write

$$\Pr\left(-z_{\alpha/2} < \frac{\hat{P} - p}{\sqrt{pq/n}} < z_{\alpha/2}\right) = 1 - \alpha.$$

Multiplying each term of the inequality by $\sqrt{pq/n}$, and then subtracting $\hat{P}$ and multiplying by -1, we obtain

$$\Pr\left(\hat{P} - z_{\alpha/2}\sqrt{\frac{pq}{n}} < p < \hat{P} + z_{\alpha/2}\sqrt{\frac{pq}{n}}\right) = 1 - \alpha.$$

It is difficult to manipulate the inequalities to obtain a random interval whose end points are independent of p, the unknown parameter. When n is large, very little error is introduced by substituting the point estimate $\hat{p} = x/n$ for the p under the radical sign. Then we can write

$$\Pr\left(\hat{P} - z_{\alpha/2}\sqrt{\frac{\hat{p}\hat{q}}{n}} < p < \hat{P} + z_{\alpha/2}\sqrt{\frac{\hat{p}\hat{q}}{n}}\right) \simeq 1 - \alpha.$$

For our particular random sample of size n, the sample proportion $\hat{p} = x/n$ is computed, and the approximate $(1 - \alpha)100\%$ confidence interval for p is given by

$$\hat{p} - z_{\alpha/2}\sqrt{\frac{\hat{p}\hat{q}}{n}} < p < \hat{p} + z_{\alpha/2}\sqrt{\frac{\hat{p}\hat{q}}{n}}.$$

CONFIDENCE INTERVAL FOR p; $n \geq 30$. *A* $(1 - \alpha)100\%$ *confidence interval for the binomial parameter p is approximately*

$$\hat{p} - z_{\alpha/2}\sqrt{\frac{\hat{p}\hat{q}}{n}} < p < \hat{p} + z_{\alpha/2}\sqrt{\frac{\hat{p}\hat{q}}{n}},$$

where $\hat{p}$ is the proportion of successes in a random sample of size n, $\hat{q} = 1 - \hat{p}$, and $z_{\alpha/2}$ is the value of the standard normal curve leaving an area of $\alpha/2$ to the right.

The method for finding a confidence interval for the binomial parameter p is also applicable when the binomial distribution is being used to approximate the hypergeometric distribution, that is, when n is small relative to N, as illustrated in Example 8.8.

Example 8.8 In a random sample of $n = 500$ families owning television sets in the city of Hamilton, Canada, it was found that $x = 160$ owned color sets. Find a 95% confidence interval for the actual proportion of families in this city with color sets.

Solution. The point estimate of p is $\hat{p} = 160/500 = 0.32$. Using Table A.4, we find $z_{0.025} = 1.96$. Therefore, substituting in the formula

$$\hat{p} - z_{\alpha/2}\sqrt{\frac{\hat{p}\hat{q}}{n}} < p < \hat{p} + z_{\alpha/2}\sqrt{\frac{\hat{p}\hat{q}}{n}},$$

we obtain the 95% confidence interval

$$0.32 - 1.96\sqrt{\frac{(0.32)(0.68)}{500}} < p < 0.32 + 1.96\sqrt{\frac{(0.32)(0.68)}{500}},$$

which simplifies to

$$0.28 < p < 0.36.$$

If p is the center value of a $(1 - \alpha)100\%$ confidence interval, then $\hat{p}$ estimates p without error. Most of the time, however, $\hat{p}$ will not be exactly equal to p and the point estimate is in error. The size of this error will be the difference between p and $\hat{p}$, and we can be $(1 - \alpha)100\%$ confident that this difference will be less than $z_{\alpha/2}\sqrt{\hat{p}\hat{q}/n}$. We can readily see this if we draw a diagram of the confidence interval as in Figure 8.6.

Error

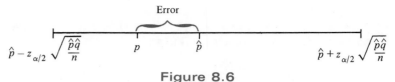

$$\hat{p} - z_{\alpha/2} \sqrt{\frac{\hat{p}\hat{q}}{n}} \qquad p \qquad \hat{p} \qquad\qquad\qquad \hat{p} + z_{\alpha/2} \sqrt{\frac{\hat{p}\hat{q}}{n}}$$

Figure 8.6

Error in estimating p by $\hat{p}$.

THEOREM 8.3 *If $\hat{p}$ is used as an estimate of p, we can be $(1 - \alpha)100\%$ confident that the error will be less than $z_{\alpha/2}\sqrt{\hat{p}\hat{q}/n}$.*

In Example 8.8 we are 95% confident that the sample proportion $\hat{p} = 0.32$ differs from the true proportion p by an amount less than 0.04.

Let us now determine how large a sample is necessary to ensure that the error in estimating p will be less than a specified amount e. By Theorem 8.3 this means we must choose n such that $z_{\alpha/2}\sqrt{\hat{p}\hat{q}/n} = e.$

THEOREM 8.4 *If $\hat{p}$ is used as an estimate of p, we can be $(1 - \alpha)100\%$ confident that the error will be less than a specified amount e when the sample size is*

$$n = \frac{z_{\alpha/2}^2 \, \hat{p}\hat{q}}{e^2}.$$

Theorem 8.4 is somewhat misleading in that we must use $\hat{p}$ to determine the sample size n, but $\hat{p}$ is computed from the sample. If a crude estimate of p can be made without taking a sample, we could use this value for $\hat{p}$ and then determine n. Lacking such an estimate, we could take a preliminary sample of size $n \geq 30$ to provide an estimate of p. Then, using Theorem 8.4, we could determine approximately how many observations are needed to provide the desired degree of accuracy.

An upper bound for n can be established for any degree of confidence by noting that $\hat{p}\hat{q} = \hat{p}(1 - \hat{p})$, which must be at most equal to $\frac{1}{4}$, since $\hat{p}$ must lie between zero and 1. This fact may be verified by completing the square. Hence

$$\hat{p}(1 - \hat{p}) = -(\hat{p}^2 - \hat{p}) = \tfrac{1}{4} - (\hat{p}^2 - \hat{p} + \tfrac{1}{4})$$
$$= \tfrac{1}{4} - (\hat{p} - \tfrac{1}{2})^2,$$

which is always less than $\frac{1}{4}$ except when $\hat{p} = \frac{1}{2}$ and then $\hat{p}\hat{q} = \frac{1}{4}$.

THEOREM 8.5 *If $\hat{p}$ is used as an estimate of p, we can be at least $(1 - \alpha)$ 100% confident that the error will be less than a specified amount e when the sample size is*

$$n = \frac{z_{\alpha/2}^2}{4e^2}.$$

Example 8.9 How large a sample is required in Example 8.8 if we want to be (a) 95% confident that our estimate of p is off by less than 0.02, (b) at least 95% confident?

Solution. (a) Let us treat the 500 families as a preliminary sample providing an estimate $\hat{p} = 0.32$. Then, by Theorem 8.4,

$$n = \frac{(1.96)^2(0.32)(0.68)}{(0.02)^2} = 2090.$$

Therefore, if we base our estimate of p on a random sample of size 2090 we can be 95% confident that our sample proportion will not differ from the true proportion by more than 0.02.

(b) According to Theorem 8.5 we can be at least 95% confident that our sample proportion will not differ from the true proportion by more than 0.02 if we choose a sample of size

$$n = \frac{(1.96)^2}{4(0.02)^2} = 2401.$$

Comparing the two parts of Example 8.9, we see that information concerning p, provided by a preliminary sample or perhaps from past experience, enables us to choose a smaller sample.

8.6 Estimating the Difference Between Two Proportions

Consider independent samples of size n_1 and n_2 selected at random from two binomial populations with means $n_1 p_1$ and $n_2 p_2$ and variances $n_1 p_1 q_1$ and $n_2 p_2 q_2$, respectively. We denote the proportion of successes in each sample by $\hat{p}_1$ and $\hat{p}_2$. A point estimator of the difference between the two proportions, $p_1 - p_2$, is given by the statistic $\hat{P}_1 - \hat{P}_2$.

A confidence interval for $p_1 - p_2$ can be established by considering the sampling distribution of $\hat{P}_1 - \hat{P}_2$. From the preceding section, for n_1 and n_2 sufficiently large, we know that $\hat{P}_1$ and $\hat{P}_2$ are each approximately normally distributed, with means p_1 and p_2 and variances $p_1 q_1 / n_1$ and $p_2 q_2 / n_2$, respectively. By choosing independent samples from the two populations, the variables $\hat{P}_1$ and $\hat{P}_2$ will be independent, and then by the reproductive property of the normal distribution established in Theorem 7.6, we conclude that $\hat{P}_1 - \hat{P}_2$ is approximately normally distributed with mean

$$\mu_{\hat{P}_1 - \hat{P}_2} = p_1 - p_2$$

and variance

$$\sigma^2_{\hat{P}_1 - \hat{P}_2} = \frac{p_1 q_1}{n_1} + \frac{p_2 q_2}{n_2}.$$

Therefore we can assert that

$$\Pr(-z_{\alpha/2} < Z < z_{\alpha/2}) = 1 - \alpha,$$

where

$$Z = \frac{(\hat{P}_1 - \hat{P}_2) - (p_1 - p_2)}{\sqrt{(p_1 q_1/n_1) + (p_2 q_2/n_2)}}$$

and $z_{\alpha/2}$ is a value of the standard normal curve above which we find an area of $\alpha/2$. Substituting for Z, we write

$$\Pr\left[-z_{\alpha/2} < \frac{(\hat{P}_1 - \hat{P}_2) - (p_1 - p_2)}{\sqrt{(p_1 q_1/n_1) + (p_2 q_2/n_2)}} < z_{\alpha/2}\right] = 1 - \alpha.$$

Multiplying each term of the inequality by $\sqrt{(p_1 q_1/n_1) + (p_2 q_2/n_2)}$, and then subtracting $\hat{P}_1 - \hat{P}_2$ and multiplying by -1, we obtain

$$\Pr\left[(\hat{P}_1 - \hat{P}_2) - z_{\alpha/2}\sqrt{\frac{p_1 q_1}{n_1} + \frac{p_2 q_2}{n_2}} < p_1 - p_2\right.$$

$$\left. < (\hat{P}_1 - \hat{P}_2) + z_{\alpha/2}\sqrt{\frac{p_1 q_1}{n_1} + \frac{p_2 q_2}{n_2}}\right]$$

$$= 1 - \alpha.$$

If n_1 and n_2 are both large, we replace p_1 and p_2 under the radical sign by their estimates $\hat{p}_1 = x_1/n_1$ and $\hat{p}_2 = x_2/n_2$. Then

$$\Pr\left[(\hat{P}_1 - \hat{P}_2) - z_{\alpha/2}\sqrt{\frac{\hat{p}_1 \hat{q}_1}{n_1} + \frac{\hat{p}_2 \hat{q}_2}{n_2}} < p_1 - p_2\right.$$

$$\left. < (\hat{P}_1 - \hat{P}_2) + z_{\alpha/2}\sqrt{\frac{\hat{p}_1 \hat{q}_1}{n_1} + \frac{\hat{p}_2 \hat{q}_2}{n_2}}\right]$$

$$\simeq 1 - \alpha.$$

For any two independent random samples of size n_1 and n_2, selected from two binomial populations, the difference of the sample proportions, $\hat{p}_1 - \hat{p}_2$, is computed and the $(1 - \alpha)100\%$ confidence interval is given by

$$(\hat{p}_1 - \hat{p}_2) - z_{\alpha/2}\sqrt{\frac{\hat{p}_1\hat{q}_1}{n_1} + \frac{\hat{p}_2\hat{q}_2}{n_2}} < p_1 - p_2$$

$$< (\hat{p}_1 - \hat{p}_2) + z_{\alpha/2}\sqrt{\frac{\hat{p}_1\hat{q}_1}{n_1} + \frac{\hat{p}_2\hat{q}_2}{n_2}}.$$

> **CONFIDENCE INTERVAL FOR $p_1 - p_2$; n_1 AND $n_2 \geq 30$** $A\,(1 - \alpha)100\%$ *confidence interval for the difference of two binomial parameters, $p_1 - p_2$, is approximately*
>
> $$(\hat{p}_1 - \hat{p}_2) - z_{\alpha/2}\sqrt{\frac{\hat{p}_1\hat{q}_1}{n_1} + \frac{\hat{p}_2\hat{q}_2}{n_2}} < p_1 - p_2$$
>
> $$< (\hat{p}_1 - \hat{p}_2) + z_{\alpha/2}\sqrt{\frac{\hat{p}_1\hat{q}_1}{n_1} + \frac{\hat{p}_2\hat{q}_2}{n_2}},$$
>
> *where $\hat{p}_1$ and $\hat{p}_2$ are the proportion of successes in random samples of size n_1 and n_2, respectively, $\hat{q}_1 = 1 - \hat{p}_1$ and $\hat{q}_2 = 1 - \hat{p}_2$, and $z_{\alpha/2}$ is the value of the standard normal curve leaving an area of $\alpha/2$ to the right.*

Example 8.10 A poll is taken among the residents of a city and the surrounding county to determine the feasibility of a proposal to construct a civic center. If 2400 of 5000 city residents favor the proposal and 1200 of 2000 county residents favor it, find a 90% confidence interval for the true difference in the fractions favoring the proposal to construct the civic center.

Solution. Let p_1 and p_2 be the true proportions of residents in the city and county, respectively, favoring the proposal. Hence $\hat{p}_1 = 2400/5000 = 0.48$, $\hat{p}_2 = 1200/2000 = 0.60$, and the point estimate of $p_1 - p_2$ is $\hat{p}_1 - \hat{p}_2 = 0.48 - 0.60 = -0.12$. Using Table A.4, we find $z_{0.05} = 1.645$. Therefore, substituting in the formula

$$(\hat{p}_1 - \hat{p}_2) - z_{\alpha/2}\sqrt{\frac{\hat{p}_1\hat{q}_1}{n_1} + \frac{\hat{p}_2\hat{q}_2}{n_2}}$$

$$< p_1 - p_2 < (\hat{p}_1 - \hat{p}_2)$$

$$+ z_{\alpha/2}\sqrt{\frac{\hat{p}_1\hat{q}_1}{n_1} + \frac{\hat{p}_2\hat{q}_2}{n_2}},$$

$\hat{q}_1 = 1 - \hat{p}_1$

$\hat{p}_1 = \dfrac{p_1}{m_1}$

look up z .9500 in table + get side + top values

or look up $z_{.05}$ + get neg number + change sign

we obtain the 90% confidence interval

$$-0.12 - 1.645\sqrt{\frac{(0.48)(0.52)}{5000} + \frac{(0.60)(0.40)}{2000}}$$

$$< p_1 - p_2 < -0.12$$

$$+ 1.645\sqrt{\frac{(0.48)(0.52)}{5000} + \frac{(0.60)(0.40)}{2000}},$$

which simplifies to

$$-0.1414 < p_1 - p_2 < -0.0986.$$

Since both end points of the interval are negative, we can also conclude that the proportion of county residents favoring the proposal is greater than the proportion of city residents favoring the proposal.

8.7 Estimating the Variance

A point estimate of the population variance σ^2 is provided by the sample variance s^2. Hence the statistic S^2 is called an *estimator* of σ^2.

An interval estimate of σ^2 can be established by using the statistic

$$X^2 = \frac{(n-1)S^2}{\sigma^2}.$$

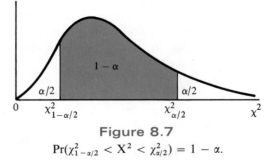

Figure 8.7
$\Pr(\chi^2_{1-\alpha/2} < X^2 < \chi^2_{\alpha/2}) = 1 - \alpha.$

According to Theorem 7.8 the statistic X^2 has a chi-square distribution with $n - 1$ degrees of freedom when samples are chosen from a normal population. We may write (see Figure 8.7)

$$\Pr(\chi^2_{1-\alpha/2} < X^2 < \chi^2_{\alpha/2}) = 1 - \alpha,$$

where $\chi^2_{1-\alpha/2}$ and $\chi^2_{\alpha/2}$ are values of the chi-square distribution with $n-1$ degrees of freedom, leaving areas of $1 - \alpha/2$ and $\alpha/2$, respectively, to the right. Substituting for X^2, we write

$$\Pr\left(\chi^2_{1-\alpha/2} < \frac{(n-1)S^2}{\sigma^2} < \chi^2_{\alpha/2}\right) = 1 - \alpha.$$

Dividing each term in the inequality by $(n-1)S^2$, and then inverting each term (thereby changing the sense of the inequalities), we obtain

$$\Pr\left[\frac{(n-1)S^2}{\chi^2_{\alpha/2}} < \sigma^2 < \frac{(n-1)S^2}{\chi^2_{1-\alpha/2}}\right] = 1 - \alpha.$$

For our particular sample of size n, the sample variance s^2 is computed, and the $(1-\alpha)100\%$ confidence interval is given by

$$\frac{(n-1)s^2}{\chi^2_{\alpha/2}} < \sigma^2 < \frac{(n-1)s^2}{\chi^2_{1-\alpha/2}}.$$

CONFIDENCE INTERVAL FOR σ^2 *A* $(1-\alpha)100\%$ *confidence interval for the variance* σ^2 *of a normal population is*

$$\frac{(n-1)s^2}{\chi^2_{\alpha/2}} < \sigma^2 < \frac{(n-1)s^2}{\chi^2_{1-\alpha/2}},$$

where s^2 *is the variance of a random sample of size n and* $\chi^2_{\alpha/2}$ *and* $\chi^2_{1-\alpha/2}$ *are the values of a chi-square distribution with* $v = n-1$ *degrees of freedom, leaving areas of* $\alpha/2$ *and* $1-\alpha/2$, *respectively, to the right.*

Example 8.11 The following are the weights, in ounces, of 10 cans of peaches distributed by a certain company: 16.4, 16.1, 15.8, 17.0, 16.1, 15.9, 15.8, 16.9, 15.2, and 16.0. Find a 95% confidence interval for the variance of all such cans of peaches distributed by this company.

Solution. First we find the sample variance s^2. After subtracting 16 from each observation,

$$s^2 = \frac{n\sum\limits_{i=1}^{n} x_i^2 - \left(\sum\limits_{i=1}^{n} x_i\right)^2}{n(n-1)} = \frac{(10)(2.72) - (1.2)^2}{(10)(9)} = 0.286.$$

To obtain a 95% confidence interval we choose $\alpha = 0.05$.

Then using Table A.6 with $v = 9$ degrees of freedom, we find $\chi^2_{0.025} = 19.023$ and $\chi^2_{0.975} = 2.700$. Substituting in the formula

$$\frac{(n-1)s^2}{\chi^2_{\alpha/2}} < \sigma^2 < \frac{(n-1)s^2}{\chi^2_{1-\alpha/2}},$$

we obtain the 95% confidence interval

$$\frac{(9)(0.286)}{19.023} < \sigma^2 < \frac{(9)(0.286)}{2.700},$$

or simply

$$0.135 < \sigma^2 < 0.953.$$

8.8 Estimating the Ratio of Two Variances

A point estimate of the ratio of two population variances σ_1^2/σ_2^2 is given by the ratio s_1^2/s_2^2 of the sample variances. Hence the statistic S_1^2/S_2^2 is the estimator of σ_1^2/σ_2^2.

If σ_1^2 and σ_2^2 are the variances of normal populations, we can establish an interval estimate of σ_1^2/σ_2^2 by using the statistic

$$F = \frac{\sigma_2^2 S_1^2}{\sigma_1^2 S_2^2}.$$

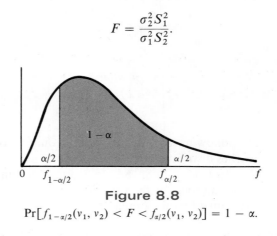

Figure 8.8

$$\Pr[f_{1-\alpha/2}(v_1, v_2) < F < f_{\alpha/2}(v_1, v_2)] = 1 - \alpha.$$

According to Theorem 7.9, the random variable F has an F distribution with $v_1 = n_1 - 1$ and $v_2 = n_2 - 1$ degrees of freedom. Therefore we may write (see Figure 8.8)

$$\Pr[f_{1-\alpha/2}(v_1, v_2) < F < f_{\alpha/2}(v_1, v_2)] = 1 - \alpha,$$

where $f_{1-\alpha/2}(v_1, v_2)$ and $f_{\alpha/2}(v_1, v_2)$ are the values of the F distribution with v_1

and v_2 degrees of freedom, leaving areas of $1 - \alpha/2$ and $\alpha/2$, respectively, to the right. Substituting for F, we write

$$\Pr\left[f_{1-\alpha/2}(v_1, v_2) < \frac{\sigma_2^2 S_1^2}{\sigma_1^2 S_2^2} < f_{\alpha/2}(v_1, v_2) \right] = 1 - \alpha.$$

Multiplying each term in the inequality by S_2^2/S_1^2, and then inverting each term (again changing the sense of the inequalities), we obtain

$$\Pr\left[\frac{S_1^2}{S_2^2} \frac{1}{f_{\alpha/2}(v_1, v_2)} < \frac{\sigma_1^2}{\sigma_2^2} < \frac{S_1^2}{S_2^2} \frac{1}{f_{1-\alpha/2}(v_1, v_2)} \right] = 1 - \alpha.$$

The results of Theorem 7.14 enable us to replace $f_{1-\alpha/2}(v_1, v_2)$ by $1/f_{\alpha/2}(v_2, v_1)$. Therefore,

$$\Pr\left[\frac{S_1^2}{S_2^2} \frac{1}{f_{\alpha/2}(v_1, v_2)} < \frac{\sigma_1^2}{\sigma_2^2} < \frac{S_1^2}{S_2^2} f_{\alpha/2}(v_2, v_1) \right] = 1 - \alpha.$$

For any two independent random samples of size n_1 and n_2 selected from two normal populations, the ratio of the sample variances, s_1^2/s_2^2, is computed and the $(1 - \alpha)100\%$ confidence interval for σ_1^2/σ_2^2 is

$$\frac{s_1^2}{s_2^2} \frac{1}{f_{\alpha/2}(v_1, v_2)} < \frac{\sigma_1^2}{\sigma_2^2} < \frac{s_1^2}{s_2^2} f_{\alpha/2}(v_2, v_1).$$

CONFIDENCE INTERVAL FOR σ_1^2/σ_2^2 *A $(1 - \alpha)100\%$ confidence interval for the ratio σ_1^2/σ_2^2 is*

$$\frac{s_1^2}{s_2^2} \frac{1}{f_{\alpha/2}(v_1, v_2)} < \frac{\sigma_1^2}{\sigma_2^2} < \frac{s_1^2}{s_2^2} f_{\alpha/2}(v_2, v_1),$$

where s_1^2 and s_2^2 are the variances of independent samples of size n_1 and n_2, respectively, from normal populations, $f_{\alpha/2}(v_1, v_2)$ is a value of the F distribution with $v_1 = n_1 - 1$ and $v_2 = n_2 - 1$ degrees of freedom leaving an area of $\alpha/2$ to the right, and $f_{\alpha/2}(v_2, v_1)$ is a similar f value with $v_2 = n_2 - 1$ and $v_1 = n_1 - 1$ degrees of freedom.

Example 8.12 A standardized placement test in mathematics was given to 25 boys and 16 girls. The boys made an average grade of 82 with a standard deviation of 8, while the girls made an average grade of 78 with a standard deviation of 7. Find a 98% confidence interval for σ_1^2/σ_2^2 and σ_1/σ_2, where σ_1^2 and

σ_2^2 are the variances of the populations of grades for all boys and girls, respectively, who at some time have taken or will take this test.

Solution. We have $n_1 = 25$, $n_2 = 16$, $s_1 = 8$, and $s_2 = 7$. For a 98% confidence interval, $\alpha = 0.02$. Using Table A.7, we find that $f_{0.01}(24, 15) = 3.29$, and $f_{0.01}(15, 24) = 2.89$. Substituting in the formula

$$\frac{s_1^2}{s_2^2} \frac{1}{f_{\alpha/2}(v_1, v_2)} < \frac{\sigma_1^2}{\sigma_2^2} < \frac{s_1^2}{s_2^2} f_{\alpha/2}(v_2, v_1),$$

we obtain the 98% confidence interval

$$\frac{64}{49}\left(\frac{1}{3.29}\right) < \frac{\sigma_1^2}{\sigma_2^2} < \frac{64}{49}(2.89),$$

which simplifies to

$$0.397 < \frac{\sigma_1^2}{\sigma_2^2} < 3.775.$$

Upon taking square roots of the confidence limits, a 98% confidence interval for σ_1/σ_2 is

$$0.630 < \frac{\sigma_1}{\sigma_2} < 1.943.$$

8.9 Decision Theory

In our discussion on the classical approach to point estimation we adopted the criterion that selects the decision function which is most efficient; that is, we choose from all possible unbiased estimators the one with the smallest variance as our "best" estimator. In *decision theory* we also take into account the rewards for making correct decisions and the penalties for making incorrect decisions. This leads to a new criterion, which chooses the decision function $\hat{\Theta}$ that penalizes us the least when the action taken is incorrect. It is convenient now to introduce a *loss function* whose values depend on the true value of the parameter θ and the action $\hat{\theta}$. This is usually written in functional notation as $L(\hat{\Theta}; \theta)$. In many decision-making problems it is desirable to use a loss function of the form

$$L(\hat{\Theta}; \theta) = |\hat{\Theta} - \theta|$$

or perhaps

$$L(\hat{\Theta}; \theta) = (\hat{\Theta} - \theta)^2$$

in arriving at a choice between two or more decision functions.

Since θ is unknown, it must be assumed that it can equal any of several possible values. The set of all possible values that θ can assume is called the *parameter space*. For each possible value of θ in the parameter space the loss function will vary from sample to sample. We define the *risk function* for the decision function $\hat{\Theta}$ to be the expected value of the loss function when the value of the parameter is θ and denote this function by $R(\hat{\Theta}; \theta)$. Hence we have

$$R(\hat{\Theta}; \theta) = E[L(\hat{\Theta}; \theta)].$$

One method of arriving at a choice between $\hat{\Theta}_1$ and $\hat{\Theta}_2$ as an estimator for θ would be to apply the *minimax criterion*. Essentially we determine the maximum value of $R(\hat{\Theta}_1; \theta)$ and the maximum value of $R(\hat{\Theta}_2; \theta)$ in the parameter space and then choose the decision function that provided the minimum of these two maximum risks.

Example 8.13 According to the minimax criterion is $\bar{X}$ or $\tilde{X}$ a better estimator of the mean μ of a normal population with known variance σ^2, based on a random sample of size n when the loss function is of the form $L(\hat{\Theta}; \theta) = (\hat{\Theta} - \theta)^2$?

Solution. The loss function corresponding to $\bar{X}$ is given by

$$L(\bar{X}; \mu) = (\bar{X} - \mu)^2.$$

Hence the risk function is

$$R(\bar{X}; \mu) = E[(\bar{X} - \mu)^2] = \frac{\sigma^2}{n}$$

for every μ in the parameter space. Similarly one can show that the risk function corresponding to $\tilde{X}$ is given by

$$R(\tilde{X}; \mu) = E[(\tilde{X} - \mu)^2] \simeq \frac{\pi\sigma^2}{2n}$$

for every μ in the parameter space. In view of the fact that $\sigma^2/n < \pi\sigma^2/2n$, the minimax criterion selects $\bar{X}$, rather than $\tilde{X}$, as the better estimator for μ.

In some practical situations we may have additional information concerning the unknown parameter θ. For example, suppose that we wish to estimate the binomial parameter p, the proportion of defectives produced by a machine during a certain day when we know that p varies from day to day. If we can write down the probability distribution $f(p)$, then it is possible to determine the expected value of the risk function for each decision function. The expected risk corresponding to the estimator $\hat{P}$, often referred to as the *Bayes risk*, is written $B(\hat{P}) = E[R(\hat{P}; P)]$, where we are now treating the true proportion of defectives as a random variable. In general, when the unknown parameter is treated as a random variable with probability distribution given by $f(\theta)$, the Bayes risk in estimating θ by means of the estimator $\hat{\Theta}$ is given by

$$B(\hat{\Theta}) = E[R(\hat{\Theta}; \Theta)] = \sum_i R(\hat{\Theta}; \theta_i)f(\theta_i).$$

The decision function $\hat{\Theta}$ that minimizes $B(\hat{\Theta})$ is called the *Bayes estimator* of θ. We shall make no attempt in this text to derive a Bayes estimator, but instead we shall employ the Bayes risk to establish a criterion for choosing between two estimators.

BAYES' CRITERION *Let $\hat{\Theta}_1$ and $\hat{\Theta}_2$ be two estimators of the unknown parameter θ, which may be looked upon as a value of the random variable Θ with probability distribution $f(\theta)$. If $B(\hat{\Theta}_1) < B(\hat{\Theta}_2)$, then $\hat{\Theta}_1$ is selected as the better estimator for θ.*

The foregoing discussion on decision theory might better be understood if one considers the following two examples.

Example 8.14   Suppose that a friend has three similar coins except for the fact that the first one has 2 heads, the second one has 2 tails, and the third one is honest. We wish to estimate which coin our friend is flipping on the basis of 2 flips of the coin. Let θ be the number of heads on the coin. Consider two decision functions $\hat{\Theta}_1$ and $\hat{\Theta}_2$, where $\hat{\Theta}_1$ is the estimator that assigns to θ the number of heads that occur when the coin is flipped twice and $\hat{\Theta}_2$ is the estimator that assigns the value of 1 to θ no matter what the experiment yields. If the loss function is of the form $L(\hat{\Theta}; \theta) = (\hat{\Theta} - \theta)^2$, which estimator is better according to the minimax procedure?

Solution.   For the estimator $\hat{\Theta}_1$, the loss function assumes the values $L(\hat{\theta}_1; \theta) = (\hat{\theta}_1 - \theta)^2$, where $\hat{\theta}_1$ may be 0, 1, or 2, depending on the true value of θ. Clearly, if $\theta = 0$ or 2, both flips will yield all tails or all heads and our decision will be a correct one. Hence $L(0; 0) = 0$ and $L(2; 2) = 0$, from which one may easily conclude that $R(\hat{\Theta}_1; 0) = 0$ and $R(\hat{\Theta}_1; 2) = 0$. However, when $\theta = 1$ we could obtain 0, 1, or 2 heads in the 2 flips with probabilities $\frac{1}{4}$, $\frac{1}{2}$, and $\frac{1}{4}$, respectively. In this case we have $L(0; 1) = 1$, $L(1; 1) = 0$, and $L(2; 1) = 1$, from which we find

$$R(\hat{\Theta}_1; 1) = 1 \times \tfrac{1}{4} + 0 \times \tfrac{1}{2} + 1 \times \tfrac{1}{4} = \tfrac{1}{2}.$$

For the estimator $\hat{\Theta}_2$, the loss function assumes values given by $L(\hat{\theta}_2; \theta) = (\hat{\theta}_2 - \theta)^2 = (1 - \theta)^2$. Hence $L(1; 0) = 1$, $L(1; 1) = 0$, and $L(1; 2) = 1$, and the corresponding risks are $R(\hat{\Theta}_2; 0) = 1$, $R(\hat{\Theta}_2; 1) = 0$, and $R(\hat{\Theta}_2; 2) = 1$. Since the maximum risk is $\frac{1}{2}$ for the estimator Θ_1 compared to a maximum risk of 1 for $\hat{\Theta}_2$, the minimax criterion selects $\hat{\Theta}_1$ as the better of the two estimators.

Example 8.15 Let us suppose, referring to Example 8.14, that our friend flips the honest coin 80% of the time and the other 2 coins each about 10% of the time. Does the Bayes criterion select $\hat{\Theta}_1$ or $\hat{\Theta}_2$ as the better estimator?

Solution. The parameter θ may now be treated as a random variable with the following probability distribution:

θ	0	1	2
$f(\theta)$	0.1	0.8	0.1

For the estimator $\hat{\Theta}_1$ the Bayes risk is

$$B(\hat{\Theta}_1) = R(\hat{\Theta}_1; 0)f(0) + R(\hat{\Theta}_1; 1)f(1) + R(\hat{\Theta}_1; 2)f(2)$$
$$= (0)(0.1) + (\tfrac{1}{2})(0.8) + (0)(0.1) = 0.4.$$

Similarly, for the estimator $\hat{\Theta}_2$, we have

$$B(\hat{\Theta}_2) = R(\hat{\Theta}_2; 0)f(0) + R(\hat{\Theta}_2; 1)f(1) + R(\hat{\Theta}_2; 2)f(2)$$
$$= (1)(0.1) + (0)(0.8) + (1)(0.1) = 0.2.$$

Since $B(\hat{\Theta}_2) < B(\hat{\Theta}_1)$, the Bayes criterion selects $\hat{\Theta}_2$ as the better estimator for the parameter θ.

EXERCISES

1. (a) Find the parameters μ and σ^2 for the finite population 3, 5, and 2.
   (b) Set up a sampling distribution similar to Table 7.2 for $\bar{X}$ when samples of size 2 are selected at random, with replacement. Show that $\bar{X}$ is an unbiased estimator of μ.
   (c) By computing s^2 for each sample, obtain the sampling distribution of S^2 and then show that S^2 is an unbiased estimator of σ^2.

2. Consider the statistic S'^2 whose values are computed from the formula $s'^2 = \sum_{i=1}^{n} (x_i - \bar{x})^2/n$. By computing s'^2 for each sample in Exercise 1, obtain the sampling distribution of S'^2. Show that $E(S'^2) \neq \sigma^2$ and hence that S'^2 is a biased estimator for σ^2.

populations σ
not sample

3. An electrical firm manufactures light bulbs that have a length of life that is approximately normally distributed, with a standard deviation of 40 hours. If a random sample of 30 bulbs has an average life of 780 hours, find a 96% confidence interval for the population mean of all bulbs produced by this firm. *use Z table even*

4. A soft-drink machine is regulated so that the amount of drink dispensed is approximately normally distributed with a standard deviation equal to 0.5 ounce. Find a 95% confidence interval for the mean of all drinks dispensed by this machine if a random sample of 36 drinks had an average content of 7.4 ounces.

5. The heights of a random sample of 50 college students showed a mean of 68.5 inches and a standard deviation of 2.7 inches.
   (a) Construct a 98% confidence interval for the mean height of all college students.
   (b) What can we assert with 98% confidence about the possible size of our error if we estimate the mean height of all college students to be 68.5 inches?

P 160

6. A random sample of 100 automobile owners shows that an automobile is driven on the average 14,500 miles per year, in the state of Virginia, with a standard deviation of 2400 miles.
   (a) Construct a 99% confidence interval for the average number of miles an automobile is driven annually in Virginia.
   (b) What can we assert with 99% confidence about the possible size of our error if we estimate the average number of miles driven by car owners in Virginia as 14,500 miles per year?

7. How large a sample is needed in Exercise 3 if we wish to be 96% confident that our sample mean will be within 10 hours of the true mean?

P 160

8. How large a sample is needed in Exercise 4 if we wish to be 95% confident that our sample mean will be within 0.3 ounce of the true mean?

9. An efficiency expert wishes to determine the average time that it takes to drill 3 holes in a certain metal clamp. How large a sample will he need to be 95% confident that his sample mean will be within 15 seconds of the true mean? Assume that it is known from previous studies that $\sigma = 40$ seconds.

10. The weights of 10 boxes of cereal are 10.2, 9.7, 10.1, 10.3, 10.1, 9.8, 9.9, 10.4, 10.3, and 9.8 ounces. Find a 99% confidence interval for the mean of all such boxes of cereal, assuming an approximate normal distribution.

P 162

11. A random sample of size 20 from a normal distribution has a mean $\bar{x} = 32.8$ and a standard deviation $s = 4.51$. Construct a 95% confidence interval for μ.

12. A random sample of 8 cigarettes of a certain brand has an average nicotine content of 18.6 milligrams and a standard deviation of 2.4 milligrams. Construct a 99% confidence interval for the true average nicotine content of this particular brand of cigarettes, assuming an approximate normal distribution.

13. A random sample of 12 male students at a fraternity party showed an average expenditure for the evening of $8.00, with a standard deviation of $1.75. Construct a 90% confidence interval for the average amount spent by male students attending this party, assuming the expenditures to be approximately normally distributed.

14. A random sample of size $n_1 = 25$ taken from a normal population with a standard deviation $\sigma_1 = 5$ has a mean $\bar{x}_1 = 80$. A second random sample of size $n_2 = 36$, taken from a different normal population with a standard deviation $\sigma_2 = 3$, has a mean $\bar{x}_2 = 75$. Find a 94% confidence interval for $\mu_1 - \mu_2$.

15. Two varieties of corn are being compared for yield. Fifty acres of each variety are planted and grown under similar conditions. Variety A yielded, on the average, 78.3 bushels per acre with a standard deviation of 5.6 bushels per acre, while variety B yielded, on the average, 87.2 bushels per acre with a standard deviation of 6.3 bushels per acre. Construct a 95% confidence interval for the difference of the population means.

16. A study was made to estimate the difference in salaries of college professors in the private and state colleges of Virginia. A random sample of 100 professors in private colleges showed an average 9-month salary of $15,000 with a standard deviation of $1200. A random sample of 200 professors in state colleges showed an average salary of $16,000 with a standard deviation of $1400. Find a 98% confidence interval for the difference between the average salaries of professors teaching in state and private colleges of Virginia.

17. Given two random samples of size $n_1 = 9$ and $n_2 = 16$, from two independent normal populations, with $\bar{x}_1 = 64$, $\bar{x}_2 = 59$, $s_1 = 6$, and $s_2 = 5$, find a 95% confidence interval for $\mu_1 - \mu_2$, assuming that $\sigma_1 = \sigma_2$.

18. Students may choose between a 3-semester-hour course in physics without labs and a 4-semester-hour course with labs. The final written examination is the same for each section. If 12 students in the section with labs made an average examination grade of 84 with a standard deviation of 4, and 18 students in the section without labs made an average grade of 77 with a standard deviation of 6, find a 99% confidence interval for the difference between the average grades for the two courses. Assume the populations to be approximately normally distributed with equal variances.

19. A taxi company is trying to decide whether to purchase brand A or brand B tires for its fleet of taxis. To estimate the difference in the two brands, an experiment is conducted using 12 of each brand. The tires are run until they wear out. The results are:

Brand A: $\bar{x}_1 = 22,500$ miles, $s_1 = 3100$ miles.
Brand B: $\bar{x}_2 = 23,600$ miles, $s_2 = 3800$ miles.

Compute a 95% confidence interval for $\mu_1 - \mu_2$, assuming the populations to be approximately normally distributed.

20. The following data represent the running times of films produced by two motion-picture companies:

	Time, Minutes						
Company I	103	94	110	87	98		
Company II	97	82	123	92	175	88	118

Compute a 90% confidence interval for the difference between the average running times of films produced by the two companies. Assume that the running times are approximately normally distributed.

21. The government awarded grants to the agricultural departments of 9 different universities to test the yield capabilities of two new varieties of wheat. Five acres of each variety are planted at each university and the yields, in bushels per acre, recorded as follows:

	University								
	1	2	3	4	5	6	7	8	9
Variety 1	38	23	35	41	44	29	37	31	38
Variety 2	45	25	31	38	50	33	36	40	43

Find a 95% confidence interval for the mean difference between the yields of the two varieties assuming the distributions of yields to be approximately normal. Explain why pairing is necessary in this problem.

22. Referring to Exercise 19, find a 99% confidence interval for $\mu_1 - \mu_2$ if a tire from each company is assigned at random to the rear wheels of 8 taxis and the following results recorded:

Taxi	Brand A	Brand B
1	21,400	22,800
2	28,300	29,100
3	22,800	23,400
4	19,900	19,300
5	30,100	29,700
6	20,400	22,600
7	23,700	24,200
8	18,700	19,600

23. It is claimed that a new diet will reduce a person's weight by 10 pounds on the average in a period of 2 weeks. The weights of 7 women who followed this diet were recorded before and after a 2-week period:

	Woman						
	1	2	3	4	5	6	7
Weight before	129	133	136	152	141	138	125
Weight after	130	121	128	137	129	132	120

Test a manufacturer's claim by computing a 95% confidence interval for the mean difference in the weight. Assume the distributions of weights to be approximately normal.

24. (a) A random sample of 200 voters is selected and 114 are found to support an annexation suit. Find the 96% confidence interval for the fraction of the voting population favoring the suit.

(b) What can we assert with 96% confidence about the possible size of our error if we estimate the fraction of voters favoring the annexation suit to be 0.57?

25. (a) A random sample of 500 cigarette smokers is selected and 86 are found to have a preference for brand X. Find the 90% confidence interval for the fraction of the population of cigarette smokers who prefer brand X.

(b) What can we assert with 90% confidence about the possible size of our error if we estimate the fraction of cigarette smokers who prefer brand X to be 0.172?

26. In a random sample of 1000 houses in a certain city it is found that 628 own color television sets. Find the 98% confidence interval for the fraction of homes in this city that have color sets.

27. A random sample of 75 college students is selected and 16 are found to have cars on campus. Use a 95% confidence interval to estimate the fraction of students who have cars on campus.

28. A certain new rocket-launching system is being considered for deployment of small short-range launches. The existing system has $p = 0.8$ as the probability of a successful launch. A sample of 40 experimental launches is made with the new system and 34 are successful.

    (a) Give a point estimate of the probability of a successful launch using the new system.

    (b) Construct a 95% confidence interval for this probability.

    (c) Does the evidence strongly indicate that the new system is better? Explain.

29. How large a sample is needed in Exercise 24 if we wish to be 96% confident that our sample proportion will be within 0.02 of the true fraction of the voting population?

30. How large a sample is needed in Exercise 26 if we wish to be 98% confident that our sample proportion will be within 0.05 of the true proportion of houses in the city that have color television sets?

31. A study is to be made to estimate the percentage of citizens in a town who favor having their water fluoridated. How large a sample is needed if one wishes to be at least 95% confident that our estimate is within 1% of the true percentage?

32. A study is to be made to estimate the proportion of housewives who own an automatic dryer. How large a sample is needed if one wishes to be at least 99% confident that our estimate differs from the true proportion by an amount less than 0.01?

33. In a study to estimate the proportion of housewives who own an automatic dryer it is found that 52 of 100 urban residents have a dryer while only 34 of 125 suburban residents own a dryer. Find a 96% confidence interval for the difference between the fraction of urban and suburban housewives who own an automatic dryer.

34. A cigarette-manufacturing firm claims that its brand A line of cigarettes outsells its brand B line by 8%. If it is found that 42 of 200 smokers prefer brand A and 18 of 150 smokers prefer brand B, compute a 94% confidence interval for the difference between the proportions of sales of the two brands and decide if the 8% difference is a valid claim.

35. A random sample of 100 men and 100 women at a Southern college are asked if they have an automobile on campus. If 24 of the men and 13 of the women have cars, find the 99% confidence interval for the difference between the proportion of men and women who have cars on campus.

36. A study is made to determine if a cold climate results in more students being absent from school during a semester than for a warmer climate. Two groups of students are selected at random, one group from Vermont and the other group from Georgia. Of the 300 students from Vermont, 64 were absent at least 1 day during the semester, and of the 400 students from Georgia, 51 were absent 1 or more days. Find a 95% confidence interval for the difference between the fractions of the students who are absent in the two states.

37. Construct a 99% confidence interval for σ^2 in Exercise 10.

38. Construct a 95% confidence interval for σ in Exercise 11.

39. Construct a 99% confidence interval for σ in Exercise 12.

40. Construct a 90% confidence interval for σ^2 in Exercise 13.

41. Construct a 98% confidence interval for σ_1/σ_2 in Exercise 17. Were we justified in assuming that $\sigma_1 = \sigma_2$?

42. Construct a 90% confidence interval for σ_1^2/σ_2^2 in Exercise 19.

43. Construct a 90% confidence interval for σ_1^2/σ_2^2 in Exercise 20.

44. We wish to estimate the binomial parameter p by the decision function $\hat{P}$, the proportion of successes in a binomial experiment consisting of n trials. Find $R(\hat{P}; p)$ when the loss function is of the form $L(\hat{P}; p) = (\hat{P} - p)^2$.

45. Suppose that an urn contains 3 balls, of which θ are red and the remainder black, where θ can vary from 0 to 3. We wish to estimate θ by selecting two balls in succession without replacement. Let $\hat{\Theta}_1$ be the decision function that assigns to θ the value 0 if neither ball is red, the value 1 if the first ball only is red, the value 2 if the second ball only is red, and the value 3 if both balls are red. Using a loss function of the form $L(\hat{\Theta}_1; \theta) = |\hat{\Theta}_1 - \theta|$, find $R(\hat{\Theta}_1; \theta)$.

46. In Exercise 45, consider the estimator $\hat{\Theta}_2 = X(X + 1)/2$, where X is the number of red balls in our sample. Find $R(\hat{\Theta}_2; \theta)$.

47. Use the minimax criterion to determine whether the estimator $\hat{\Theta}_1$ of Exercise 45 or the estimator $\hat{\Theta}_2$ of Exercise 46 is the better estimator.

48. Use the Bayes criterion to determine whether the estimator $\hat{\Theta}_1$ of Exercise 45 or the estimator $\hat{\Theta}_2$ of Exercise 46 is the better estimator, given the following additional information:

θ	0	1	2	3
$f(\theta)$	0.1	0.5	0.1	0.3

TESTS OF HYPOTHESES CHAPTER 9

9.1 Statistical Hypotheses

The testing of statistical hypotheses is perhaps the most important area of decision theory. First let us define precisely what we mean by a statistical hypothesis.

> DEFINITION *A* statistical hypothesis *is an assumption or statement, which may or may not be true, concerning one or more populations.*

The truth or falsity of a statistical hypothesis is never known with certainty unless we examine the entire population. This, of course, would be impractical in most situations. Instead we take a random sample from the population of interest and use the information contained in this sample to decide whether the hypothesis is likely to be true or false. Evidence from the sample that is inconsistent with the stated hypothesis leads to a rejection of the hypothesis, whereas evidence supporting the hypothesis leads to its acceptance. We should make it clear at this point that the acceptance of a statistical hypothesis is a result of insufficient evidence to reject it and does not necessarily imply that it is true. For example, in tossing a coin 100 times we might test the hypothesis that the coin is balanced. In terms of population parameters, we are testing the hypothesis that the proportion of heads is $p = 0.5$ if the coin were tossed indefinitely. An outcome of 48 heads would not be surprising if the coin is balanced. Such a result would surely support the hypothesis $p = 0.5$. One might argue that such an occurrence is also consistent with the hypothesis that $p = 0.45$. Thus, in accepting the hypothesis, the only thing we can be reasonably certain about is that the true proportion of heads is not a great deal different from one half. If the 100 trials had resulted in only 35 heads, we would then have evidence to support the rejection of our hypothesis. In view of the fact that the probability of obtaining 35 or fewer heads in 100 tosses of a balanced coin is approximately 0.002, either a very rare event has occurred or we are right in concluding that $p \neq 0.5$.

Although we shall use the terms "accept" and "reject" frequently through-out this chapter, it is important to understand that the rejection of a hypothesis is to conclude that it is false, while the acceptance of a hypothesis merely implies that we have no evidence to believe otherwise. Because of this terminology, the statistician or experimenter should always state as his hypothesis that which he hopes to reject. If he is interested in a new cold vaccine, he should assume that it is no better than the vaccine now on the market and then set out to reject this contention. Similarly, to prove that one teaching technique is superior to another, we test the hypothesis that there is no difference in the two techniques.

Hypotheses that we formulate with the hope of rejecting are called *null hypotheses* and are denoted by H_0. The rejection of H_0 leads to the acceptance of an *alternative hypothesis*, denoted by H_1. Hence if H_0 is the null hypothesis $p = 0.5$ for a binomial population, the alternative hypothesis H_1 might be $p = 0.75, p > 0.5, p < 0.5$, or $p \neq 0.5$.

9.2 Type I and Type II Errors

To illustrate the concepts used in testing a statistical hypothesis about a population, consider the following example. A certain type of cold vaccine is known to be only 25% effective after a period of 2 years. In order to determine if a new and somewhat more expensive vaccine is superior in providing protection against the same virus for a longer period of time, 20 people are chosen at random and inoculated. If fewer than 12 of those receiving the new vaccine contract the virus within a 2-year period, the new vaccine will be considered superior to the one presently in use. We shall test the null hypothesis that the new vaccine is equally effective after a period of 2 years as the one now commonly used against the alternative hypothesis that the new vaccine is in fact superior. This is equivalent to testing the hypothesis that the binomial parameter for the probability of a success on a given trial is $p = \frac{1}{4}$ against the alternative that $p > \frac{1}{4}$. This is usually written as follows:

$$H_0 : p = \tfrac{1}{4},$$
$$H_1 : p > \tfrac{1}{4}.$$

The above decision could lead to either of two wrong conclusions. For instance the new vaccine may be no better than the one now in use and, for this particular randomly selected group of individuals, 9 or more surpass the 2-year period without contracting the virus. We would be committing an error by rejecting H_0 in favor of H_1 when, in fact, H_0 is true. Such an error is called a *type I error*.

> DEFINITION *A* type I error *has been committed if we reject the null hypothesis when it is true.*

A second kind of error is committed if fewer than 9 of the group surpass the 2-year period successfully, and we conclude that the new vaccine is no better when it actually is. In this case we would accept H_0 when it is false. This is called a *type II error*.

> DEFINITION *A* type II error *has been committed if we accept the null hypothesis when it is false.*

The statistic on which we base our decision is the number of individuals who receive protection from the new vaccine for a period of at least 2 years. The possible values, from 0 to 20, are divided into two groups: those numbers less than 9 and those greater than or equal to 9. All possible scores above 8.5 constitute the *critical region*, and all possible scores below 8.5 determine the *acceptance region*. The number 8.5 separating these two regions is called the *critical value*. If the statistic on which we base our decision falls in the critical region, we reject H_0 in favor of the alternative hypothesis H_1. If it falls in the acceptance region, we accept H_0.

> DEFINITION *The probability of committing a type I error is called the* level of significance *of the test and is denoted by* α.

In our example a type I error will occur when 9 or more individuals surpass the 2-year period without contracting the virus using a new vaccine that is actually equivalent to the one in use. Hence, if X is the number of individuals who remain healthy for at least 2 years,

$$\alpha = \Pr(\text{type I error})$$
$$= \Pr(X \geq 9 | p = \tfrac{1}{4})$$

$$= \sum_{x=9}^{20} b(x; 20, \tfrac{1}{4})$$

$$= 1 - \sum_{x=0}^{8} b(x; 20, \tfrac{1}{4})$$

$$= 1 - 0.9591$$

$$= 0.0409.$$

We say that the null hypothesis, $p = \tfrac{1}{4}$, is being tested at the $\alpha = 0.0409$ level of significance. Sometimes the level of significance is called the *size* of the critical region. A critical region of size 0.0409 is very small, and therefore it is unlikely that a type I error will be committed. Consequently it would be most unusual for 9 or more individuals to remain immune to a virus for a 2-year period using a new vaccine that is essentially equivalent to the one now on the market.

The probability of committing a type II error, denoted by β, is impossible to compute unless we have a specific alternative hypothesis. If we test the null hypothesis that $p = \tfrac{1}{4}$ against the alternative hypothesis that $p = \tfrac{1}{2}$, then we are able to compute the probability of accepting H_0 when it is false. We simply find the probability of obtaining fewer than 9 in the group that surpass the 2-year period when $p = \tfrac{1}{2}$. In this case

$$\beta = \text{Pr(type II error)}$$

$$= \text{Pr}(X < 9 | p = \tfrac{1}{2})$$

$$= \sum_{x=0}^{8} b(x; 20, \tfrac{1}{2})$$

$$= 0.2517.$$

This is a rather high probability, indicating a poor test procedure. It is quite likely that we will reject the new vaccine when, in fact, it is superior to that now in use. Ideally we like to use a test procedure for which both the type I and type II errors are small.

It is possible that the director of the testing program is willing to make a type II error if the more expensive vaccine is not significantly superior. The only time he wishes to guard against the type II error is when the true value of p is at least 0.7. Letting $p = 0.7$, the above test procedure gives

$$\beta = \text{Pr(type II error)}$$

$$= \text{Pr}(X < 9 | p = 0.7)$$

$$= \sum_{x=0}^{8} b(x; 20, 0.7)$$

$$= 0.0051.$$

With such a small probability of committing a type II error it is extremely unlikely that the new vaccine would be rejected when it is 70% effective after a period of 2 years. As the alternative hypothesis approaches unity, the value of β diminishes to zero.

Let us assume that the director of the testing program is unwilling to commit a type II error when the alternative hypothesis $p = \frac{1}{2}$ is true, even though we have found the probability of such an error to be $\beta = 0.2517$. A reduction in β is always possible by increasing the size of the critical region. For example consider what happens to the values of α and β when we change our critical value to 7.5, so that all scores of 8 or more fall in the critical region and those below 8 fall in the acceptance region. Now, in testing $p = \frac{1}{4}$ against the alternative hypothesis that $p = \frac{1}{2}$, we find

$$\alpha = \sum_{x=8}^{20} b(x; 20, \tfrac{1}{4})$$

$$= 1 - \sum_{x=0}^{7} b(x; 20, \tfrac{1}{4})$$

$$= 1 - 0.8982$$

$$= 0.1018,$$

$$\beta = \sum_{x=0}^{7} b(x; 20, \tfrac{1}{2})$$

$$= 0.1316.$$

By adopting a new decision procedure, we have reduced the probability of committing a type II error at the expense of increasing the probability of committing a type I error. For a fixed sample size, a decrease in the probability of one error will always result in an increase in the probability of the other error. Fortunately the probability of committing both types of errors can be reduced by increasing the sample size. Consider the same problem using a random sample of 100 individuals. If 37 or more of the group surpass the 2-year period, we reject the null hypothesis that $p = \frac{1}{4}$ and accept the alternative hypothesis that $p > \frac{1}{4}$. The critical value is now 36.5. All possible scores above 36.5 constitute the critical region and all possible scores below 36.5 fall in the acceptance region.

To determine the probability of committing a type I error, we shall use the normal-curve approximation with

$$\mu = np = (100)(\tfrac{1}{4}) = 25,$$

$$\sigma = \sqrt{npq} = \sqrt{(100)(\tfrac{1}{4})(\tfrac{3}{4})} = 4.33.$$

Referring to Figure 9.1,

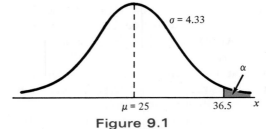

Figure 9.1

Probability of a type I error.

$$\alpha = \Pr(\text{type I error})$$
$$= \Pr(X > 36.5 \mid H_0 \text{ is true}).$$

The z value corresponding to $x = 36.5$ is

$$z = \frac{36.5 - 25}{4.33} = 2.656.$$

Therefore,

$$\alpha = \Pr(Z > 2.656)$$
$$= 1 - \Pr(Z < 2.656)$$
$$= 1 - 0.9961$$
$$= 0.0039.$$

If H_0 is false and the true value of H_1 is $p = \frac{1}{2}$, we can determine the probability of a type II error using the normal-curve approximation with

$$\mu = np = (100)(\tfrac{1}{2}) = 50,$$
$$\sigma = \sqrt{npq} = \sqrt{(100)(\tfrac{1}{2})(\tfrac{1}{2})} = 5.$$

The probability of falling in the acceptance region when H_1 is true is given by the area of the shaded region in Figure 9.2. Hence

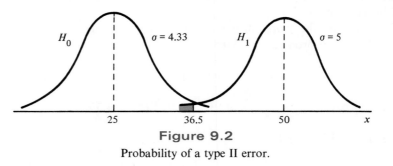

Figure 9.2

Probability of a type II error.

$$\beta = \Pr(\text{type II error})$$
$$= \Pr(X < 36.5 | H_1 \text{ is true}).$$

The z value corresponding to $x = 36.5$ is

$$z = \frac{36.5 - 50}{5} = -2.7.$$

Therefore,

$$\beta = \Pr(Z < -2.7)$$
$$= 0.0035.$$

Obviously the type I and type II errors will rarely occur if the experiment consists of 100 individuals.

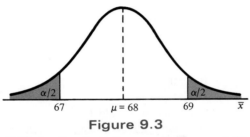

Figure 9.3

Critical region for testing $\mu = 68$ versus $\mu \neq 68$.

The concepts discussed above can easily be seen graphically when the population is continuous. Consider the null hypothesis that the average height of students in a certain college is 68 inches against the alternative hypothesis that it is unequal to 68; that is, we wish to test

$$H_0: \mu = 68,$$
$$H_1: \mu \neq 68.$$

The alternative hypothesis allows for the possibility that $\mu < 68$ or $\mu > 68$.

Assume the standard deviation of the population of heights to be $\sigma = 3.6$. Our decision statistic, based on a sample of size $n = 36$, will be $\bar{X}$, the most efficient estimator of μ. From Chapter 7 we know that the sampling distribution of $\bar{X}$ is approximately normally distributed with standard deviation $\sigma_{\bar{X}} = \sigma/\sqrt{n} = 3.6/6 = 0.6$.

A sample mean that falls close to the hypothesized value of 68 would be considered evidence in favor of H_0. On the other hand a sample mean that is considerably less than or more than 68 would be evidence inconsistent with H_0 and therefore favoring H_1. A critical region, indicated by the shaded area in Figure 9.3, is arbitrarily chosen to be $\bar{X} < 67$ and $\bar{X} > 69$. The acceptance region will therefore be $67 < \bar{X} < 69$. Hence if our sample mean $\bar{x}$ falls inside the critical region, H_0 is rejected; otherwise we accept H_0.

The probability of committing a type I error, or the level of significance of our test, is equal to the sum of the areas that have been shaded in each tail of the distribution in Figure 10.3. Therefore,

$$\alpha = \Pr(\bar{X} < 67 | H_0 \text{ is true}) + \Pr(\bar{X} > 69 | H_0 \text{ is true}).$$

The z values corresponding to $\bar{x}_1 = 67$ and $\bar{x}_2 = 69$ when H_0 is true are

$$z_1 = \frac{67 - 68}{0.6} = -1.67,$$

$$z_2 = \frac{69 - 68}{0.6} = 1.67.$$

Therefore,

$$\alpha = \Pr(Z < -1.67) + \Pr(Z > 1.67)$$
$$= 2\Pr(Z < -1.67)$$
$$= 0.0950.$$

Thus 9.5% of all samples of size 36 would lead us to reject $\mu = 68$ inches when it is true. To reduce α, we have a choice of increasing the sample size or widening the acceptance region. Suppose that we increase the sample size to $n = 64$. Then $\sigma_{\bar{X}} = 3.6/8 = 0.45$. Now

$$z_1 = \frac{67 - 68}{0.45} = -2.22,$$

$$z_2 = \frac{69 - 68}{0.45} = 2.22.$$

Hence

$$\alpha = \Pr(Z < -2.22) + \Pr(Z > 2.22)$$
$$= 2\Pr(Z < -2.22)$$
$$= 0.0264.$$

The reduction in α is not sufficient by itself to guarantee a good testing procedure. We must evaluate β for various alternative hypotheses that we feel should be accepted if true. Therefore, if it is important to reject H_0 when the true mean is some value $\mu \geq 70$ or $\mu \geq 66$, then the probability of committing a type II error should be computed and examined for the alternatives $\mu = 66$ and $\mu = 70$. As a result of symmetry it is only necessary to evaluate the probability of accepting the null hypothesis that $\mu = 68$ when the alternative $\mu = 70$ is true. A type II error will result when the sample mean $\bar{x}$ falls between 67 and 69 when H_1 is true. Therefore, referring to Figure 9.4,

$$\beta = \Pr(67 < \bar{X} < 69 | H_1 \text{ is true}).$$

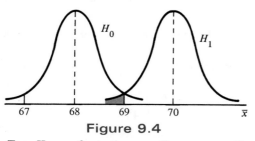

Figure 9.4

Type II error for testing $\mu = 68$ versus $\mu = 70$.

The z values corresponding to $\bar{x}_1 = 67$ and $\bar{x}_2 = 69$, when H_1 is true, are

$$z_1 = \frac{67 - 70}{0.45} = -6.67,$$

$$z_2 = \frac{69 - 70}{0.45} = -2.22.$$

Therefore,

$$\begin{aligned}
\beta &= \Pr(-6.67 < Z < -2.22) \\
&= \Pr(Z < -2.22) - \Pr(Z < -6.67) \\
&= 0.0132 - 0.0000 \\
&= 0.0132.
\end{aligned}$$

If the true value of μ is the alternative $\mu = 66$, the value of β will again be 0.0132. For all possible values of $\mu < 66$ or $\mu > 70$, the value of β will be even smaller when $n = 64$, and consequently there would be little chance of accepting H_0 when it is false.

The probability of committing a type II error increases rapidly when the true value of μ approaches, but is not equal to, the hypothesized value. Of course this is usually the situation where we do not mind making a type II error. For example, if the alternative hypothesis $\mu = 68.5$ is true, we do not mind committing a type II error by concluding that the true answer is $\mu = 68$. The probability of making such an error will be high when $n = 64$. Referring to Figure 9.5, we have

$$\beta = \Pr(67 < \bar{X} < 69 | H_1 \text{ is true}).$$

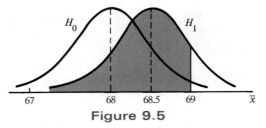

Figure 9.5

Type II error for testing $\mu = 68$ versus $\mu \neq 68.5$.

The z values corresponding to $\bar{x}_1 = 67$ and $\bar{x}_2 = 69$ when $\mu = 68.5$ are

$$z_1 = \frac{67 - 68.5}{0.45} = -3.33,$$

$$z_2 = \frac{69 - 68.5}{0.45} = 1.11.$$

Therefore,

$$\begin{aligned} \beta &= \Pr(-3.33 < Z < 1.11) \\ &= \Pr(Z < 1.11) - \Pr(Z < -3.33) \\ &= 0.8665 - 0.0004 \\ &= 0.8661. \end{aligned}$$

The preceding examples illustrate the following important properties:

1. The type I error and type II error are related. A decrease in the probability of one results in an increase in the probability of the other.
2. The size of the critical region, and therefore the probability of committing a type I error, can always be reduced by adjusting the critical value(s).
3. An increase in the sample size n will reduce α and β simultaneously.
4. If the null hypothesis is false, β is a maximum when the true value of a parameter is close to the hypothesized value. The greater the distance between the true value and the hypothesized value, the smaller β will be.

9.3 One-Tailed and Two-Tailed Tests

A test of any statistical hypothesis, where the alternative is *one-sided*, such as

$$H_0 : \theta = \theta_0,$$
$$H_1 : \theta > \theta_0,$$

or perhaps

$$H_0: \theta = \theta_0,$$
$$H_1: \theta < \theta_0,$$

is called a *one-tailed test*. The critical region for the alternative hypothesis $\theta > \theta_0$ lies entirely in the right tail of the distribution, while the critical region for the alternative hypothesis $\theta < \theta_0$ lies entirely in the left tail. A one-sided test was used in the vaccine experiment to test the hypothesis $p = \frac{1}{4}$ against the one-sided alternative $p > \frac{1}{4}$ for the binomial distribution.

A test of any statistical hypothesis where the alternative is *two-sided*, such as

$$H_0: \theta = \theta_0,$$
$$H_1: \theta \neq \theta_0,$$

is called a *two-tailed test*. The alternative hypothesis states that either $\theta < \theta_0$ or $\theta > \theta_0$. Values in both tails of the distribution constitute the critical region. A two-tailed test was used to test the null hypothesis that $\mu = 68$ inches against the two-sided alternative $\mu \neq 68$ for the continuous population of students' heights.

Whether one sets up a one-sided or a two-sided alternative hypothesis will depend on the conclusion to be drawn if H_0 is rejected. The location of the critical region can be determined only after H_1 has been stated. For example, in testing a new drug, one sets up the hypothesis that it is no better than similar drugs now on the market and tests this against the alternative hypothesis that the new drug is superior. Such an alternative hypothesis will result in a one-tailed test with the critical region in the right tail. However, if we wish to determine whether two teaching procedures are equally effective, the alternative hypothesis should allow for either procedure to be superior. Hence the test is two-tailed with the critical region divided so as to fall in the extreme left and right tails of the distribution.

In testing hypotheses about discrete populations the critical region is chosen arbitrarily and its size determined. If the size α is too large, it can be reduced by making an adjustment in the critical value. It may be necessary to increase the sample size to offset the increase that automatically occurs in β. In testing hypotheses about continuous populations, it is customary to choose the value of α to be 0.05 or 0.01 and then find the critical value(s). For example, in a two-tailed test at the 0.05 level of significance, the critical values for a statistic having a standard normal distribution will be $-z_{0.025} = -1.96$ and $z_{0.025} = 1.96$. In terms of z values, the critical region of size 0.05 will be $Z < -1.96$ and $Z > 1.96$ and the acceptance region is $-1.96 < Z < 1.96$. A test is said to be *significant* if the null hypothesis is rejected at the 0.05 level of significance and is considered *highly significant* if the null hypothesis is rejected at the 0.01 level of significance.

In the remaining sections of this chapter we shall consider several special tests of hypotheses that are frequently used by statisticians.

9.4 Tests Concerning Means and Variances

Consider the problem of testing the hypothesis that the mean μ of a population, with known variance σ^2, equals a specified value μ_0 against the two-sided alternative that the mean is not equal to μ_0; that is, we shall test

$$H_0: \mu = \mu_0,$$

$$H_1: \mu \neq \mu_0.$$

An appropriate statistic on which we base our decision criterion is the random variable $\bar{X}$. From Chapter 7 we already know that the sampling distribution of $\bar{X}$ is approximately normally distributed with mean $\mu_{\bar{X}} = \mu$ and variance $\sigma_{\bar{X}}^2 = \sigma^2/n$, where μ and σ^2 are the mean and variance of the population from which we select random samples of size n. By using a significance level of α, it is possible to find two critical values, $\bar{x}_1$ and $\bar{x}_2$, such that the interval $\bar{x}_1 < \bar{X} < \bar{x}_2$ defines the acceptance region and the two tails of the distribution, $\bar{X} < \bar{x}_1$ and $\bar{X} > \bar{x}_2$, constitute the critical region.

The critical region can be given in terms of z values by means of the transformation

$$z = \frac{\bar{x} - \mu_0}{\sigma/\sqrt{n}}.$$

Hence, for an α level of significance, the critical values of the random variable Z, corresponding to $\bar{x}_1$ and $\bar{x}_2$, are shown in Figure 9.6 to be

$$-z_{\alpha/2} = \frac{\bar{x}_1 - \mu_0}{\sigma/\sqrt{n}},$$

$$z_{\alpha/2} = \frac{\bar{x}_2 - \mu_0}{\sigma/\sqrt{n}}.$$

From the population we select a random sample of size n and compute the sample mean $\bar{x}$. If $\bar{x}$ falls in the acceptance region, $\bar{x}_1 < \bar{X} < \bar{x}_2$, then

$$z = \frac{\bar{x} - \mu_0}{\sigma/\sqrt{n}}$$

will fall in the region $-z_{\alpha/2} < Z < z_{\alpha/2}$ and we conclude that $\mu = \mu_0$; otherwise we reject H_0 and accept the alternative hypothesis that $\mu \neq \mu_0$. The critical region is usually stated in terms of Z rather than $\bar{X}$.

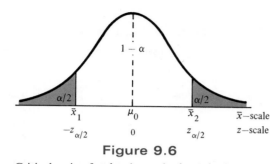

Figure 9.6

Critical region for the alternative hypothesis $\mu \neq \mu_o$.

The test procedure just described is equivalent to finding a $(1 - \alpha)100\%$ confidence interval for μ and accepting H_0 if μ_0 lies in the interval. If μ_0 lies outside the interval, we reject H_0 in favor of the alternative hypothesis H_1. Consequently, when one makes inferences about the mean μ from a population with known variance σ^2, whether it be by the construction of a confidence interval or through the testing of a statistical hypothesis, the same statistic, $Z = (\bar{X} - \mu)/(\sigma/\sqrt{n})$, is employed.

In general, if one uses a statistic to construct a confidence interval for a parameter θ, whether it be the statistic Z, T, X^2, or F, we can use that same statistic to test the hypothesis that the parameter equals some specified value θ_0 against an appropriate alternative. Of course all the underlying assumptions made in Chapter 8 relative to the use of a given statistic will apply to the tests described here. This essentially means that all our samples are selected from approximately normal populations. However, a Z statistic may be used to test hypotheses about means from nonnormal populations when $n \geq 30$.

In Table 9.1 we list the statistic used to test a specified hypothesis H_0 and give the appropriate critical regions for one- and two-sided alternative hypotheses H_1. The steps for testing a hypothesis concerning a population parameter θ against some alternative hypothesis may be summarized as follows:

1. H_0: $\theta = \theta_0$.
2. H_1: Alternatives are $\theta < \theta_0$, $\theta > \theta_0$, or $\theta \neq \theta_0$.
3. Choose a level of significance equal to α.
4. Select the appropriate test statistic and establish the critical region.
5. Compute the value of the statistic from a random sample of size n.
6. Conclusion: Reject H_0 if the statistic has a value in the critical region; otherwise accept H_0.

Example 9.1 A manufacturer of sports equipment has developed a new synthetic fishing line, which he claims has a mean breaking strength of 15 pounds with a standard deviation of 0.5 pound. Test the hypothesis that $\mu = 15$ pounds against the

normal
pop $\to \sigma$
estimated
Standard Dev $\to S$

Table 9.1
Tests Concerning Means and Variances

$S(p128)$

H_0	Test Statistic	H_1	Critical Region
$\mu = \mu_0$	$Z = \dfrac{\bar{X} - \mu_0}{\sigma/\sqrt{n}}$; σ known	$\mu < \mu_0$ $\mu > \mu_0$ $\mu \neq \mu_0$	$Z < -z_\alpha$ $Z > z_\alpha$ $Z < -z_{\alpha/2}$ and $Z > z_{\alpha/2}$
$\mu = \mu_0$	$T = \dfrac{\bar{X} - \mu_0}{S/\sqrt{n}}$; $v = n - 1$, σ unknown	$\mu < \mu_0$ $\mu > \mu_0$ $\mu \neq \mu_0$	$T < -t_\alpha$ $T > t_\alpha$ $T < -t_{\alpha/2}$ and $T > t_{\alpha/2}$
$\mu_1 - \mu_2 = d_0$	$Z = \dfrac{(\bar{X}_1 - \bar{X}_2) - d_0}{\sqrt{(\sigma_1^2/n_1) + (\sigma_2^2/n_2)}}$; σ_1 and σ_2 known	$\mu_1 - \mu_2 < d_0$ $\mu_1 - \mu_2 > d_0$ $\mu_1 - \mu_2 \neq d_0$	$Z < -z_\alpha$ $Z > z_\alpha$ $Z < -z_{\alpha/2}$ and $Z > z_{\alpha/2}$
$\mu_1 - \mu_2 = d_0$	$T = \dfrac{(\bar{X}_1 - \bar{X}_2) - d_0}{S_p\sqrt{(1/n_1) + (1/n_2)}}$; $v = n_1 + n_2 - 2$, $\sigma_1 = \sigma_2$ but unknown, $S_p^2 = \dfrac{(n_1 - 1)S_1^2 + (n_2 - 1)S_2^2}{n_1 + n_2 - 2}$	$\mu_1 - \mu_2 < d_0$ $\mu_1 - \mu_2 > d_0$ $\mu_1 - \mu_2 \neq d_0$	$T < -t_\alpha$ $T > t_\alpha$ $T < -t_{\alpha/2}$ and $T > t_{\alpha/2}$
$\mu_1 - \mu_2 = d_0$	$T' = \dfrac{(\bar{X}_1 - \bar{X}_2) - d_0}{\sqrt{(S_1^2/n_1) + (S_2^2/n_2)}}$; $v = \dfrac{(s_1^2/n_1 + s_2^2/n_2)^2}{\dfrac{(s_1^2/n_1)^2}{n_1 - 1} + \dfrac{(s_2^2/n_2)^2}{n_2 - 1}}$; $\sigma_1 \neq \sigma_2$ and unknown	$\mu_1 - \mu_2 < d_0$ $\mu_1 - \mu_2 > d_0$ $\mu_1 - \mu_2 \neq d_0$	$T' < -t_\alpha$ $T' > t_\alpha$ $T' < -t_{\alpha/2}$ and $T' > t_{\alpha/2}$
$\mu_D = d_0$	$T = \dfrac{\bar{D} - d_0}{S_d/\sqrt{n}}$; $v = n - 1$, paired observations	$\mu_D < d_0$ $\mu_D > d_0$ $\mu_D \neq d_0$	$T < -t_\alpha$ $T > t_\alpha$ $T < -t_{\alpha/2}$ and $T > t_{\alpha/2}$
$\sigma^2 = \sigma_0^2$	$X^2 = \dfrac{(n - 1)S^2}{\sigma_0^2}$; $v = n - 1$	$\sigma^2 < \sigma_0^2$ $\sigma^2 > \sigma_0^2$ $\sigma^2 \neq \sigma_0^2$	$X^2 < \chi_{1-\alpha}^2$ $X^2 > \chi_\alpha^2$ $X^2 < \chi_{1-\alpha/2}^2$ and $X^2 > \chi_{\alpha/2}^2$
$\sigma_1^2 = \sigma_2^2$	$F = \dfrac{S_1^2}{S_2^2}$; $v_1 = n_1 - 1$ and $v_2 = n_2 - 1$	$\sigma_1^2 < \sigma_2^2$ $\sigma_1^2 > \sigma_2^2$ $\sigma_1^2 \neq \sigma_2^2$	$F < f_{1-\alpha}(v_1, v_2)$ $F > f_\alpha(v_1, v_2)$ $F < f_{1-\alpha/2}(v_1, v_2)$ and $F > f_{\alpha/2}(v_1, v_2)$

alternative that $\mu \neq 15$ pounds if a random sample of 50 lines is tested and found to have a mean breaking strength of 14.8 pounds Use a 0.01 level of significance.

Solution.   1. H_0: $\mu = 15$ pounds.
2. H_1: $\mu \neq 15$ pounds.
3. $\alpha = 0.01$.
4. Critical region: $Z < -2.58$ and $Z > 2.58$, where

$Z_{\alpha/2} = 2.58 = Z_{.005}$

$\alpha/2 = .01$

$\alpha = .005$

in table

$$Z = \frac{\bar{X} - \mu_0}{\sigma/\sqrt{n}}.$$

5. Computations:

$$\bar{x} = 14.8 \text{ pounds}, \qquad n = 50,$$

$$z = \frac{14.8 - 15}{0.5/\sqrt{50}} = -2.828.$$

inside critical range

-2.58

6. Conclusion: Reject H_0 and conclude that the average breaking strength is not equal to 15 but is in fact less than 15 pounds.

Example 9.2 The average length of time for students to register for fall classes at a certain college has been 50 minutes with a standard deviation of 10 minutes. A new registration procedure using modern computing machines is being tried. If a random sample of 12 students had an average registration time of 42 minutes with a standard deviation of 11.9 minutes under the new system, test the hypothesis that the population mean is now less than 50, using a level of significance of (a) 0.05, (b) 0.01. Assume the population of times to be normal.

Solution.   1. H_0: $\mu = 50$ minutes.
2. H_1: $\mu < 50$ minutes.
3. (a) $\alpha = 0.05$, (b) $\alpha = 0.01$.
4. Critical region: (a) $T < -1.796$, (b) $T < -2.718$, where $T = (\bar{X} - \mu_0)/(S/\sqrt{n})$ with $v = 11$ degrees of freedom.
5. Computations: $\bar{x} = 42$ minutes, $s = 11.9$ minutes, $n = 12$. Hence

$$t = \frac{42 - 50}{11.9/\sqrt{12}} = -2.33.$$

6. Conclusion: Reject H_0 at the 0.05 level of significance

need to take more samples

X | but not at the 0.01 level. This essentially means that the true mean is likely to be less than 50 minutes but does not differ sufficiently to warrant the high cost that would be required to operate a computer.

Example 9.3 A course in mathematics is taught to 12 students by the conventional classroom procedure. A second group of 10 students was given the same course by means of programed materials. At the end of the semester the same examination was given to each group. The 12 students meeting in the classroom made an average grade of 85 with a standard deviation of 4, while the 10 students using programed materials made an average of 81 with a standard deviation of 5. Test the hypothesis that the two methods of learning are equal using a 0.10 level of significance. Assume the populations to be approximately normal with equal variances.

Solution. Let μ_1 and μ_2 represent the average grades of all students that might take this course by the classroom and programed presentations, respectively. Using the six-step procedure, we have $\sigma_1 = \sigma_2$ *but unknown* ?

1. $H_0: \mu_1 = \mu_2$ or $\mu_1 - \mu_2 = 0$.
2. $H_1: \mu_1 \neq \mu_2$ or $\mu_1 - \mu_2 \neq 0$.
3. $\alpha = 0.10$. $v = n_1 + n_2 - 2 = 12 + 10 - 2 = 20$
4. Critical region: $T < -1.725$ and $T > 1.725$.
5. Computations:

$$\bar{x}_1 = 85, \qquad s_1 = 4, \qquad n_1 = 12,$$
$$\bar{x}_2 = 81, \qquad s_2 = 5, \qquad n_2 = 10.$$

Hence

$$s_p = \sqrt{\frac{(11)(16) + (9)(25)}{12 + 10 - 2}} = 4.478,$$

$$t = \frac{(85 - 81) - 0}{4.478\sqrt{\frac{1}{12} + \frac{1}{10}}} = 2.07.$$

6. Conclusion: Reject H_0 and conclude that the two methods of learning are not equal. Since the computed t value falls in the part of the critical region in the right tail of the distribution, we can conclude that the classroom procedure is superior to the method using programed materials. Note that we arrived at this same conclusion in Example 8.5 by means of a 90% confidence interval.

Example 9.4 To determine whether membership in a fraternity is beneficial or detrimental to one's grades, the following grade-point averages were collected over a period of 5 years:

	Year				
	1	2	3	4	5
Fraternity	2.0	2.0	2.3	2.1	2.4
Nonfraternity	2.2	1.9	2.5	2.3	2.4

Assuming the populations normal, test at the 0.025 level of significance whether membership in a fraternity is detrimental to one's grades.

Solution. Let μ_1 and μ_2 be the average grades of fraternity and non-fraternity students, respectively. We proceed by the six-step rule:

1. $H_0: \mu_1 = \mu_2$ or $\mu_D = 0$.
2. $H_1: \mu_1 < \mu_2$ or $\mu_D < 0$.
3. $\alpha = 0.025$.
4. Critical region: $T < -2.776$.
5. Computations:

Fraternity	Nonfraternity	d_i	d_i^2
2.0	2.2	−0.2	0.04
2.0	1.9	0.1	0.01
2.3	2.5	−0.2	0.04
2.1	2.3	−0.2	0.04
2.4	2.4	0.0	0.00
		−0.5	0.13

We find $\bar{d} = -0.5/5 = -0.1$ and

$$s_d^2 = \frac{(5)(0.13) - (-0.5)^2}{(5)(4)} = 0.02.$$

Taking the square root, we have $s_d = 0.14142$. Hence

$$t = \frac{-0.1 - 0}{0.14142/\sqrt{5}} = -1.6.$$

6. Conclusion: Accept H_0 and conclude that membership in a fraternity does not significantly affect one's grades.

Example 9.5 A manufacturer of car batteries claims that the life of his batteries have a standard deviation equal to 0.9 year. If a random sample of 10 of these batteries have a standard deviation of 1.2 years, do you think that $\sigma > 0.9$ year? Use a 0.05 level of significance.

Solution. 1. $H_0: \sigma^2 = 0.81$.
2. $H_1: \sigma^2 > 0.81$.
3. $\alpha = 0.05$.
4. Critical region: $X^2 > 16.919$.
5. Computations: $s^2 = 1.44$, $n = 10$, and

$$\chi^2 = \frac{(9)(1.44)}{0.81} = 16.0.$$

6. Conclusion: Accept H_0 and conclude that there is no reason to doubt that the standard deviation is 0.9 year.

9.5 Wilcoxon Two-Sample Test

The procedures discussed in the previous section for testing hypotheses about the difference between two means are valid only if the populations are approximately normal or if the samples are large. How then does one make such a test when small independent samples are selected from non-normal populations? In 1945 Wilcoxon proposed a very simple test which assumes no knowledge about the distribution and parameters of the population. A test performed without this information is called a *nonparametric* or *distribution-free* test. The particular nonparametric test which we shall consider here is referred to as the *Wilcoxon two-sample test*.

We shall test the null hypothesis H_0 that $\mu_1 = \mu_2$ against some suitable alternative. First we select a random sample from each of the populations. Let n_1 be the number of observations in the smaller sample and n_2 the number of observations in the larger sample. When the samples are of equal size, n_1 and n_2 may be randomly assigned. Arrange the $n_1 + n_2$ observations of the combined samples in ascending order and substitute a rank of $1, 2, \ldots,$ $n_1 + n_2$ for each observation. In the case of ties (identical observations) we replace the observations by the mean of the ranks that the observations would have if they were distinguishable. For example, if the seventh and eighth observations are identical, we would assign a rank of 7.5 to each of the two observations.

The sum of the ranks corresponding to the n_1 observations in the smaller sample is denoted by w_1. Similarly the value w_2 represents the sum of the n_2 ranks corresponding to the larger sample. The total $w_1 + w_2$ depends only on the number of observations in the two samples and is in no way affected by the results of the experiment. Hence, if $n_1 = 3$ and $n_2 = 4$, then $w_1 + w_2 = 1 + 2 + \cdots + 7 = 28$, regardless of the numerical values of the observations. In general

$$w_1 + w_2 = \frac{(n_1 + n_2)(n_1 + n_2 + 1)}{2},$$

the arithmetic sum of the integers $1, 2, \ldots, n_1 + n_2$. Once we have determined w_1, it may be easier to find w_2 by the formula

$$w_2 = \frac{(n_1 + n_2)(n_1 + n_2 + 1)}{2} - w_1.$$

In choosing repeated samples of size n_1 and n_2, we would expect w_1, and therefore w_2, to vary. Thus we may think of w_1 and w_2 as values of the random variables W_1 and W_2, respectively. The null hypothesis $\mu_1 = \mu_2$ will be rejected in favor of the alternative $\mu_1 < \mu_2$ only if w_1 is small and w_2 is large. Likewise the alternative $\mu_1 > \mu_2$ can be accepted only if w_1 is large and w_2 is small. For a two-tailed test we may reject H_0 in favor of H_1 if w_1 is small and w_2 is large or if w_1 is large and w_2 is small. Owing to symmetry in the distributions of W_1 and W_2 upper tail probabilities may be obtained from the lower tail probabilities. Hence, no matter what the alternative hypothesis may be, we reject the null hypothesis when the smaller of w_1 and w_2 is sufficiently small. Suppose for a given experiment that $w_1 < w_2$. Knowing the distribution of W_1, we can determine the $\Pr(W_1 \le w_1 | H_0$ is true). If this probability is less than or equal to 0.05, our test is significant and we would reject H_0 in favor of the appropriate one-sided alternative. When the probability does not exceed 0.01, the test is highly significant. In the case of a two-tailed test, symmetry permits us to base our decision on the value of $2\Pr(W_1 \le w_1 | H_0$ is true). Therefore, when $2\Pr(W_1 \le w_1 | H_0$ is true) < 0.05, the test is significant and we conclude that $\mu_1 \ne \mu_2$.

The distribution of W_1, when H_0 is true, is based on the fact that all the observations in the smaller sample could be assigned ranks at random as long as their sum is less than or equal to w_1. The total number of ways of assigning $n_1 + n_2$ ranks to n_1 observations so that the sum of the ranks does not exceed w_1 is denoted by $n(W_1 \le w_1)$. There are $\binom{n_1 + n_2}{n_1}$ equally likely ways to assign the $n_1 + n_2$ ranks to n_1 observations giving all possible values of W_1. Hence

$$\Pr(W_1 \le w_1 | H_0 \text{ is true}) = \frac{n(W_1 \le w_1)}{\binom{n_1 + n_2}{n_1}}, \quad \text{for } n_1 \le n_2.$$

It is possible to find $n(W_1 \leq w_1)$ for any given test by listing all the cases and counting them. Thus, when $n_1 = 3$ and $n_2 = 5$, the number of cases where the sum of the ranks in the smaller sample is less than or equal to 8 may be listed as follows:

$$1 + 2 + 3 = 6,$$
$$1 + 2 + 4 = 7,$$
$$1 + 3 + 4 = 8,$$
$$1 + 2 + 5 = 8.$$

Therefore there are 4 favorable cases out of a possible $\binom{8}{3} = 56$ equally likely cases. Hence

$$\Pr(W_1 \leq 8 | H_0 \text{ is true}) = \tfrac{4}{56} = 0.0714.$$

For the case where $w_2 < w_1$, we could proceed as above to determine $\Pr(W_2 \leq w_2 | H_0 \text{ is true})$. However, in either case it is usually easier to find the desired probability by using Table A.8 when n_2 does not exceed eight. Table A.8 is based upon the statistic U, the minimum of U_1 and U_2, where

$$U_1 = W_1 - \frac{n_1(n_1 + 1)}{2}$$

and

$$U_2 = W_2 - \frac{n_2(n_2 + 1)}{2}.$$

If $\Pr(U \leq u | H_0 \text{ is true}) \leq \alpha$, our test is significant and we reject H_0 in favor of the appropriate one-sided alternative. For a two-tailed test our test is significant when $2\Pr(U \leq u | H_0 \text{ is true}) \leq \alpha$, in which case we accept the alternative hypothesis that $\mu_1 \neq \mu_2$.

In the preceding illustration, where we had $n_1 = 3$, $n_2 = 5$, and $w_1 = 8$, we find $w_2 = [(8)(9)/2] - 8 = 28$, and then

$$u_1 = 8 - \left[\frac{(3)(4)}{2}\right] = 2,$$

$$u_2 = 28 - \left[\frac{(5)(6)}{2}\right] = 13.$$

Using Table A.8, with $u = 2$, we have

$$\Pr(U \leq 2 | H_0 \text{ is true}) = 0.071,$$

which agrees with the previous answer. If, for the same illustration, $w_1 = 7$ so that $u = 1$, we find

$$\Pr(U \leq 1 | H_0 \text{ is true}) = 0.036,$$

which is significant for a one-tailed test at the 0.05 level but not at 0.01 level. For a two-tailed test the probability that the sample means differ by an amount as great as or greater than that observed is

$$2\Pr(U \leq 1 | H_0 \text{ is true}) = (2)(0.036) = 0.072,$$

from which we conclude that H_0 is true.

When n_2 is between 9 and 20, Table A.9 may be used. If the observed value of U is less than or equal to the tabled value, the null hypothesis may be rejected at the level of significance indicated by the table. Table A.9 gives critical values of U for levels of significance equal to 0.001, 0.01, 0.025, and 0.05 for a one-tailed test. In the case of a two-tailed test the critical values of U correspond to the 0.002, 0.02, 0.05, and 0.1 levels of significance. When n_1 and n_2 increase in size, the sampling distribution of U approaches the normal distribution with mean

$$\mu_U = \frac{n_1 n_2}{2}$$

and variance

$$\sigma_U^2 = \frac{n_1 n_2 (n_1 + n_2 + 1)}{12}.$$

Consequently, when n_2 is greater than 20, one could use the statistic $Z = (U - \mu_U)/\sigma_U$ for our test, with the critical region falling in either or both tails of the standard normal distribution, depending on the form of H_1.

To test the null hypothesis that the means of two nonnormal populations are equal when only small independent samples are available, we proceed by the following steps:

1. $H_0: \mu_1 = \mu_2$.
2. $H_1:$ Alternatives are $\mu_1 < \mu_2$, $\mu_1 > \mu_2$, or $\mu_1 \neq \mu_2$.
3. Choose a level of significance equal to α.
4. Critical region:
   (a) All u values for which $\Pr(U \leq u | H_0 \text{ is true}) < \alpha$ when $n_2 \leq 8$ and the test is one-tailed.
   (b) All u values for which $2\Pr(U \leq u | H_0 \text{ is true}) < \alpha$ when $n_2 \leq 8$ and the test is two-tailed.
   (c) All u values less than or equal to the appropriate critical value in Table A.9 when $9 \leq n_2 \leq 20$.

5. Compute w_1, w_2, u_1, and u_2 from independent samples of size n_1 and n_2 where $n_1 \leq n_2$. Using the smaller of u_1 and u_2 for u, determine whether u falls in the acceptance or critical region.

6. Conclusion: Reject H_0 if u falls in the critical region; otherwise accept H_0.

Example 9.6 To find out whether a new serum will arrest leukemia, 9 mice, which have all reached an advanced stage of the disease, are selected. Five mice receive the treatment and four do not. The survival times, in years, from the time the experiment commenced are as follows:

Treatment	2.1	5.3	1.4	4.6	0.9
No Treatment	1.9	0.5	2.8	3.1	

At the 0.05 level of significance can the serum be said to be effective?

Solution. We follow the six-step procedure above with $n_1 = 4$ and $n_2 = 5$.

1. $H_0: \mu_1 = \mu_2$.
2. $H_1: \mu_1 < \mu_2$.
3. $\alpha = 0.05$.
4. Critical region: All u values for which $\Pr(U \leq u | H_0$ is true$) < 0.05$.
5. Computations: The observations are arranged in ascending order and ranks from 1 to 9 assigned.

Original Data	0.5	0.9	1.4	1.9	2.1	2.8	3.1	4.6	5.3
Ranks	1	2	3	4	5	6	7	8	9

The treatment observations are underscored for identification purposes. Now,

$$w_1 = 1 + 4 + 6 + 7 = 18$$

and

$$w_2 = \left[\frac{(9)(10)}{2} \right] - 18 = 27.$$

Therefore,

$$u_1 = 18 - \left[\frac{(4)(5)}{2}\right] = 8,$$

$$u_2 = 27 - \left[\frac{(5)(6)}{2}\right] = 12,$$

so that $u = 8$. Since $\Pr(U \leq 8 | H_0$ is true$) = 0.365 >$ 0.05, the value $u = 8$ falls in the acceptance region.

6. Conclusion: Accept H_0 and conclude that the serum does not prolong life by arresting leukemia.

Example 9.7 The nicotine content of two brands of cigarettes, measured in milligrams, was found to be as follows:

Brand A	22.1	24.0	26.3	25.4	24.8	23.7	26.1	23.3	
Brand B	24.1	20.6	23.1	22.5	24.0	26.2	21.6	22.2	21.9	25.4

Test the hypothesis, at the 0.05 level of significance, that the average nicotine contents of the two brands are equal against the alternative that they are unequal.

Solution. We proceed by the six-step rule with $n_1 = 8$ and $n_2 = 10$.

1. $H_0: \mu_1 = \mu_2$.
2. $H_1: \mu_1 \neq \mu_2$.
3. $\alpha = 0.05$.
4. Critical region: $U \leq 17$ (from Table A.9).
5. Computations: The observations are arranged in ascending order and ranks from 1 to 18 assigned.

Original Data	Ranks
20.6	1
21.6	2
21.9	3
22.1	4
22.2	5
22.5	6
23.1	7
23.3	8
23.7	9
24.0	10.5
24.0	10.5
24.1	12
24.8	13
25.4	14.5
25.4	14.5
26.1	16
26.2	17
26.3	18

The ranks of the observations belonging to the smaller sample are underscored. Now,

$$w_1 = 4 + 8 + 9 + 10.5 + 13 + 14.5 + 16 + 18 = 93$$

and

$$w_2 = \left[\frac{(18)(19)}{2} \right] - 93 = 78.$$

Therefore,

$$u_1 = 93 - \left[\frac{(8)(9)}{2} \right] = 57,$$

$$u_2 = 78 - \left[\frac{(10)(11)}{2} \right] = 23,$$

so that $u = 23$.

6. Conclusion: Accept H_0 and conclude that there is no difference in the average nicotine contents of the two brands of cigarettes.

The use of the Wilcoxon two-sample test is not restricted to nonnormal populations. It can be used in place of the t test when the populations are normal, although the probability of committing a type II error will be larger. The Wilcoxon two-sample test is always superior to the t test for nonnormal populations.

9.6 Wilcoxon Test for Paired Observations

Assume that n pairs of observations are selected from two nonnormal populations. For large n the distribution of the mean of the differences in repeated sampling is approximately normal and tests of hypotheses concerning the means may be carried out using the statistic $T = (\bar{D} - d_0)/(S_d/\sqrt{n})$ given in Table 9.1. However, if n is small and the population of differences is decidedly nonnormal, we must resort to a nonparametric test. Perhaps the easiest and quickest to perform is a test called the *sign test*. In testing the null hypothesis that $\mu_1 = \mu_2$ or $\mu_D = 0$, each difference d_i is assigned a plus or minus sign, depending on whether d_i is positive or negative. If the null hypothesis is true, the sum of the plus signs should be approximately equal to the sum of the minus signs. When one sign appears more frequently than it should, based on chance alone, we would reject the hypothesis that the populations means are equal.

The sign test shows, by the assigned plus or minus sign, which member of a pair of observations is the larger, but it does not indicate the magnitude of the difference. A test utilizing both direction and magnitude was proposed in 1945 by Wilcoxon and is now commonly referred to as the *Wilcoxon test for paired observations*. Wilcoxon's test is more sensitive than the sign test in detecting a difference in the population means and therefore will be considered in detail.

To test the hypothesis that $\mu_1 = \mu_2$ by the Wilcoxon test, first discard all differences equal to zero and then rank the remaining d_i's without regard to sign. A rank of 1 is assigned to the smallest d_i in absolute value, a rank of 2 to the next smallest, and so on. When the absolute value of two or more differences is the same, assign to each the average of the ranks that would have been assigned if the differences were distinguishable. If there is no difference between the two population means, the total of the ranks corresponding to the positive differences should be almost equal to the total of the ranks corresponding to the negative differences. Let us represent these totals by w_+ and w_-, respectively. We shall designate the smaller of the w_+ and w_- by w and find the probability of obtaining, by chance alone, a value less than or equal to w when H_0 is true.

In selecting repeated samples of paired observations, we would expect w to vary. Thus we may think of w as a value of some random variable W. Once the distribution of W is known, we can determine $\Pr(W \leq w | H_0$ is true). For a level of significance equal to α, we reject H_0 when

$$\Pr(W \leq w | H_0 \text{ is true}) < \alpha$$

and accept the appropriate one-sided alternative. In the case of a two-tailed test, we reject H_0 at the α level of significance in favor of the two-sided alternative hypothesis $\mu_1 \neq \mu_2$ when

$$2\Pr(W \leq w | H_0 \text{ is true}) < \alpha.$$

If we assume that there is no difference in the population means, each d_i is just as likely to be positive as it is to be negative. Thus there are two equally likely ways for a given rank to receive a sign. For n differences there are 2^n equally likely ways for the n ranks to receive signs. Let $n(W \leq w)$ be the number of the 2^n ways of assigning signs to the n ranks such that the value of W does not exceed w. Then

$$\Pr(W \leq w | H_0 \text{ is true}) = \frac{n(W \leq w)}{2^n}.$$

Consider, for example, the case of $n = 6$ matched pairs that yield a value $w = 5$. What is the probability that $W \leq 5$ when the two population means are equal? The sets of ranks whose total does not exceed 5 may be listed as follows:

Value of W	Sets of Ranks Totaling W
0	$\varnothing$
1	{1}
2	{2}
3	{3}, {1, 2}
4	{4}, {1, 3}
5	{5}, {1, 4}, {2, 3}

Therefore $n(W \leq 5) = 10$ out of a possible $2^6 = 64$ equally likely cases. Hence

$$\Pr(W \leq 5 | H_0 \text{ is true}) = \tfrac{10}{64} = 0.1563,$$

a result that is quite likely to occur when $\mu_1 = \mu_2$.

When $5 \leq n \leq 30$, Table A.10 gives approximate critical values of W for levels of significance equal to 0.01, 0.025, and 0.05 for a one-tailed test, and equal to 0.02, 0.05, and 0.10 for a two-tailed test. In the preceding example for which $n = 6$, Table A.10 shows that a value of $W \leq 2$ is required for a one-tailed test to be significant at the 0.05 level. When $n > 30$, the sampling distribution of W approaches the normal distribution with mean

$$\mu_W = \frac{n(n + 1)}{4}$$

and variance

$$\sigma_W^2 = \frac{n(n + 1)(2n + 1)}{24}.$$

In this case the statistic $Z = (W - \mu_W)/\sigma_W$ can be used to determine the critical region for our test.

The Wilcoxon test for paired observations may also be used to test the null hypothesis that $\mu_1 - \mu_2 = \mu_D = d_0$. We simply apply the same procedure as above after each d_i is adjusted by subtracting d_0. Therefore, to test a hypothesis about the difference between the means of two populations, whose distributions are unknown, where the observations occur in pairs and the sample size is small, we proceed by the following six steps:

1. $H_0: \mu_1 - \mu_2 = \mu_D = d_0$.
2. H_1: Alternatives are $\mu_1 - \mu_2 < d_0$, $\mu_1 - \mu_2 > d_0$, or $\mu_1 - \mu_2 \neq d_0$.
3. Choose a level of significance equal to α.
4. Critical region:
   (a) All w values for which $\Pr(W \leq w | H_0 \text{ is true}) < \alpha$ when $n < 5$ and the test is one-tailed.
   (b) All w values for which $2\Pr(W \leq w | H_0 \text{ is true}) < \alpha$ when $n < 5$ and the test is two-tailed.

(c) All w values less than or equal to the appropriate critical value in Table A.10 when $5 \leq n \leq 30$.

5. Rank the n differences, $d_i - d_0$, without regard to sign, and then compute w.

6. Conclusion: Reject H_0 if w falls in the critical region; otherwise accept H_0.

Example 9.8 It is claimed that a college senior can increase his score in the major field area of the graduate record examination by at least 50 points if he is provided sample problems in advance. To test this claim, 20 college seniors were divided into 10 pairs such that each matched pair had almost the same overall quality grade-point average for their first 3 years in college. Sample problems and answers were provided at random to 1 member of each pair 1 week prior to the examination. The following examination scores were recorded:

	Pair									
	1	2	3	4	5	6	7	8	9	10
With sample problems	531	621	663	579	451	660	591	719	543	575
Without sample problems	509	540	688	502	424	683	568	748	530	524

Test the null hypothesis at the 0.05 level of significance that sample problems increase the scores by 50 points against the alternative hypothesis that the increase is less than 50 points.

Solution. Let μ_1 and μ_2 represent the mean score of all students taking the test in question with and without sample problems, respectively. We follow the six-step procedure:

1. $H_0: \mu_1 - \mu_2 = 50$.
2. $H_1: \mu_1 - \mu_2 < 50$.
3. $\alpha = 0.05$.
4. Critical region: Since $n = 10$, Table A.10 shows the critical region to be $W \leq 11$.
5. Computations:

	Pair									
	1	2	3	4	5	6	7	8	9	10
d_i	22	81	-25	77	27	-23	23	-29	13	51
$d_i - d_0$	-28	31	-75	27	-23	-73	-27	-79	-37	1
Ranks	5	6	9	3.5	2	8	3.5	10	7	1

Now $w_+ = 10.5$ and $w_- = 44.5$, so that $w = 10.5$, the smaller of w_+ and w_-.

6. Conclusion: Reject H_0 and conclude that the sample problems do not, on the average, increase one's graduate record score by as much as 50 points.

9.7 Tests Concerning Proportions

go to Binomial table

Tests of hypotheses concerning proportions are required in many areas. The politician is certainly interested in knowing what fraction of the voters will favor him in the next election. All manufacturing firms are concerned about the proportion of defectives when a shipment is made. The gambler depends upon a knowledge of the proportion of outcomes that he considers favorable.

We shall consider the problem of testing the hypothesis that the proportion of successes in a binomial experiment equals some specified value. That is, we are testing the null hypothesis H_0 that $p = p_0$, where p is the parameter of the binomial distribution. The alternative hypothesis may be one of the usual one-sided or two-sided alternatives: $p < p_0, p > p_0$, or $p \neq p_0$.

The appropriate statistic on which we base our decision criterion is the binomial random variable X, although we could just as well use the statistic $\hat{P} = X/n$. Values of X that are far from the mean $\mu = np_0$ will lead to the rejection of the null hypothesis. To test the hypothesis

$$H_0: p = p_0,$$
$$H_1: p < p_0,$$

we use the binomial distribution with $p = p_0$ and $q = 1 - p_0$ to determine $\Pr(X \leq x | H_0 \text{ is true})$. The value x is the number of successes in our sample of size n. If $\Pr(X \leq x | H_0 \text{ is true}) < \alpha$, our test is significant at the α level and we reject H_0 in favor of H_1. Similarly, to test the hypothesis

$$H_0: p = p_0,$$
$$H_1: p > p_0,$$

we find $\Pr(X \geq x | H_0 \text{ is true})$ and reject H_0 in favor of H_1 if this probability is less than α. Finally, to test the hypothesis

$$H_0: p = p_0,$$
$$H_1: p \neq p_0,$$

at the α level of significance, we reject H_0 when $x < np_0$ and $\Pr(X \leq x | H_0 \text{ is true}) < \alpha/2$ or when $x > np_0$ and $\Pr(X \geq x | H_0 \text{ is true}) < \alpha/2$.

The steps for testing a hypothesis about a proportion against various alternatives are:

1. $H_0: p = p_0$.
2. H_1: Alternatives are $p < p_0$, $p > p_0$, or $p \neq p_0$.
3. Choose a level of significance equal to α.
4. Critical region:
   (a) All x values such that $\Pr(X \leq x | H_0$ is true$) < \alpha$ for the alternative $p < p_0$.
   (b) All x values such that $\Pr(X \geq x | H_0$ is true$) < \alpha$ for the alternative $p > p_0$.
   (c) All x values such that $\Pr(X \leq x | H_0$ is true$) < \alpha/2$ when $x < np_0$ and all x values such that $\Pr(X \geq x | H_0$ is true $< \alpha/2$ when $x > np_0$, for the alternative $p \neq p_0$.
5. Computations: Find x and compute the appropriate probability.
6. Conclusion: Reject H_0 if x falls in the critical region; otherwise accept H_0.

Example 9.9 A pheasant hunter claims that he hits 80% of the birds he shoots at. Would you agree with this claim if on a given day he brings down 9 of the 15 pheasants he shoots at? Use a 0.05 level of significance.

Solution.   1. $H_0: p = 0.8$.
                   2. $H_1: p \neq 0.8$.
                   3. $\alpha = 0.05$.
                   4. Critical region: All x values such that $\Pr(X \leq x | H_0$ is true$) < 0.025$.
                   5. Computations: We have $x = 9$ and $n = 15$. Therefore, using Table A.2,

$$\Pr(X \leq 9 | p = 0.8) = \sum_{x=0}^{9} b(x; 15, 0.8)$$

$$= 0.0611 > 0.025.$$

                   6. Conclusion: Accept H_0 and conclude that there is no reason to doubt the hunter's claim.

In Section 6.3 we saw that binomial probabilities were obtainable from the actual binomial formula or from Table A.2 when n is small. For large n, approximation procedures are required. When the hypothesized value p_0 is very close to zero or 1, the Poisson distribution, with parameter $\mu = np_0$, may be used. The normal-curve approximation is usually preferred for large n and is very accurate as long as p_0 is not extremely close to zero or to 1.

Using the normal approximation, we base our decision criterion on

$$z = \frac{\hat{p} - p_0}{\sqrt{p_0 q_0/n}} = \frac{x - np_0}{\sqrt{np_0 q_0}},$$

which is a value of the standard normal variable Z. Hence for a two-tailed test at the α level of significance, the critical region is $Z < -z_{\alpha/2}$ and $Z > z_{\alpha/2}$. For the one-sided alternative $p < p_0$, the critical region is $Z < -z_\alpha$, and for the alternative $p > p_0$, the critical region is $Z > z_\alpha$.

To test a hypothesis about a proportion using the normal-curve approximation, we proceed as follows:

1. $H_0: p = p_0$.
2. H_1: Alternatives are $p < p_0$, $p > p_0$, or $p \neq p_0$.
3. Choose a level of significance equal to α.
4. Critical region:

$$Z < -z_\alpha \text{ for the alternative } p < p_0,$$

$$Z > z_\alpha \text{ for the alternative } p > p_0,$$

$$Z < -z_{\alpha/2} \text{ and } Z > z_{\alpha/2} \text{ for the alternative } p \neq p_0.$$

5. Computations: Find x from a sample of size n, and then compute

$$z = \frac{x - np_0}{\sqrt{np_0 q_0}}.$$

6. Conclusion: Reject H_0 if z falls in the critical region; otherwise accept H_0.

Example 9.10 A basketball player has hit on 60% of his shots from the floor. If on the next 100 shots he makes 70 baskets, would you say his shooting has improved? Use a 0.05 level of significance.

Solution.
1. $H_0: p = 0.6$.
2. $H_1: p > 0.6$.
3. $\alpha = 0.05$.
4. Critical region: $Z > 1.645$.
5. Computations: $x = 70$, $n = 100$, $np_0 = (100)(0.6) = 60$, and

$$z = \frac{70 - 60}{\sqrt{(100)(0.6)(0.4)}} = 2.04.$$

6. Conclusion: Reject H_0 and conclude that his shooting percentage has improved.

9.8 Testing the Difference Between Two Proportions

Situations often arise in which we wish to test the hypothesis that two proportions are equal. For example we might try to prove that the proportion of doctors who are pediatricians in one state is greater than the proportion of pediatricians in another state. A person may decide to give up smoking only if he is convinced that the proportion of smokers with lung cancer exceeds the proportion of nonsmokers with lung cancer.

In general we wish to test the null hypothesis

$$H_0: p_1 = p_2 = p$$

against some suitable alternative. The parameters p_1 and p_2 are the two population proportions of the attribute under investigation. The statistic on which we base our decision criterion is the random variable $\hat{P}_1 - \hat{P}_2$. Independent samples of size n_1 and n_2 are selected at random from two binomial populations and the proportion of successes $\hat{P}_1$ and $\hat{P}_2$, for the two samples are computed. From Section 8.6 we know that

$$z = \frac{\hat{p}_1 - \hat{p}_2}{\sqrt{(p_1 q_1/n_1) + (p_2 q_2/n_2)}} = \frac{\hat{p}_1 - \hat{p}_2}{\sqrt{pq[(1/n_1) + (1/n_2)]}}$$

is a value of the standard normal variable Z when H_0 is true and n_1 and n_2 are large. To compute z, we must estimate the value of p appearing in the radical. Pooling the data from both samples, we write

$$\hat{p} = \frac{x_1 + x_2}{n_1 + n_2},$$

where x_1 and x_2 are the number of successes in each of the two samples. The value of the statistic Z becomes

$$z = \frac{\hat{p}_1 - \hat{p}_2}{\sqrt{\hat{p}\hat{q}[(1/n_1) + (1/n_2)]}},$$

where $\hat{q} = 1 - \hat{p}$. The critical regions for the appropriate alternative hypotheses are set up as before using the critical points of the standard normal curve.

To test the hypothesis that two proportions are equal, when the samples are large, we proceed by the following six steps:

1. $H_0: p_1 = p_2$.
2. H_1: Alternatives are $p_1 < p_2, p_1 > p_2$, or $p_1 \neq p_2$.
3. Choose a level of significance equal to α.

4. Critical region:

$Z < -z_\alpha$ for the alternative $p_1 < p_2$,

$Z > z_\alpha$ for the alternative $p_1 > p_2$,

$Z < -z_{\alpha/2}$ and $Z > z_{\alpha/2}$ for the alternative $p_1 \neq p_2$.

5. Computations: Compute $\hat{p}_1 = x_1/n_1$, $\hat{p}_2 = x_2/n_2$, $\hat{p} = (x_1 + x_2)/(n_1 + n_2)$ and then find

$$z = \frac{\hat{p}_1 - \hat{p}_2}{\sqrt{\hat{p}\hat{q}[(1/n_1) + (1/n_2)]}}.$$

6. Conclusion: Reject H_0 if z falls in the critical region; otherwise accept H_0.

Example 9.11 A vote is to be taken among the residents of a town and the surrounding country to determine whether a civic center will be constructed. The proposed construction site is within the town limits and for this reason many voters in the county feel that the proposal will pass, because of the large proportion of town voters who favor the construction. To determine if there is a significant difference in the proportion of town voters and county voters favoring the proposal, a poll is taken. If 120 of 200 town voters favor the proposal and 240 of 500 county residents favor it, would you agree that the proportion of town voters favoring the proposal is higher than the proportion of county voters? Use a 0.025 level of significance.

Solution. Let p_1 and p_2 be the true proportion of voters in the town and county, respectively, favoring the proposal. We now follow the six-step procedure:

1. $H_0: p_1 = p_2$.
2. $H_1: p_1 > p_2$.
3. $\alpha = 0.025$.
4. Critical region: $Z > 1.96$.
5. Computations:

$$\hat{p}_1 = x_1/n_1 = 120/200 = 0.60,$$
$$\hat{p}_2 = x_2/n_2 = 240/500 = 0.48,$$
$$\hat{p} = (x_1 + x_2)/(n_1 + n_2)$$
$$= (120 + 240)/(200 + 500) = 0.51.$$

Therefore,

$$z = \frac{0.60 - 0.48}{\sqrt{(0.51)(0.49)[(1/200) + (1/500)]}} = 2.9.$$

6. Conclusion: Reject H_0 and agree that the proportion of town voters favoring the proposal is higher than the proportion of county voters.

skip

9.9 Goodness-of-Fit Test

Throughout this chapter we have been concerned with the testing of statistical hypotheses about single population parameters such as μ, σ^2, and p. Now we shall consider a test to determine if a population has a specified theoretical distribution. The test is based upon how good a fit we have between the frequency of occurrence of observations in an observed sample and the expected frequencies obtained from the hypothesized distribution.

Table 9.2
Observed and Expected Frequencies of 120 Tosses of a Die

	Faces					
	1	2	3	4	5	6
Observed	20	22	17	18	19	24
Expected	20	20	20	20	20	20

To illustrate, consider the tossing of a die. We hypothesize that the die is honest, which is equivalent to testing the hypothesis that the distribution of outcomes is uniform. Suppose that the die is tossed 120 times and that each outcome is recorded. Theoretically, if the die is balanced, we would expect each face to occur 20 times. The results are given in Table 9.2. By comparing the observed frequencies with the corresponding expected frequencies we must decide whether these discrepancies are likely to occur due to sampling fluctuations and the die is balanced, or the die is not honest and the distribution of outcomes is not uniform. It is common practice to refer to each possible outcome of an experiment as a cell. Hence in our illustration we have six cells. The appropriate statistic on which we base our decision criterion for an experiment involving k cells is defined in the following theorem.

THEOREM 9.1 *A* goodness-of-fit test *between observed and expected frequencies is based on the quantity*

$$\chi^2 = \sum_{i=1}^{k} \frac{(o_i - e_i)^2}{e_i},$$

where χ^2 is a value of the random variable X^2 whose sampling distribution is approximated very closely by the chi-square distribution. The symbols o_i and e_i represent the observed and expected frequencies, respectively, for the ith cell.

If the observed frequencies are close to the corresponding expected frequencies, the χ^2 value will be small, indicating a good fit. If the observed frequencies differ considerably from the expected frequencies, the χ^2 value will be large and the fit is poor. A good fit leads to the acceptance of H_0, whereas a poor fit leads to its rejection. The critical region will, therefore, fall in the right tail of the chi-square distribution. For a level of significance equal to α we find the critical value χ_α^2 from Table A.6, and then $X^2 > \chi_\alpha^2$ constitutes the critical region. The decision criterion described here should not be used unless each of the expected frequencies is at least equal to 5.

The number of degrees of freedom associated with the chi-square distribution used here depends on two factors: the number of cells in the experiment, and the number of quantities obtained from the observed data that are necessary in the calculation of the expected frequencies. We arrive at this number by the following theorem.

THEOREM 9.2 *The number of degrees of freedom in a chi-square goodness-of-fit test is equal to the number of cells minus the number of quantities obtained from the observed data, which are used in the calculations of the expected frequencies.*

The only quantity provided by the observed data, in computing expected frequencies for the outcome when a die is tossed, is the total frequency. Hence, according to our definition, the computed χ^2 value has $6 - 1 = 5$ degrees of freedom.

From Table 9.2 we find the χ^2 value to be

$$\chi^2 = \frac{(20 - 20)^2}{20} + \frac{(22 - 20)^2}{20} + \frac{(17 - 20)^2}{20} + \frac{(18 - 20)^2}{20}$$
$$+ \frac{(19 - 20)^2}{20} + \frac{(24 - 20)^2}{20}$$
$$= 1.7.$$

Using Table A.6, we find $\chi_{0.05}^2 = 11.070$ for $v = 5$ degrees of freedom. Since 1.7 is less than the critical value, we fail to reject H_0 and conclude that the distribution is uniform. In other words, the die is balanced.

As a second illustration let us test the hypothesis that the frequency distribution of battery lives given in Table 3.2 may be approximated by the normal distribution. The expected frequencies for each class (cell), listed in Table 9.3, are obtained from a normal curve having the same mean and standard deviation as our sample. From the data of Table 3.1 we find that the sample of 40 batteries has a mean $\bar{x} = 3.4125$ and a standard deviation $s = 0.703$. These values will be used for μ and σ in computing z values

Table 9.3

Observed and Expected Frequencies of Battery
Lives Assuming Normality

Class Boundaries	o_i		e_i	
1.45–1.95	2		0.6	
1.95–2.45	1	7	2.7	10.1
2.45–2.95	4		6.8	
2.95–3.45	15		10.6	
3.45–3.95	10		10.3	
3.95–4.45	5	8	6.1	8.3
4.45–4.95	3		2.2	

corresponding to the class boundaries. The z value corresponding to the
boundaries of the fourth class, for example, are

$$z_1 = \frac{2.95 - 3.4125}{0.703} = -0.658,$$

$$z_2 = \frac{3.45 - 3.4125}{0.703} = 0.053.$$

From Table A.4 we find the area between $z_1 = -0.658$ and $z_2 = 0.053$ to be

$$\text{Area} = \Pr(-0.658 < Z < 0.053)$$
$$= \Pr(Z < 0.053) - \Pr(Z < -0.658)$$
$$= 0.5211 - 0.2552$$
$$= 0.2659.$$

Hence the expected frequency for the fourth class is

$$e_4 = (0.2659)(40) = 10.6.$$

The expected frequency for the first class interval is obtained by using the
total area under the normal curve to the left of the boundary 1.95. For the
last class interval we use the total area to the right of the boundary 4.45. All
other expected frequencies are determined by the method described above
for the fourth class. Note that we have combined adjacent classes in Table 9.3
where the expected frequencies are less than 5. Consequently the total number
of intervals is reduced from 7 to 4. The χ^2 value is then given by

$$\chi^2 = \frac{(7 - 10.1)^2}{10.1} + \frac{(15 - 10.6)^2}{10.6} + \frac{(10 - 10.3)^2}{10.3} + \frac{(8 - 8.3)^2}{8.3}$$

$$= 2.797.$$

The number of degrees of freedom for this test will be $4 - 3 = 1$, since three quantities—the total frequency, mean, and standard deviation—of the observed data were required to find the expected frequencies. Since the computed χ^2 value is less than $\chi^2_{0.05} = 3.841$ for 1 degree of freedom, we have no reason to reject the null hypothesis and conclude that the normal distribution provides a good fit for the distribution of battery lives.

9.10 Test for Independence

The chi-square test procedure discussed in Section 9.9 can also be used to test the hypothesis of independence of two variables. Suppose that we wish to study the relationship between religious affiliation and geographical region. Two groups of people are chosen at random, one from the East coast and one from the West coast of the United States, and each person is classified as Protestant, Catholic, or Jewish. The observed frequencies are presented in Table 9.4, which is known as a *contingency table*.

Table 9.4
2×3 Contingency Table

	Protestant	Catholic	Jewish	Total
East coast	182	215	203	600
West coast	154	136	110	400
Total	336	351	313	1000

A contingency table with r rows and c columns is referred to as an $r \times c$ table. The term "$r \times c$" is read "r by c." The row and column totals in Table 9.4 are called *marginal frequencies*. To test the null hypothesis, H_0, of independence between a person's religious faith and the region where he lives, we must first find the expected frequencies for each cell of Table 9.4 under the assumption that H_0 is true.

Let us define the following events:

P: An individual selected from our sample is Protestant.
C: An individual selected from our sample is Catholic.
J: An individual selected from our sample is Jewish.
E: An individual selected from our sample lives on the East coast.
W: An individual selected from our sample lives on the West coast.

Using the marginal frequencies, we can list the following probabilities:

$$\Pr(P) = \frac{336}{1000}, \qquad \Pr(C) = \frac{351}{1000},$$

$$\Pr(J) = \frac{313}{1000}, \qquad \Pr(E) = \frac{600}{1000},$$

$$\Pr(W) = \frac{400}{1000}.$$

Now, if H_0 is true and the two variables are independent, we should have

$$\Pr(P \cap E) = \Pr(P)\Pr(E) = \left(\frac{336}{1000}\right)\left(\frac{600}{1000}\right),$$

$$\Pr(P \cap W) = \Pr(P)\Pr(W) = \left(\frac{336}{1000}\right)\left(\frac{400}{1000}\right),$$

$$\Pr(C \cap E) = \Pr(C)\Pr(E) = \left(\frac{351}{1000}\right)\left(\frac{600}{1000}\right),$$

$$\Pr(C \cap W) = \Pr(C)\Pr(W) = \left(\frac{351}{1000}\right)\left(\frac{400}{1000}\right),$$

$$\Pr(J \cap E) = \Pr(J)\Pr(E) = \left(\frac{313}{1000}\right)\left(\frac{600}{1000}\right),$$

$$\Pr(J \cap W) = \Pr(J)\Pr(W) = \left(\frac{313}{1000}\right)\left(\frac{400}{1000}\right).$$

The expected frequencies are obtained by multiplying each cell probability by the total number of observations. Thus the expected number of Protestants living on the East coast in our sample will be

$$\left(\frac{336}{1000}\right)\left(\frac{600}{1000}\right)(1000) = \frac{(336)(600)}{1000} = 202$$

when H_0 is true. The general formula for obtaining the expected frequency of any cell is given by

$$e = \frac{RC}{T},$$

where R and C are the corresponding row and column totals and T is the grand total of all the observed frequencies. The expected frequencies for each cell are recorded in parentheses beside the actual observed value in Table 9.5. Note that the expected frequencies in any row or column add up to the appropriate marginal total. In our example we need to compute only the two expected frequencies in the top row of Table 9.5 and then find the others by subtraction. By using three marginal totals and the grand total to arrive at the expected frequencies, we have lost 4 degrees of freedom, leaving

Expected values

Table 9.5
Observed and Expected Frequencies

	Protestant	Catholic	Jewish	Total
East coast	182 (202)	215 (211)	203 (187)	600
West coast	154 (134)	136 (140)	110 (126)	400
Total	336	351	313	1000

a total of 2. A simple formula providing the correct number of degrees of freedom is given by $v = (r - 1)(c - 1)$. Hence, for our example, $v = (2 - 1)(3 - 1) = 2$ degrees of freedom.

To test the null hypothesis of independence, we use the following decision criterion.

> **TEST FOR INDEPENDENCE** *Calculate*
>
> $$\chi^2 = \sum_i \frac{(o_i - e_i)^2}{e_i},$$
>
> *where the summation extends over all cells in the $r \times c$ contingency table. If $\chi^2 > \chi^2_\alpha$, reject the null hypothesis of independence at the α level of significance; otherwise accept the null hypothesis. The number of degrees of freedom is*
>
> $$v = (r - 1)(c - 1).$$

Applying this criterion to our example, we find that

$$\chi^2 = \frac{(182 - 202)^2}{202} + \frac{(215 - 211)^2}{211} + \frac{(203 - 187)^2}{187}$$
$$+ \frac{(154 - 134)^2}{134} + \frac{(136 - 140)^2}{140} + \frac{(110 - 126)^2}{126}$$
$$= 8.556.$$

From Table A.6 we find that $\chi^2_{0.05} = 5.991$ for $v = (2 - 1)(3 - 1) = 2$ degrees of freedom. The null hypothesis is rejected at the 0.05 level of significance, and we conclude that religious faith and the region where one lives are not independent.

The chi-square statistic for testing independence is also applicable when testing the hypothesis that k binomial populations have the same parameter p. This is, therefore, an extension of the test presented in Section 9.8 for the difference between two proportions to the differences among k proportions.

Hence we are interested in testing the hypothesis

$$H_0 : p_1 = p_2 = \cdots = p_k = p$$

against the alternative hypothesis that the population proportions are not all equal, which is equivalent to testing that the number of successes or failures is independent of the sample chosen. To perform this test, we first select independent random samples of size $n_1, n_2, \ldots, n_k$ from the k populations and arrange the data as in the $2 \times k$ contingency table, Table 9.6. The expected cell frequencies are calculated as before and substituted together with the observed frequencies into the above chi-square formula for independence with $v = k - 1$ degrees of freedom. By selecting an appropriate critical region, one can now reach a conclusion concerning H_0.

Table 9.6

k Independent Binomial Samples

	Sample			
	1	2	$\cdots$	k
Successes	x_1	x_2	$\cdots$	x_k
Failures	$n_1 - x_1$	$n_2 - x_2$	$\cdots$	$n_k - x_k$

It is important to remember that the statistic on which we base our decision has a distribution that is only approximated by the chi-square distribution. The computed χ^2 values depend on the cell frequencies and consequently are discrete. The continuous chi-square distribution seems to approximate the discrete sampling distribution of X^2 very well, provided that the number of degrees of freedom is greater than 1. In a 2×2 contingency table, where we have only 1 degree of freedom, a correction called *Yates' correction for continuity* is applied. The corrected formula then becomes

$$\chi^2(\text{corrected}) = \sum_i \frac{(|o_i - e_i| - 0.5)^2}{e_i}.$$

If the expected cell frequencies are large, the corrected and uncorrected results are almost the same. When the expected frequencies are between 5 and 10, Yates' correction should be applied. For expected frequencies less than 5 the Fisher–Irwin exact test should be used. A discussion of this test may be found in *Basic Concepts of Probability and Statistics* by Hodges and Lehmann (see the References). The Fisher–Irwin test may be avoided, however, by choosing a larger sample.

EXERCISES

1. The proportion of adults living in a small town who are college graduates is estimated to be $p = 0.3$. To test this hypothesis, a random sample of 15 adults is selected. If the number of college graduates in our sample is anywhere from 2 to 7, we shall accept the null hypothesis that $p = 0.3$; otherwise we shall conclude that $p \neq 0.3$. Evaluate α assuming $p = 0.3$. Evaluate β for the alternatives $p = 0.2$ and $p = 0.4$. Is this a good test procedure?

2. The proportion of families buying milk from company A in a certain city is believed to be $p = 0.6$. If a random sample of 10 families shows that 3 or less buy milk from company A, we shall reject the hypothesis that $p = 0.6$ in favor of the alternative $p < 0.6$. Find the probability of committing a type I error if the true proportion is $p = 0.6$. Evaluate the probability of committing a type II error for the alternatives $p = 0.3, p = 0.4$, and $p = 0.5$.

3. A random sample of 400 voters in a certain city are asked if they favor a new 3% sales tax. If more than 220 but less than 260 favor the sales tax, we shall conclude that 60% of the voters are for it. Find the probability of committing a type I error if 60% of the voters favor the tax. What is the probability of committing a type II error using this test procedure if actually only 48% of the voters are in favor of the new sales tax?

4. A manufacturer has developed a new fishing line, which he claims has a mean breaking strength of 15 pounds with a standard deviation of 0.5 pound. To test the hypothesis that $\mu = 15$ pounds against the alternative that $\mu < 15$ pounds, a random sample of 50 lines will be tested. The critical region is defined to be $\bar{X} < 14.9$. Find the probability of committing a type I error when H_0 is true. Evaluate β for the alternatives $\mu = 14.8$ and $\mu = 14.9$.

5. An electrical firm manufactures light bulbs that have a length of life that is approximately normally distributed with a mean of 800 hours and a standard deviation of 40 hours. Test the hypothesis that $\mu = 800$ hours against the alternative $\mu \neq 800$ hours if a random sample of 30 bulbs has an average life of 788 hours. Use a 0.04 level of significance.

6. A random sample of 36 drinks from a soft-drink machine has an average content of 7.4 ounces with a standard deviation of 0.48 ounce. Test the hypothesis that $\mu = 7.5$ ounces against the alternative hypothesis $\mu < 7.5$ at the 0.05 level of significance.

7. The average height of males in the freshman class of a certain college has been 68.5 inches, with a standard deviation of 2.7 inches. Is there reason to believe that there has been a change in the average height if a random sample of 50 males in the present freshman class have an average height of 69.7 inches? Use a 0.02 level of significance.

8. It is claimed that an automobile is driven on the average less than 12,000 miles per year. To test this claim, a random sample of 100 automobile owners are asked to keep a record of the miles they travel. Would you agree with this claim if the random sample showed an average of 14,500 miles and a standard deviation of 2400 miles? Use a 0.01 level of significance.

9. Test the hypothesis that the average weight of containers of a particular lubricant is 10 ounces if the weights of a random sample of 10 containers are 10.2, 9.7, 10.1,

10.3, 10.1, 9.8, 9.9, 10.4, 10.3, and 9.8 ounces. Use a 0.01 level of significance and assume that the distribution of weights is normal.

10. A random sample of size 20 from a normal distribution has a mean $\bar{x} = 32.8$ and a standard deviation $s = 4.51$. Does this suggest, at the 0.05 level of significance, that the population mean is greater than 30?

11. A random sample of 8 cigarettes of a certain brand has an average nicotine content of 18.6 milligrams and a standard deviation of 2.4 milligrams. Is this in line with the manufacturer's claim that the average nicotine content does not exceed 17.5 milligrams? Use a 0.01 level of significance and assume the distribution of nicotine contents to be normal.

12. A male student will spend, on the average, $8 for a Saturday evening fraternity party. Test the hypothesis at the 0.1 level of significance that $\mu = \$8$ against the alternative $\mu \neq \$8$ if a random sample of 12 male students attending a homecoming party showed an average expenditure of $8.90 with a standard deviation of $1.75. Assume that the expenses are approximately normally distributed.

13. A random sample of size $n_1 = 25$, taken from a normal population with a standard deviation $\sigma_1 = 5.2$, has a mean $\bar{x}_1 = 81$. A second random sample of size $n_2 = 36$, taken from a different normal population with a standard deviation $\sigma_2 = 3.4$, has a mean $\bar{x}_2 = 76$. Test the hypothesis at the 0.06 level of significance that $\mu_1 = \mu_2$ against the alternative $\mu_1 \neq \mu_2$.

14. A farmer claims that the average yield of corn of variety A exceeds the average yield of variety B by at least 12 bushels per acre. To test this claim, 50 acres of each variety are planted and grown under similar conditions. Variety A yielded, on the average, 86.7 bushels per acre with a standard deviation of 6.28 bushels per acre, while variety B yielded, on the average, 77.8 bushels per acre with a standard deviation of 5.61 bushels per acre. Test the farmer's claim using a 0.05 level of significance.

15. A study was made to estimate the difference in salaries of college professors in the private and state colleges of Virginia. A random sample of 100 professors in private colleges showed an average 9-month salary of $16,000 with a standard deviation of $1300. A random sample of 200 professors in state colleges showed an average salary of $16,900 with a standard deviation of $1400. Test the hypothesis that the average salary for professors teaching in state colleges does not exceed the average salary for professors teaching in private colleges by more than $500. Use a 0.02 level of significance.

16. Given two random samples of size $n_1 = 11$ and $n_2 = 14$, from two independent normal populations, with $\bar{x}_1 = 75$, $\bar{x}_2 = 60$, $s_1 = 6.1$, and $s_2 = 5.3$, test the hypothesis at the 0.05 level of significance that $\mu_1 = \mu_2$ against the alternative that $\mu_1 \neq \mu_2$. Assume that the population variances are equal.

17. A study is made to see if increasing the substrate concentration has an appreciable effect on the velocity of a chemical reaction. With the substrate concentration of 1.5 moles per liter, the reaction was run 15 times with an average velocity of 7.5 micromoles per 30 minutes and a standard deviation of 1.5. With a substrate concentration of 2.0 moles per liter, 12 runs were made yielding an average velocity of 8.8 micromoles per 30 minutes and a sample standard deviation of 1.2. Would you say that the increase in substrate concentration increases the mean velocity by as much as 0.5 micromole per 30 minutes? Use a 0.01 level of significance and assume the populations to be approximately normally distributed with equal variances.

18. A study was made to determine if the subject matter in a physics course is better understood when a lab constitutes part of the course. Students were allowed to choose between a 3-semester-hour course without labs and a 4-semester-hour course with labs. In the section with labs 11 students made an average grade of 85 with a standard deviation of 4.7, and in the section without labs 17 students made an average grade of 79 with a standard deviation of 6.1. Would you say that the laboratory course increases the average grade by as much as 8 points? Use a 0.01 level of significance and assume the populations to be approximately normally distributed with equal variances.

19. A taxi company is trying to decide whether to purchase brand A or brand B tires for its fleet of taxis. To help arrive at a decision an experiment is conducted using 12 of each brand. The tires are run until they wear out. The results are

$$\text{Brand } A: \bar{x}_1 = 23{,}600 \text{ miles, } s_1 = 3{,}200 \text{ miles,}$$
$$\text{Brand } B: \bar{x}_2 = 24{,}800 \text{ miles, } s_2 = 3{,}700 \text{ miles.}$$

Test the hypothesis at the 0.05 level of significance that there is no difference in the two brands of tires. Assume the populations to be approximately normally distributed.

20. The following data represent the running times of films produced by two different motion-picture companies:

			Time, Minutes				
Company 1	102	86	98	109	92		
Company 2	81	165	97	134	92	87	114

Test the hypothesis that the average running time of films produced by company 2 exceeds the average running time of films produced by company 1 by 10 minutes against the one-sided alternative that the difference is more than 10 minutes. Use a 0.1 level of significance and assume the distributions of times to be approximately normal.

21. In Exercise 21, Chapter 8, use the t distribution to test the hypothesis, at the 0.05 level of significance, that the average yields of the two varieties of wheat are equal against the alternative hypothesis that they are unequal.

22. In Exercise 22, Chapter 8, use the t distribution to test the hypothesis, at the 0.01 level of significance, that $\mu_1 = \mu_2$ against the alternative hypothesis that $\mu_1 < \mu_2$.

23. In Exercise 23, Chapter 8, use the t distribution to test the hypothesis, at the 0.05 level of significance, that the diet reduces a person's weight by 10 pounds on the average against the alternative hypothesis that the mean difference in weight is less than 10 pounds.

24. Test the hypothesis that $\sigma^2 = 0.03$ against the alternative hypothesis that $\sigma^2 \neq 0.03$ in Exercise 9. Use a 0.01 level of significance.

25. Test the hypothesis that $\sigma = 6$ against the alternative that $\sigma < 6$ in Exercise 10. Use a 0.05 level of significance.

26. Test the hypothesis that $\sigma^2 = 2.3$ against the alternative that $\sigma^2 \neq 2.3$ in Exercise 11. Use a 0.05 level of significance.

27. Test the hypothesis that $\sigma = 1.40$ against the alternative that $\sigma > 1.40$ in Exercise 12. Use a 0.01 level of significance.

28. Test the hypothesis that $\sigma_1^2 = \sigma_2^2$ against the alternative that $\sigma_1^2 > \sigma_2^2$ in Exercise 16. Use a 0.01 level of significance.

29. Test the hypothesis that $\sigma_1 = \sigma_2$ against the alternative that $\sigma_1 < \sigma_2$ in Exercise 19. Use a 0.05 level of significance.

30. Test the hypothesis that $\sigma_1^2 = \sigma_2^2$ against the alternative that $\sigma_1^2 \neq \sigma_2^2$ in Exercise 20. Use a 0.10 level of significance.

31. The following data represent the weights of personal luggage carried on a large aircraft by the members of two baseball clubs:

Club A	34	39	41	28	33	
Club B	36	40	35	31	39	36

Use the Wilcoxon two-sample test with $\alpha = 0.05$ to test the hypothesis that the two clubs carry the same amount of luggage on the average against the alternative hypothesis that the average weight of luggage for club B is greater than that of club A.

32. A fishing line is being manufactured by two processes. To determine if there is a difference in the mean breaking strength of the lines, 10 pieces by each process are selected and tested for breaking strength. The results are as follows:

Process 1	10.4	9.8	11.5	10.0	9.9	9.6	10.9	11.8	9.3	10.7
Process 2	8.7	11.2	9.8	10.1	10.8	9.5	11.0	9.8	10.5	9.9

Use the Wilcoxon two-sample test with $\alpha = 0.1$ to determine if there is a difference between the mean breaking strengths of the lines manufactured by the two processes.

33. From a mathematics class of 12 equally capable students using programed materials, 5 are selected at random and given additional instruction by the teacher. The results on the final examination were as follows:

	Grade						
Additional instruction	87	69	78	91	80		
No additional instruction	75	88	64	82	93	79	67

Use the Wilcoxon two-sample test with $\alpha = 0.05$ to determine if the additional instruction affects the average grade.

34. The weights of 4 people before they stopped smoking and 5 weeks after they stopped smoking are as follows:

	Individual			
	1	2	3	4
Before	148	176	153	116
After	154	179	151	121

Use the Wilcoxon test for paired observations to test the hypothesis, at the 0.05 level of significance, that giving up smoking has no effect on a person's weight against the alternative that one's weight increases if he quits smoking.

35. In Exercise 21, Chapter 8, use the Wilcoxon test for paired observations to test the hypothesis, at the 0.05 level of significance, that the average yields of the two varieties of wheat are equal against the alternative hypothesis that they are unequal. Compare your conclusion with that of Exercise 21.

36. In Exercise 22, Chapter 8, use the Wilcoxon test for paired observations to test the hypothesis, at the 0.01 level of significance, that $\mu_1 = \mu_2$ against the alternative hypothesis that $\mu_1 < \mu_2$. Compare your conclusions with that of Exercise 22.

37. In Exercise 23, Chapter 8, use the Wilcoxon test for paired observations to test the hypothesis, at the 0.05 level of significance, that the diet reduces a person's weight by 10 pounds on the average against the alternative hypothesis that the mean difference in weight is less than 10 pounds. Compare your conclusion with that of Exercise 23.

38. It is believed that at least 60% of the residents in a certain area favor an annexation suit by a neighboring city. What conclusion would you draw if only 110 in a sample of 200 voters favor the suit? Use a 0.04 level of significance.

39. A manufacturer of cigarettes claims that 20% of the cigarette smokers prefer brand X. To test this claim a random sample of 20 cigarette smokers are selected and asked what brand they prefer. If 6 of the 20 smokers prefer brand X, what conclusion do we draw? Use a 0.01 level of significance.

40. In a random sample of 1000 houses in a certain city, it is found that 618 own color television sets. Is this sufficient evidence to conclude that $\frac{2}{3}$ of the houses in this city have color television sets? Use a 0.02 level of significance.

41. At a certain college it is estimated that fewer than 25% of the students have cars on campus. Does this seem to be a valid estimate if, in a random sample of 90 college students, 28 are found to have cars? Use a 0.05 level of significance.

42. In a study to estimate the proportion of housewives who own an automatic dryer, it is found that 63 of 100 urban residents have a dryer and 59 of 125 suburban residents own a dryer. Is there a significant difference between the proportion of urban and suburban housewives who own an automatic dryer? Use a 0.04 level of significance.

43. A cigarette-manufacturing firm distributes two brands of cigarettes. If it is found that 56 of 200 smokers prefer brand A and that 29 of 150 smokers prefer brand B, can we conclude at the 0.06 level of significance that brand A outsells brand B?

44. A random sample of 100 men and 100 women at a Southern college are asked if they have an automobile on campus. If 31 of the men and 24 of the women have cars, can we conclude that more men than women have cars on campus? Use a 0.01 level of significance.

45. A study is made to determine if a cold climate contributes more to absenteeism from school during a semester than a warmer climate. Two groups of students are selected at random, one group from Maine and the other from Alabama. Of the 300 students from Maine, 72 were absent at least 1 day during the semester, and of the 400 students from Alabama, 70 were absent 1 or more days. Can we conclude that a colder climate results in a greater number of students being absent from school at least 1 day during the semester? Use a 0.05 level of significance.

46. A die is tossed 180 times with the following results:

x	1	2	3	4	5	6
f	28	36	36	30	27	23

Is this a balanced die? Use a 0.01 level of significance.

47. In 100 tosses of a coin, 63 heads and 37 tails are observed. Is this a balanced coin? Use a 0.05 level of significance.

48. Three marbles are selected from an urn containing 5 red marbles and 3 green marbles. After recording the number X of red marbles, the marbles are replaced in the urn and the experiment repeated 112 times. The results obtained are as follows:

x	0	1	2	3
f	1	31	55	25

Test the hypothesis at the 0.05 level of significance that the recorded data may be fitted by the hypergeometric distribution $h(x; 8, 3, 5)$, $x = 0, 1, 2, 3$.

49. Three cards are drawn from an ordinary deck of playing cards, with replacement, and the number Y of spades is recorded. After repeating the experiment 64 times, the following outcomes were recorded:

y	0	1	2	3
f	21	31	12	0

Test the hypothesis at the 0.01 level of significance that the recorded data may be fitted by the binomial distribution $b(y; 3, \frac{1}{4})$, $y = 0, 1, 2, 3$.

50. The grades in a statistics course for a particular semester were as follows:

Grade	A	B	C	D	F
f	14	18	32	20	16

Test the hypothesis, at the 0.05 level of significance, that the distribution of grades is uniform.

51. A coin is thrown until a head occurs and the number X of tosses recorded. After repeating the experiment 256 times, we obtained the following results:

x	1	2	3	4	5	6	7	8
f	136	60	34	12	9	1	3	1

Test the hypothesis at the 0.05 level of significance that the observed distribution of X may be fitted by the geometric distribution $g(x; \frac{1}{2})$, $x = 1, 2, 3, \ldots$.

52. Repeat Exercise 49 using a new set of data obtained by actually carrying out the described experiment 64 times.

53. Repeat Exercise 51 using a new set of data obtained by performing the described experiment 256 times.

54. In Exercise 9, Chapter 3, test the goodness of fit between the observed frequencies and the expected normal frequencies, using a 0.05 level of significance.

55. In Exercise 10, Chapter 3, test the goodness of fit between the observed frequencies and the expected normal frequencies, using a 0.01 level of significance.

56. In an experiment to study the dependence of hypertension on smoking habits, the following data were taken on 180 individuals:

	Nonsmokers	Moderate Smokers	Heavy Smokers
Hypertension	21	36	30
No hypertension	48	26	19

Test the hypothesis that the presence or absence of hypertension is independent of smoking habits. Use a 0.05 level of significance.

57. A random sample of 200 college students are classified according to class status and drinking habits.

	Freshmen	Sophomore	Junior	Senior
Drinkers	21	33	25	20
Nondrinkers	47	26	19	9

Test the hypothesis that class status and drinking habits are independent. Use a 0.05 level of significance.

58. A random sample of 30 adults are classified according to sex and the number of hours they watch television during a week.

	Male	Female
Over 25 hours	5	9
Under 25 hours	9	7

Using a 0.01 level of significance, test the hypothesis that a person's sex and time watching television are independent.

59. A random sample of 200 married men, all retired, were classified according to education and number of children.

Education	Number of Children		
	0–1	2–3	Over 3
Elementary	14	37	32
Secondary	19	42	17
College	12	17	10

Test the hypothesis, at the 0.05 level of significance, that the size of a family is independent of the level of education attained by the father.

60. In a shop study, a set of data was collected to determine whether or not the proportion of defectives produced by workers was the same for the day, evening, or night shift worked. The following data were collected on the items produced:

	Shift		
	Day	Evening	Midnight
Defective	45	55	70
Nondefective	905	890	870

What is your conclusion? Use an $\alpha = 0.025$ level of significance.

REGRESSION
AND CORRELATION CHAPTER **10**

10.1 Linear Regression

The problem we shall consider in this section is that of estimating or predicting the value of a dependent random variable Y on the basis of a known measurement of an independent *controlled* variable X. Problems of this type are known as *regression problems*. Scientists, economists, psychologists, and sociologists have always been concerned with the problems of prediction. Meterologists are constantly analyzing data in hopes of predicting or forecasting with a high degree of accuracy. Measurements from meteorological data are used extensively today to predict impact areas for missiles fired under various atmospheric conditions. Various tests are taken by freshmen entering college for the purpose of predicting success in later years.

Suppose we wish to predict a student's grade in freshman chemistry based upon his score on an intelligence test administered prior to his attending college. To make such a prediction, we must examine the distribution of chemistry grades achieved in the past by students having this same intelligence test score. Denoting an individual's chemistry grade by Y and his intelligence test score by X, then (x, y) represents the results of any student in the population. A random sample of size n from the population might be represented by the set $\{(x_i, y_i); i = 1, 2, \ldots, n\}$. If additional samples were taken using exactly the same values of x, we would expect the values of y to vary. Hence the value y_i in the ordered pair (x_i, y_i) is a value of some random variable Y_i.

For convenience we define $Y|x$ to be the random variable Y corresponding to a fixed value x and denote its probability distribution by $f(y|x)$. Clearly then, if $x = x_i$, the symbol $Y|x_i$ represents the random variable Y_i. We are interested in the distributions of the set of random variables $\{Y_i;\, i = 1, 2, \ldots, n\}$, all of which are assumed to be independent. For the purpose of constructing confidence intervals and making tests of hypotheses, we shall also require $Y_1, Y_2, \ldots, Y_n$ to be normally distributed.

In our prediction problem above, we define $\mu_{Y|x} = E(Y|x)$ to be the mean of the distribution of all chemistry grades for a given or known intelligence test score x, and define $\sigma^2_{Y|x}$ to be the variance of this distribution. We shall assume the variances equal, that is, $\sigma^2_{Y|x} = \sigma^2$, for all x. The parameter $\mu_{Y|x}$ is constant for any fixed x but may vary for different intelligence test scores.

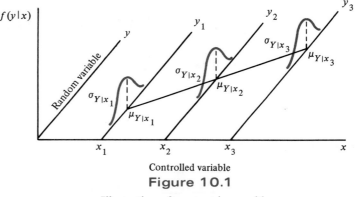

Controlled variable

Figure 10.1

Illustration of a regression problem.

The above discussion is illustrated graphically in Figure 10.1 for three intelligence test scores x_1, x_2, and x_3. The curve connecting the means of all the distributions is called the *regression curve*. If the means $\mu_{Y|x}$ fall on a straight line as is the case in Figure 10.1, the regression is linear and may be represented by the equation

$$\mu_{Y|x} = \alpha + \beta x,$$

where α and β are parameters to be estimated from the sample data. Denoting their estimates by a and b, respectively, we can then estimate $\mu_{Y|x}$ by $\bar{y}_x$ from the sample regression line

$$\bar{y}_x = a + bx.$$

Let us consider the distributions of chemistry grades corresponding to the intelligence test scores $x_1 = 50, x_2 = 55, x_3 = 65$, and $x_4 = 70$. The chemistry grades for a sample of 12 freshmen having these intelligence test scores are presented in Table 10.1. The data of Table 10.1 have been plotted in Figure

Table 10.1

Intelligence Test Scores and Freshmen Chemistry Grades

Student	Test Score x	Chemistry Grade y
1	65	85
2	50	74
3	55	76
4	65	90
5	55	85
6	70	87
7	65	94
8	70	98
9	55	81
10	70	91
11	50	76
12	55	74

10.2 to give a *scatter diagram*. The assumption of linearity, that is, the assumption that all the $\mu_{Y|x}$ lie on a straight line, appears reasonable.

The sample line and a hypothetical population line have been drawn on the scatter diagram of Figure 10.2. The agreement between the sample line and the unknown population line should be good if we have a large number of chemistry grades for each intelligence test score. Using the sample line in Figure 10.2, we would predict a chemistry grade of 84 for a student whose intelligence test score was 60.

The sample line is completely determined once the point estimates a and b, of the parameters α and β, are evaluated from the sample data. We shall now consider the problem of deriving computational formulas for these point estimates and then provide a measure of their reliability by constructing confidence intervals for the parameters α and β.

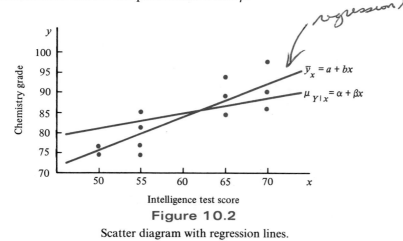

Figure 10.2

Scatter diagram with regression lines.

10.2 Estimation of Parameters

Assuming that all the means, $\mu_{Y|x}$, fall on a straight line, we can write the population regression line in the form

$$\mu_{Y|x} = \alpha + \beta x.$$

The random variable $Y_i = Y|x_i$ may then be written

$$Y_i = \alpha + \beta x_i + E_i,$$

where the random variable E_i, from previous assumptions on Y_i, must necessarily have a mean of zero and a variance $\sigma^2_{Y|x_i} = \sigma^2$. Each observation (x_i, y_i) in our sample satisfies the relation

$$y_i = \alpha + \beta x_i + \varepsilon_i.$$

For a random sample of n observations $\{(x_i, y_i); i = 1, 2, \ldots, n\}$, the regression line, which is an estimate of $\mu_{Y|x}$, is given by

$$\bar{y}_x = a + bx,$$

and each pair of observations satisfies the relation

$$y_i = a + bx_i + e_i,$$

where e_i is called the *residual*. The difference between e_i and ε_i is clearly shown in Figure 10.3.

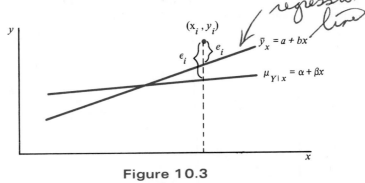

Figure 10.3

Comparing ϵ_i with the residual e_i.

We shall find a and b, the estimates of α and β, so that the sum of the squares of the residuals is a minimum. The residual sum of squares is often called the *sum of squares of the errors* about the regression line and is denoted

by SSE. This minimization procedure for estimating parameters is called the
method of least squares. Hence we shall find a and b so as to minimize

$$\text{SSE} = \sum_{i=1}^{n} e_i^2 = \sum_{i=1}^{n} (y_i - a - bx_i)^2.$$

The determination of a and b so as to minimize SSE is most easily accomplished by means of differential calculus. We shall omit the details and state the final formulas.

ESTIMATION OF THE PARAMETERS α AND β *The regression line*

$$\mu_{Y|x} = \alpha + \beta x$$

is estimated from a sample of size n by the line

$$\bar{y}_x = a + bx,$$

where

$$b = \frac{n \sum_{i=1}^{n} x_i y_i - \left(\sum_{i=1}^{n} x_i \right) \left(\sum_{i=1}^{n} y_i \right)}{n \sum_{i=1}^{n} x_i^2 - \left(\sum_{i=1}^{n} x_i \right)^2},$$

$$a = \bar{y} - b\bar{x}.$$

The calculations of a and b using the data of Table 10.1 are illustrated by the following example.

Example 10.1 Find the regression line for the data of Table 10.1.

Solution. Using a desk calculator, we find that

$$\sum_{i=1}^{12} x_i = 725, \qquad \sum_{i=1}^{12} y_i = 1011, \qquad \sum_{i=1}^{12} x_i y_i = 61,685,$$

$$\sum_{i=1}^{12} x_i^2 = 44,475, \qquad \bar{x} = 60.417, \qquad \bar{y} = 84.250.$$

Therefore,

$$b = \frac{(12)(61,685) - (725)(1011)}{(12)(44,475) - (725)^2} = 0.897,$$

$$a = 84.250 - (0.897)(60.417) = 30.056.$$

The regression line is then given by

$$\bar{y}_x = 30.056 + 0.897x.$$

By substituting any two of the given values of x into this equation, say $x_1 = 50$ and $x_2 = 70$, we obtain the ordinates $\bar{y}_{50} = 74.9$ and $\bar{y}_{70} = 92.8$. The regression line in Figure 10.2 was drawn by connecting these two points with a straight line.

Before we can construct confidence intervals and perform tests of hypotheses on α and β, the unknown variance σ^2 must be estimated from the data. An unbiased estimate of σ^2 with $n - 2$ degrees of freedom is given by the formula

$$s^2 = \frac{\text{SSE}}{n - 2},$$

where

$$\text{SSE} = \sum_{i=1}^{n} (y_i - a - bx_i)^2$$
$$= \sum_{i=1}^{n} (y_i - \bar{y}_{x_i})^2.$$

The value $\bar{y}_{x_i}$ is the estimated value of the random variable Y when $x = x_i$, and is obtained from the regression line $\bar{y}_x = a + bx$. Although this formula for SSE appears to have a simple form, the computations involved can be tedious. An equivalent and preferred formula is given by

$$\text{SSE} = (n - 1)(s_y^2 - b^2 s_x^2),$$

where s_x^2 and s_y^2 are the variances of the selected values of X and the observed values of Y, respectively. Hence we may estimate σ^2 by means of the formula given in the following theorem.

THEOREM 10.1 *An unbiased estimate of σ^2 with $n - 2$ degrees of freedom is given by the formula*

$$s^2 = \frac{n - 1}{n - 2}(s_y^2 - b^2 s_x^2).$$

The calculation of s^2 for the data of Table 10.1 is illustrated in the following example.

Example 10.2 Calculate s^2 for the data of Table 10.1.

Solution. In Example 10.1 we found that

$$\sum_{i=1}^{12} x_i = 725, \qquad \sum_{i=1}^{12} x_i^2 = 44,475, \qquad \sum_{i=1}^{12} y_i = 1011.$$

Referring to the data of Table 10.1, we now find $\sum_{i=1}^{12} y_i^2 = 85,905$. Therefore,

$$s_x^2 = \frac{(12)(44,475) - (725)^2}{(12)(11)} = 61.174,$$

$$\frac{m\sum x^2 - (\sum x)^2}{m(m-1)}$$

$$s_y^2 = \frac{(12)(85,905) - (1011)^2}{(12)(11)} = 66.205.$$

Recall that $b = 0.897$. Hence *Smooth thm 10.1*

$$s^2 = \tfrac{11}{10}[66.205 - (0.805)(61.174)] = 18.656.$$

$$S = 4.3$$

It is important to remember that our values of a and b are only estimates of the true parameters α and β based upon a given sample of n observations. The different estimates of α and β that could be computed by drawing several samples of size n may be thought of as the values assumed by the random variables A and B.

Since the values of x remain fixed, the values of A and B depend on the variations in the values of y or, more precisely, on the values of the random variables $Y_1, Y_2, \ldots, Y_n$. If we assume $Y_1, Y_2, \ldots, Y_n$ to be independent and normally distributed, then it can be shown that the random variable A is also normally distributed with mean

$$\mu_A = \alpha$$

and variance

$$\sigma_A^2 = \left[\frac{\sum_{i=1}^{n} x_i^2}{n(n-1)s_x^2} \right] \sigma^2.$$

Using the Z-transformation of Section 6.2, we can write

$$Z = \frac{A - \alpha}{\sigma \sqrt{\sum_{i=1}^{n} x_i^2} \Big/ s_x \sqrt{n(n-1)}} = \frac{(A - \alpha)s_x \sqrt{n(n-1)}}{\sigma \sqrt{\sum_{i=1}^{n} x_i^2}},$$

which is a random variable having the standard normal distribution.

The standard deviation, σ, is usually unknown and is replaced by its estimator S to give

$$T = \frac{(A - \alpha)s_x\sqrt{n(n-1)}}{S\sqrt{\sum\limits_{i=1}^{n} x_i^2}},$$

which is a random variable having the t distribution with $n - 2$ degrees of freedom. This quantity can be used to establish a $(1 - \alpha)100\%$ confidence interval for the parameter α by noting that

$$\Pr\left[-t_{\alpha/2} < \frac{(A - \alpha)s_x\sqrt{n(n-1)}}{S\sqrt{\sum\limits_{i=1}^{n} x_i^2}} < t_{\alpha/2}\right] = 1 - \alpha.$$

The algebraic techniques that were used repeatedly in Chapter 8 allow us to write

$$\Pr\left[A - \frac{t_{\alpha/2}S\sqrt{\sum\limits_{i=1}^{n} x_i^2}}{s_x\sqrt{n(n-1)}} < \alpha < A + \frac{t_{\alpha/2}S\sqrt{\sum\limits_{i=1}^{n} x_i^2}}{s_x\sqrt{n(n-1)}}\right] = 1 - \alpha,$$

and then for a given sample we compute a and s to give the $(1 - \alpha)100\%$ confidence interval

$$a - \frac{t_{\alpha/2}s\sqrt{\sum\limits_{i=1}^{n} x_i^2}}{s_x\sqrt{n(n-1)}} < \alpha < a + \frac{t_{\alpha/2}s\sqrt{\sum\limits_{i=1}^{n} x_i^2}}{s_x\sqrt{n(n-1)}}.$$

> **CONFIDENCE INTERVAL FOR α** *A $(1 - \alpha)100\%$ confidence interval for the parameter α in the regression line $\mu_{Y|x} = \alpha + \beta x$ is*
>
> $$a - \frac{t_{\alpha/2}s\sqrt{\sum\limits_{i=1}^{n} x_i^2}}{s_x\sqrt{n(n-1)}} < \alpha < a + \frac{t_{\alpha/2}s\sqrt{\sum\limits_{i=1}^{n} x_i^2}}{s_x\sqrt{n(n-1)}}.$$

Example 10.3 Find a 95% confidence interval for α in the regression line $\mu_{Y|x} = \alpha + \beta x$, based on the data in Table 10.1.

Solution. In Example 10.2 we found that $s_x^2 = 61.174$ and $s^2 = 18.656$. Therefore, taking square roots we obtain $s_x = 7.8$ and $s = 4.3$. From Example 10.1 we had $\sum\limits_{i=1}^{n} x_i^2 = 44{,}475$ and

$a = 30.056$. Using Table A.5, we find $t_{0.025} = 2.228$ for 10 degrees of freedom. Therefore a 95% confidence interval for α is given by

$$30.056 - \frac{(2.228)(4.3)\sqrt{44{,}475}}{7.8\sqrt{(12)(11)}} < \alpha < 30.056$$

$$+ \frac{(2.228)(4.3)\sqrt{44{,}475}}{7.8\sqrt{(12)(11)}},$$

which simplifies to

$$7.510 < \alpha < 52.602.$$

To test the null hypothesis H_0 that $\alpha = \alpha_0$ against a suitable alternative, we can use the t distribution with $n - 2$ degrees of freedom to establish a critical region and then base our decision on the value of

$$t = \frac{(a - \alpha_0)s_x\sqrt{n(n - 1)}}{s\sqrt{\sum_{i=1}^{n} x_i^2}}.$$

Example 10.4 Using the estimated value $a = 30.056$ in Example 10.1, test the hypothesis that $\alpha = 35$ at the 0.05 level of significance.

Solution.   1. $H_0: \alpha = 35$.
2. $H_1: \alpha \neq 35$.
3. Choose a 0.05 level of significance.
4. Critical region: $T < -2.228$ and $T > 2.228$.
5. Computations:

$$t = \frac{(30.056 - 35)(7.8)\sqrt{(12)(11)}}{4.3\sqrt{44{,}475}} = -0.489.$$

6. Conclusion: Accept H_0.

The random variable B may also be shown to have a normal distribution with mean

$$\mu_B = \beta$$

and variance

$$\sigma_B^2 = \frac{\sigma^2}{(n - 1)s_x^2}.$$

Therefore,

$$Z = \frac{B - \beta}{\sigma/s_x\sqrt{n-1}} = \frac{s_x\sqrt{n-1}(B-\beta)}{\sigma}$$

is a random variable having the standard normal distribution. Replacing σ by its estimator S, we obtain the random variable

$$T = \frac{s_x\sqrt{n-1}(B-\beta)}{S},$$

which has a t distribution with $n - 2$ degrees of freedom.

The statistic T can be used to construct a $(1 - \alpha)100\%$ confidence interval for the parameter β.

CONFIDENCE INTERVAL FOR β *A $(1 - \alpha)100\%$ confidence interval for the parameter β in the regression line $\mu_{Y|x} = \alpha + \beta x$ is*

$$b - \frac{t_{\alpha/2}\,s}{s_x\sqrt{n-1}} < \beta < b + \frac{t_{\alpha/2}\,s}{s_x\sqrt{n-1}}.$$

Example 10.5 Find a 95% confidence interval for β in the regression line $\mu_{Y|x} = \alpha + \beta x$, based on the data in Table 10.1.

Solution. From Example 10.3 we note that $s_x = 7.8$, $s = 4.3$ and $t_{0.025} = 2.228$ for 10 degrees of freedom. In Example 10.1 the slope was calculated to be $b = 0.897$. Therefore a 95% confidence interval for β is given by

$$0.897 - \frac{(2.228)(4.3)}{(7.8)\sqrt{11}} < \beta < 0.897 + \frac{(2.228)(4.3)}{(7.8)\sqrt{11}},$$

which simplifies to

$$0.527 < \beta < 1.267.$$

To test the null hypothesis H_0 that $\beta = \beta_0$, against a suitable alternative, we again use the t distribution with $n - 2$ degrees of freedom to establish a critical region and then base our decision on the value of

$$t = \frac{s_x\sqrt{n-1}(b-\beta_0)}{s}.$$

The method is illustrated in the following example.

Example 10.6 Using the estimated value $b = 0.897$ in Example 10.1, test the hypothesis that $\beta = 0$ at the 0.01 level of significance against the alternative that $\beta > 0$.

Solution.   1. $H_0: \beta = 0$.
2. $H_1: \beta > 0$.
3. Choose a 0.01 level of significance.
4. Critical region: $T > 2.764$.
5. Computations:

$$t = \frac{(7.8)\sqrt{11}(0.897 - 0)}{4.3} = 5.396.$$

6. Conclusion: Reject H_0 and conclude that $\beta > 0$.

10.3 Prediction

The equation $\bar{y}_x = a + bx$ may be used to predict the value of $\mu_{Y|x'}$, where x' is not necessarily one of the prechosen values, or it may be used to predict a single value $y_{x'}$ of the variable $Y|x'$. We would expect the error of prediction to be higher in the case of a single predicted value than in the case where a mean is predicted. This, then, will affect the width of our confidence intervals for the parameters being estimated.

If the experimenter wishes to construct a confidence interval for $\mu_{Y|x'}$, it is necessary for him to examine the sampling distribution of the differences between the ordinates $\bar{y}_{x'}$, obtained from the computed regression line in repeated sampling, and the corresponding ordinate $\mu_{Y|x'}$ of the true regression line. It is not difficult to show that such a sampling distribution of differences is normal with mean

$$\mu_{\bar{Y}_{x'} - \mu_{Y|x'}} = E(\bar{Y}_{x'} - \mu_{Y|x'}) = 0$$

and variance

$$\sigma^2_{\bar{Y}_{x'} - \mu_{Y|x'}} = \left[\frac{1}{n} + \frac{(x' - \bar{x})^2}{(n - 1)s_x^2} \right] \sigma^2.$$

In practice we replace σ^2 by s^2, a value of the estimator S^2. Therefore the statistic

$$T = \frac{\bar{Y}_{x'} - \mu_{Y|x'}}{S\sqrt{(1/n) + [(x' - \bar{x})^2/(n - 1)s_x^2]}}$$

has a t distribution with $n - 2$ degrees of freedom. A $(1 - \alpha)100\%$ confidence interval for $\mu_{Y|x'}$ can now be constructed based on the t distribution.

CONFIDENCE INTERVAL FOR $\mu_{Y|x'}$ *A* $(1 - \alpha)100\%$ *confidence interval for* $\mu_{Y|x'}$ *is given by*

$$\bar{y}_{x'} - t_{\alpha/2} s \sqrt{\frac{1}{n} + \frac{(x' - \bar{x})^2}{(n-1)s_x^2}} < \mu_{Y|x'} < \bar{y}_{x'} + t_{\alpha/2} s \sqrt{\frac{1}{n} + \frac{(x' - \bar{x})^2}{(n-1)s_x^2}}.$$

Example 10.7 Using the data in Table 10.1, construct a 95% confidence interval for $\mu_{Y|60}$.

Solution. From the regression equation we find

$$\bar{y}_{60} = 30.056 + (0.897)(60) = 83.876.$$

$n - 2$

$t_{\alpha/2} =$ Previously we had $\bar{x} = 60.417$, $s_x^2 = 61.174$, $s = 4.3$, and $t_{0.025} = 2.228$ for 10 degrees of freedom. Therefore a 95% confidence interval for $\mu_{Y|60}$ is given by

$$83.876 - (2.228)(4.3)\sqrt{\frac{1}{12} + \frac{(60 - 60.417)^2}{(11)(61.174)}}$$

$$< \mu_{Y|60} < 83.876$$

$$+ (2.228)(4.3)\sqrt{\frac{1}{12} - \frac{(60 - 60.417)^2}{(11)(61.174)}},$$

or simply

$$81.106 < \mu_{Y|60} < 86.646.$$

To obtain a confidence interval for any single value $y_{x'}$ of the variable $Y|x'$, it is necessary to estimate the variance of the differences between the ordinates $\bar{y}_{x'}$, obtained from the computed regression line in repeated sampling, and the corresponding true ordinate $y_{x'}$. The sampling distribution of these differences can also be shown to be normal with mean

$$\mu_{\bar{Y}_{x'} - Y|x'} = E(\bar{Y}_{x'} - Y|x') = 0$$

and variance

$$\sigma^2_{\bar{Y}_{x'} - Y|x'} = \left[1 + \frac{1}{n} + \frac{(x' - \bar{x})^2}{(n-1)s_x^2}\right]\sigma^2.$$

An estimator for $\sigma^2_{\bar{Y}_{x'} - Y|x'}$ is obtained by substituting S^2 for σ^2 in the preceding formula. The statistic

$$T = \frac{\bar{Y}_{x'} - Y|x'}{S\sqrt{1 + (1/n) + [(x' - \bar{x})^2/(n - 1)s_x^2]}},$$

which has a t distribution with $n - 2$ degrees of freedom, is then used to construct a confidence interval for $y_{x'}$.

> **CONFIDENCE INTERVAL FOR $y_{x'}$** *A* $(1 - \alpha)100\%$ *confidence interval for $y_{x'}$ is given by*
>
> $$\bar{y}_{x'} - t_{\alpha/2}s\sqrt{1 + \frac{1}{n} + \frac{(x' - \bar{x})^2}{(n + 1)s_x^2}} < y_{x'} < \bar{y}_{x'} + t_{\alpha/2}s\sqrt{1 + \frac{1}{n} + \frac{(x' - \bar{x})^2}{(n - 1)s_x^2}}.$$

Example 10.8 Using the data in Table 10.1, construct a 95% confidence interval for y_{60}.

suppose I know a person has a IQ = 60 what is his chemistry grade

Solution. We have $n = 12$, $x' = 60$, $\bar{x} = 60.417$, $\bar{y}_{x'} = 83.876$, $s_x^2 = 61.174$, $s = 4.3$, and $t_{0.025} = 2.228$ for 10 degrees of freedom. Therefore a 95% confidence interval for y_x is given by

→ from p2 43 eg 10.1

n - 2

$t_{\alpha/2} = t_{.025}$
$v = m - 2 \quad v = 10$
$t_{\alpha/2} = t_{.05\%}$

83.876

$\bar{Y}_{x'} \pm$ *error*

$$83.876 - (2.228)(4.3)\sqrt{1 + \frac{1}{12} + \frac{(60 - 60.417)^2}{(11)(61.174)}}$$

$$< y_{60} < 83.876$$

$$+ (2.228)(4.3)\sqrt{1 + \frac{1}{12} + \frac{(60 - 60.417)^2}{(11)(61.174)}},$$

which simplifies to

$$73.903 < y_{60} < 93.849. \quad = \quad 83.876 \pm 9.973$$

10.4 Test for Linearity of Regression

In Section 10.1 we defined the regression to be linear when all the $\mu_{Y|x}$ fall on a straight line. For any given problem we either assume the regression is linear and proceed with the estimation of parameters as in Section 10.2, or we conclude that the regression is nonlinear and resort to the methods discussed in Sections 10.5 and 10.6. To avoid laborious calculations, a linear regression equation is always preferred over a nonlinear regression curve if the assumption of linearity can be justified. Fortunately a test for linearity of regression does exist and will now be presented.

Let us select a random sample of n observations using k distinct values of x, say $x_1, x_2, \ldots, x_k$, such that the sample contains n_1 observed values of the random variable Y_1 corresponding to x_1, n_2 observed values of Y_2 corresponding to $x_2, \ldots, n_k$ observed values of Y_k corresponding to x_k. Of necessity $n = \sum\limits_{i=1}^{k} n_i$. We define

$$y_{ij} = \text{the } j\text{th value of the random variable } Y_i,$$

$$y_{i.} = \text{the sum of the values of } Y_i \text{ in our sample.}$$

Hence if $n_4 = 3$ measurements of Y are made corresponding to $x = x_4$, we could indicate these observations by y_{41}, y_{42}, and y_{43}. Then $y_{4.} = y_{41} + y_{42} + y_{43}$. Now the computed value

$$f = \frac{\chi_1^2/(k-2)}{\chi_2^2/(n-k)},$$

where

$$\chi_1^2 = \sum \frac{y_{i.}^2}{n_i} - \frac{\left(\sum y_{ij}\right)^2}{n} - b^2(n-1)s_x^2,$$

and

$$\chi_2^2 = \sum y_{ij}^2 - \sum \frac{y_{i.}^2}{n_i},$$

is a value of the random variable F, having an F distribution with $k-2$ and $n-k$ degrees of freedom when the $\mu_{Y|x}$ fall on a straight line and therefore may be used to test the hypothesis H_0 for linearity of regression.

When H_0 is true, $\chi_1^2/(k-2)$ and $\chi_2^2/(n-k)$ are independent estimates of σ^2. However, if H_0 is false, $\chi_1^2/(k-2)$ overestimates σ^2. Hence we reject the hypothesis of linearity of regression at the α level of significance when our f value falls in a critical region of size α located in the upper tail of the F distribution.

Example 10.9   Use the data in Table 10.1 to test the hypothesis that the regression is linear.

      Solution.   1. H_0: the regression is linear.
                     2. H_1: the regression is nonlinear.
                     3. Choose a 0.05 level of significance.
                     4. Critical region: $F > 4.46$.
                     5. Computations: From Table 10.1 we have

$$x_1 = 50, \quad n_1 = 2, \quad y_{1.} = 150,$$
$$x_2 = 55, \quad n_2 = 4, \quad y_{2.} = 316,$$
$$x_3 = 65, \quad n_3 = 3, \quad y_{3.} = 269,$$
$$x_4 = 70, \quad n_4 = 3, \quad y_{4.} = 276.$$

Therefore,

$$\chi_1^2 = \left(\frac{150^2}{2} + \frac{316^2}{4} + \frac{269^2}{3} + \frac{276^2}{3} \right) - \frac{1011^2}{12}$$
$$- (0.897)^2(11)(61.174)$$
$$= 8.1506,$$

$$\chi_2^2 = 85,905 - \left(\frac{150^2}{2} + \frac{316^2}{4} + \frac{269^2}{3} + \frac{276^2}{3} \right)$$
$$= 178.6667.$$

Hence

$$f = \frac{8.1506/2}{178.6667/8} = 0.182.$$

6. Conclusion: Accept H_0 and conclude that the data may be fitted by a linear equation.

10.5 Exponential Regression

If a set of data appears to be best represented by a nonlinear regression curve, we must then try to determine the form of the curve and estimate the parameters. Sometimes a scatter diagram indicates that the means $\mu_{Y|x}$ will probably be best represented by an exponential curve of the form

$$\mu_{Y|x} = \gamma \, \delta^x,$$

where γ and δ are parameters to be estimated from the data. Denoting these estimates by c and d, respectively, we can estimate $\mu_{Y|x}$ by $\bar{y}_x$ from the sample regression curve

$$\bar{y}_x = cd^x.$$

Taking logarithms to the base 10, we obtain the regression curve,

$$\log \bar{y}_x = \log c + (\log d)x,$$

and each pair of observations in the sample satisfies the relation

$$\log y_i = \log c + (\log d)x_i + e_i$$
$$= a + bx_i + e_i,$$

where $a = \log c$ and $b = \log d$. Therefore it is possible to find a and b by the formulas in Section 10.2, using the points $(x_i, \log y_i)$, and then determine c and d by taking antilogarithms.

The least-squares procedure for fitting an exponential curve to a set of data is illustrated in the following example.

Example 10.10 The following data represent the enrollments at a small liberal arts college during the past 7 years:

x (years)	1	2	3	4	5	6	7
y (enrollment)	304	341	393	457	548	670	882

Use the method of least squares to estimate a curve of the form $\mu_{Y|x} = \gamma \, \delta^x$ and predict the enrollment 5 years from now.

Solution. The logarithms of the y values are, respectively, 2.483, 2.533, 2.594, 2.660, 2.739, 2.826, and 2.945. To compute the estimates a and b we need the following:

$$\sum_{i=1}^{7} x_i = 28, \qquad \sum_{i=1}^{7} \log y_i = 18.780, \qquad \sum_{i=1}^{7} x_i^2 = 140,$$

$$\sum_{i=1}^{7} x_i \log y_i = 77.237, \qquad \bar{x} = 4, \quad \overline{\log y} = 2.683.$$

Substituting in the formulas for a and b of Section 10.2, we have

$$b = \frac{(7)(77.237) - (28)(18.780)}{(7)(140) - (28)^2} = 0.076$$

and then

$$a = 2.683 - (0.076)(4) = 2.379.$$

Hence

$$c = 10^{2.379} = 239,$$
$$d = 10^{0.076} = 1.19,$$

and our least-squares regression curve is

$$\bar{y}_x = (239)(1.19)^x.$$

Based on this rate of growth, we should expect the enrollment 5 years from now ($x = 12$) to be

$$\bar{y}_{12} = (239)(1.19)^{12} = 1954.$$

10.6 Multiple Regression

We shall now consider the problem of estimating or predicting the value of a dependent variable Y on the basis of a set of measurements taken on several independent variables $X_1, X_2, \ldots, X_r$. As in the case of simple linear regression the values of each dependent variable are selected by the experimenter and remain fixed. We may, for example, wish to estimate the speed of wind as a function of height, temperature, and pressure. The prediction equation is obtained by using a least-squares procedure on data collected at predetermined heights, temperatures, and pressures, to evaluate the necessary coefficients in the assumed equation.

A random sample of size n from the population might be represented by the set $\{(x_{1i}, x_{2i}, \ldots, x_{ri}, y_i); i = 1, 2, \ldots, n\}$. The value y_i is again a value of some random variable Y_i. We shall assume a theoretical equation of the form

$$\mu_{Y|x_1, x_2, \ldots, x_r} = \beta_0 + \beta_1 x_1 + \beta_2 x_2 + \cdots + \beta_r x_r,$$

where $\beta_0, \beta_1, \ldots, \beta_r$ are parameters to be estimated from the data. Denoting these estimates by $b_0, b_1, \ldots, b_r$, respectively, we can write the sample regression equation in the form

$$\bar{y}_{x_1, x_2, \ldots, x_r} = b_0 + b_1 x_1 + b_2 x_2 + \cdots + b_r x_r.$$

In what follows we shall restrict our attention to the case of two independent variables, X_1 and X_2. The results can then be generalized to the case of several independent variables. In dealing with more than two independent variables, a knowledge of matrix theory can facilitate the mathematical manipulations considerably. For a matrix presentation the reader is referred to more advanced texts on the subject.

With only two independent variables the sample regression equation reduces to the form

$$\bar{y}_{x_1, x_2} = b_0 + b_1 x_1 + b_2 x_2,$$

and each set of observations satisfies the relation

$$y_i = b_0 + b_1 x_{1i} + b_2 x_{2i} + e_i.$$

The least-squares estimates of b_0, b_1, and b_2 are obtained by solving the simultaneous linear equations

$$n b_0 + b_1 \sum_{i=1}^{n} x_{1i} + b_2 \sum_{i=1}^{n} x_{2i} = \sum_{i=1}^{n} y_i,$$

$$b_0 \sum_{i=1}^{n} x_{1i} + b_1 \sum_{i=1}^{n} x_{1i}^2 + b_2 \sum_{i=1}^{n} x_{1i} x_{2i} = \sum_{i=1}^{n} x_{1i} y_i,$$

$$b_0 \sum_{i=1}^{n} x_{2i} + b_1 \sum_{i=1}^{n} x_{1i}x_{2i} + b_2 \sum_{i=1}^{n} x_{2i}^2 = \sum_{i=1}^{n} x_{2i}y_i.$$

These equations can be solved for b_1 and b_2 by any appropriate method, such as Cramer's rule, and then b_0 can be obtained from the first of the three equations by observing that

$$b_0 = \bar{y} - b_1\bar{x}_1 - b_2\bar{x}_2.$$

Example 10.11 Given the data

y	2	5	7	8	5
x_1	8	8	6	5	3
x_2	0	1	1	3	4

fit a regression equation of the form $\mu_{Y|x_1,x_2} = \beta_0 + \beta_1 x_1 + \beta_2 x_2$.

Solution. From the given data we find

$$\sum_{i=1}^{5} x_{1i} = 30, \qquad \sum_{i=1}^{5} x_{2i} = 9, \qquad \sum_{i=1}^{5} x_{1i}x_{2i} = 41,$$

$$\sum_{i=1}^{5} x_{1i}^2 = 198, \qquad \sum_{i=1}^{5} x_{2i}^2 = 27, \qquad \sum_{i=1}^{5} y_i = 27,$$

$$\sum_{i=1}^{5} x_{1i}y_i = 153, \qquad \sum_{i=1}^{5} x_{2i}y_i = 56.$$

Inserting these values in the simultaneous equations above, we obtain

$$5b_0 + 30b_1 + 9b_2 = 27,$$
$$30b_0 + 198b_1 + 41b_2 = 153,$$
$$9b_0 + 41b_1 + 27b_2 = 56.$$

The solution of this set of equations yields the unique estimates $b_0 = 4.486, b_1 = -0.039,$ and $b_2 = 0.638$. Therefore our regression equation is

$$\bar{y}_{x_1,x_2} = 4.486 - 0.039x_1 + 0.638x_2.$$

If it is desired to fit a regression equation using successive powers of one or more selected independent variables, then the methods in this section can be applied. For example suppose we wish to estimate the parameters of the regression equation

$$\mu_{Y|x} = \beta_0 + \beta_1 x + \beta_2 x^2.$$

We are actually fitting a polynomial of the form

$$\mu_{Y|x_1,x_2} = \beta_0 + \beta_1 x_1 + \beta_2 x_2,$$

where we set $x_1 = x$ and $x_2 = x^2$ in the simultaneous equations listed above.

When the values of x are equally spaced, a convenient method of curve fitting by means of orthogonal polynomials can be used. The reader may refer to the text *Probability and Statistics for Engineers and Scientists* by Walpole and Myers (see the References) for a full explanation on the use of orthogonal polynomials.

10.7 Correlation

A correlation problem differs from a regression problem in that we are concerned with a measure of the relationship between two or more variables rather than predicting one variable from a knowledge of the independent variables. Recall that the values of the independent variables were fixed in our regression study. Such is not the case now. We define the *linear correlation coefficient* to be a measure of the linear relationship between two random variables X and Y, and denote it by ρ. The variables X and Y are assumed to have a joint probability distribution $f(x, y)$. Measures of the degree of association among several variables are given by the *multiple* and *partial correlation coefficients*. In this text we shall consider only the linear correlation coefficient and refer the reader to the book *Methods of Correlation Analysis* by Ezekiel (see the References) for a detailed discussion of the multiple and partial correlation coefficients.

To estimate a linear correlation coefficient, we first choose a random sample of n pairs of measurements (x, y). By constructing a scatter diagram for the (x, y) values (see Figure 10.4), we are able to draw certain conclusions. Should the points follow closely a straight line of positive slope, we have a high positive correlation between the two variables. On the other hand, if the points follow closely a straight line of negative slope, we have a high negative correlation between the two variables. The correlation between the two variables decreases numerically as the scattering of points from a straight line increases. If the points follow a strictly random pattern as in Figure 10.4(c), we have zero correlation and conclude that no relationship exists between X and Y.

It is important to remember that the correlation coefficient between two variables is a measure of their linear relationship, and a value of $\rho = 0$ implies a lack of linearity and not a lack of association. Hence, if a strong quadratic relationship exists between X and Y as indicated in Figure 10.4(d), we will still obtain a zero correlation to indicate a nonlinear relationship.

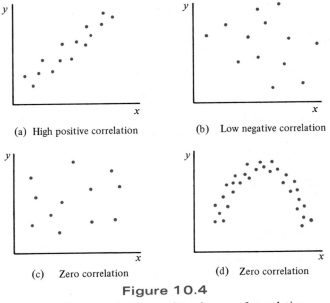

(a) High positive correlation   (b) Low negative correlation

(c) Zero correlation   (d) Zero correlation

Figure 10.4

Scatter diagrams showing various degrees of correlation.

The most commonly used measure of linear correlation between two variables is called the *Pearson product-moment correlation coefficient*, or simply the *sample correlation coefficient*.

CORRELATION COEFFICIENT *The measure ρ of linear relationship between two variables X and Y is estimated by the* sample correlation coefficient r, *where*

$$r = \frac{n \sum_{i=1}^{n} x_i y_i - \left(\sum_{i=1}^{n} x_i \right)\left(\sum_{i=1}^{n} y_i \right)}{\sqrt{\left[n \sum_{i=1}^{n} x_i^2 - \left(\sum_{i=1}^{n} x_i \right)^2 \right]\left[n \sum_{i=1}^{n} y_i^2 - \left(\sum_{i=1}^{n} y_i \right)^2 \right]}} = b\frac{s_x}{s_y}.$$

In Section 10.2 we saw that

$$\text{SSE} = (n - 1)(s_y^2 - b^2 s_x^2).$$

Dividing both sides of this equation by $(n - 1)s_y^2$, we obtain the relation

$$r^2 = 1 - \frac{\text{SSE}}{(n - 1)s_y^2},$$

from which we conclude that r^2 must be between zero and 1. Consequently r must range from -1 to $+1$. A value of -1 or $+1$ will occur when SSE $= 0$, but this is the case where all the points lie in a straight line. Hence a perfect relationship exists between X and Y when $r = \pm 1$. If r is close to zero, the linear relationship between X and Y is weak.

One must be careful in interpreting r beyond what has been stated above. For example, values of r equal to 0.3 and 0.6 only mean that we have two positive correlations, one somewhat stronger than the other. It is wrong to conclude that $r = 0.6$ indicates a relationship twice as good as that indicated by the value $r = 0.3$. On the other hand, if we consider r^2, then $100 \times r^2\%$ of the variation in the values of the variable Y may be accounted for by the linear relationship with the variable X. Thus a correlation of 0.6 means that 36% of the variation of the random variable Y is accounted for by differences in the variable X.

Example 10.12   Compute and interpret the correlation coefficient for the following data:

x (height)	12	10	14	11	12	9
y (weight)	18	17	23	19	20	15

Solution.   From the data we find

$$\sum_{i=1}^{6} x_i = 68, \qquad \sum_{i=1}^{6} y_i = 112, \qquad \sum_{i=1}^{6} x_i y_i = 1292,$$

$$\sum_{i=1}^{6} x_i^2 = 786, \qquad \sum_{i=1}^{6} y_i^2 = 2128.$$

Therefore,

$$r = \frac{(6)(1292) - (68)(112)}{\sqrt{[(6)(786) - (68)^2][(6)(2128) - (112)^2]}}$$

$$= 0.947.$$

A correlation coefficient of 0.947 indicates a very good linear relationship between X and Y. Since $r^2 = 0.90$, we can say that 90% of the variation in the values of Y is accounted for by a linear relationship with X.

A test of the hypothesis

$$H_0: \rho = \rho_0,$$
$$H_1: \rho \neq \rho_0$$

is easily conducted from the sample information by using the quantity

$$\frac{1}{2} \ln \left(\frac{1 + r}{1 - r}\right),$$

which is a value of a random variable that follows approximately the normal distribution with mean $(\frac{1}{2}) \ln [(1 + \rho)/(1 - \rho)]$ and variance $1/(n - 3)$. Thus the test procedure is to compute

$$z = \frac{\sqrt{n - 3}}{2} \left[\ln \left(\frac{1 + r}{1 - r}\right) - \ln \left(\frac{1 + \rho_0}{1 - \rho_0}\right) \right]$$

$$= \frac{\sqrt{n - 3}}{2} \ln \left[\frac{(1 + r)(1 - \rho_0)}{(1 - r)(1 + \rho_0)} \right]$$

and compare to the critical points of the standard normal distribution.

Example 10.13 For the data of Example 10.12 test the null hypothesis that there is no linear relationship between the variables. Use a 0.05 level of significance.

Solution.   1. $H_0: \rho = 0$.
2. $H_1: \rho \neq 0$.
3. $\alpha = 0.05$.
4. Critical region: $Z < -1.96$ and $Z > 1.96$.
5. Computations:

$$z = \frac{\sqrt{3}}{2} \ln \left(\frac{1.947}{0.053}\right) = 3.12.$$

6. Conclusion: Reject the hypothesis of no linear relationship.

EXERCISES

1. Consider the following data:

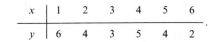

x	1	2	3	4	5	6
y	6	4	3	5	4	2

(a) Find the equation of the regression line.
(b) Graph the line on a scatter diagram.
(c) Find a point estimate of $\mu_{Y|4}$.

2. The grades of a class of 9 students on a midterm report (x) and on the final examination (y) are as follows:

x	77	50	71	72	81	94	96	99	67
y	82	66	78	34	47	85	99	99	68

.

(a) Find the equation of the regression line.
(b) Estimate the final examination grade of a student who received a grade of 85 on the midterm report but was ill at the time of the final examination.

3. A study was made on the amount of converted sugar in a certain process at various temperatures. The data were coded and recorded as follows:

Temperature, x	Converted Sugar, y
1.0	8.1
1.1	7.8
1.2	8.5
1.3	9.8
1.4	9.5
1.5	8.9
1.6	8.6
1.7	10.2
1.8	9.3
1.9	9.2
2.0	10.5

(a) Estimate the linear regression line.
(b) Estimate the amount of converted sugar produced when the coded temperature is 1.75.

4. The amounts of a chemical compound y, which dissolved in 100 grams of water at various temperatures, x, were recorded as follows:

x, °C	y, grams		
0	8	6	8
15	12	10	14
30	25	21	24
45	31	33	28
60	44	39	42
75	48	51	44

(a) Find the equation of the regression line.
(b) Graph the line on a scatter diagram.
(c) Estimate the amount of chemical that will dissolve in 100 grams of water at 50°C.

5. A mathematics placement test is given to all entering freshmen at a small college. A student who receives a grade below 35 is denied admission to the regular mathematics course and placed in a remedial class. The placement test scores and the final grades for 20 students who took the regular course were recorded as follows:

Placement Test	Course Grade	Placement Test	Course Grade
50	53	90	54
35	41	80	91
35	61	60	48
40	56	60	71
55	68	60	71
65	36	40	47
35	11	55	53
60	70	50	68
90	79	65	57
35	59	50	79

(a) Plot a scatter diagram.

(b) Find the equation of the regression line to predict course grades from placement test scores.

(c) Graph the line on the scatter diagram.

(d) If 60 is the minimum passing grade, below which placement test score should students in the future be denied admission to this course?

6. A study was made by a retail merchant to determine the relation between weekly advertising expenditures and sales. The following data were recorded:

Advertising Costs, $	Sales, $
40	385
20	400
25	395
20	365
30	475
50	440
40	490
20	420
50	560
40	525
25	480
50	510

(a) Plot a scatter diagram.

(b) Find the equation of the regression line to predict weekly sales from advertising expenditures.

7. Test the hypothesis that $\beta = 0$ in Exercise 1, using a 0.05 level of significance, against the alternative that $\beta \neq 0$.

8. Test the hypothesis that $\alpha = 6$ in Exercise 1, using a 0.01 level of significance, against the alternative that $\alpha \neq 6$.

9. Construct 95% confidence intervals for α and β in Exercise 2.

10. Construct 95% confidence intervals for α and β in Exercise 3.

11. Construct 99% confidence intervals for α and β in Exercise 4.

12. Construct a 95% confidence interval for α in Exercise 5.

13. Construct a 90% confidence interval for β in Exercise 6.

14. Construct a 95% confidence interval for $\mu_{Y|80}$ in Exercise 2.

15. Construct a 95% confidence interval for the amount of converted sugar corresponding to $x = 1.6$ in Exercise 3.

16. Construct 99% confidence intervals for $\mu_{Y|50}$ and y_{50} in Exercise 4.

17. Construct 95% confidence intervals for $\mu_{Y|35}$ and y_{35} in Exercise 5.

18. Construct 90% confidence intervals for $\mu_{Y|45}$ and y_{45} in Exercise 6.

19. Test for linearity of regression in Exercise 4.

20. Test for linearity of regression in Exercise 5.

21. The following data are the selling prices, y, of a certain make and model of used car x years old:

x, Years	y, $
1	2350
2	1695
2	1750
3	1395
5	985
5	895

(a) Fit an exponential curve of the form $\mu_{Y|x} = \gamma\delta^x$.
(b) Estimate the selling price of such a car when it is 4 years old.

22. The pressure P of a gas corresponding to various volumes V was recorded as follows:

V, in.^{3}	50	60	70	90	100
P, lb/in.^{2}	64.7	51.3	40.5	25.9	7.8

The ideal gas law is given by the equation $PV^\gamma = C$, where γ and C are constants.
(a) Find the least-squares estimates of γ and C from the given data.
(b) Estimate P when $V = 80$ cubic inches.

23. The following data resulted from 15 experimental runs made on four independent variables and a single response y:

y	x_1	x_2
14.8	11.5	6.3
12.1	14.3	7.4
19.0	9.4	5.9
14.5	15.2	8.7
16.6	8.8	9.1
17.2	9.8	5.6
17.5	11.2	6.8
14.1	10.9	7.4
13.8	14.7	8.2
14.7	15.1	9.2
17.7	8.7	4.7
17.0	8.6	5.5
17.6	9.3	6.6
16.3	10.8	8.7
18.2	11.9	5.4

(a) Fit a regression equation of the form $\mu_{Y|x_1, x_2} = \beta_0 + \beta_1 x_1 + \beta_2 x_2$.

(b) Predict the response y when $x_1 = 10.3$ and $x_2 = 5.8$.

24. The following data were collected to determine the relationship between the college entrance aptitude examination score, high school rank in class, and grade-point average at the end of the freshman year.

Grade-Point Average, y	Aptitude Score, x_1	Decile Rank, x_2
1.93	565	3
2.55	525	2
1.72	477	1
2.48	555	1
2.87	502	1
1.87	469	3
1.34	517	4
3.03	555	1
2.54	576	2
2.34	559	2
1.40	574	8
1.45	578	4
1.72	548	8
3.80	656	1
2.13	688	5
1.81	465	6
2.33	661	1
2.53	477	1
2.04	490	2
3.20	524	2

(a) Fit a regression equation of the form $\mu_{Y|x_1, x_2} = \beta_0 + \beta_1 x_1 + \beta_2 x_2$.

(b) Predict the grade-point average for an entering freshman who has an aptitude score of 575 and ranks in the third decile of his graduating class.

25. Given the data

x	0	1	2	3	4	5	6	7	8	9
y	9.1	7.3	3.2	4.6	4.8	2.9	5.7	7.1	8.8	10.2

(a) Fit a regression curve of the form $\mu_{Y|x} = \beta_0 + \beta_1 x + \beta_2 x^2$.

(b) Estimate Y when $x = 2$.

26. An experiment was conducted on a new model of a particular make of automobile to determine the stopping distance at various speeds. The following data were recorded:

Speed v, miles/hr	20	30	40	50	60	70
Stopping distance d, ft	51	86	135	202	290	392

(a) Fit a regression curve of the form $\mu_{D|v} = \beta_0 + \beta_1 v + \beta_2 v^2$.

(b) Estimate the stopping distance when the car is traveling at 45 miles/hr.

27. Compute and interpret the correlation coefficient for the following data:

x	4	5	9	14	18	22	24
y	16	22	11	16	7	3	17

28. Compute and interpret the correlation coefficient for the aptitude scores and grade-point averages in Exercise 24.

29. Compute and interpret the correlation coefficient for the following grades of 6 students selected at random:

Math Grade	70	92	80	74	65	83
English Grade	74	84	63	87	78	90

30. Test the hypothesis that $\rho = -0.8$ in Exercise 27, using a 0.05 level of significance.

31. Test the hypothesis that $\rho = 0$ in Exercise 29, using a 0.05 level of significance.

ANALYSIS OF VARIANCE CHAPTER **11**

11.1 Introduction

Consider an experiment in which three varieties of wheat are planted on several acre lots and their yields per acre recorded. We are interested in testing the null hypothesis that the three varieties of wheat produce equal yields on the average. To test whether a particular two of the three varieties are significantly different, we use the appropriate tests described in Chapter 9. However, to test for the equality of several means simultaneously, a new technique, called the *analysis of variance*, is required.

The analysis of variance is a method for splitting the total variation of our data into meaningful components that measure different sources of variation. In our experiment we obtain two components, the first measuring the variation due to experimental error and the second measuring variation due to experimental error plus any variation due to the different varieties of wheat. If the null hypothesis is true and the three varieties of wheat produce equal yields on the average, then both components provide independent estimates of the experimental error. Hence we base our test on a comparison of these two components by means of the F distribution.

The yields may vary in part due to different types of soil on which the wheat is planted. This source of variation may be eliminated if the experiment is planned so that all plots of ground have the same soil composition. Failure to eliminate large sources of variation by means of controlled experimentation will result in an inflated estimate of the experimental error and thereby increase the probability of committing a type II error.

Let us now complicate the experiment by considering the yields of three varieties of wheat using four different kinds of fertilizer. We are now interested in testing whether the variation in the yields is caused by differences in the varieties of wheat, differences in the types of fertilizer, or perhaps differences in both. In this case, the analysis of variance provides a means of partitioning the total variation of the yields into three components, the first measuring experimental error only, the second measuring experimental error plus any variation due to the different varieties of wheat, and the third measuring experimental error plus any variation due to the different fertilizers. Hence a comparison of the second component with the first will provide a test of the hypothesis that the varieties of wheat produce equal yields on the average. Similarly we can test for no differences in the mean yields due to different fertilizers by comparing the third and first components.

The classification of observations on the basis of a single criterion, such as variety of wheat, is called a *one-way classification*. If the observations are classified according to two criteria, such as variety of wheat and type of fertilizer, we have what is called a *two-way classification*. The analysis-of-variance procedure will be considered in detail for each of these classifications in Sections 11.2, 11.5, and 11.6. Methods for the analysis of variance of multiway classifications are not difficult but are beyond the scope of this text.

11.2 One-Way Classification

Random samples of size n are selected from each of k populations. It will be assumed that the k populations are independent and normally distributed with means $\mu_1, \mu_2, \ldots, \mu_k$ and common variance σ^2. We wish to derive appropriate methods for testing the hypothesis

$$H_0: \mu_1 = \mu_2 = \cdots = \mu_k,$$

H_1: At least two of the means are not equal.

Let x_{ij} denote the jth observation from the ith population and arrange the data as in Table 11.1. Here T_i is the total of all observations in the sample from the ith population, $\bar{x}_{i.}$ the mean of all observations in the sample from the ith population, $T..$ the total of all nk observations, and $\bar{x}..$ the mean of all nk observations. Each observation may be written in the form

$$x_{ij} = \mu_i + \varepsilon_{ij},$$

where ε_{ij} measures the deviation of the jth observation of the ith sample from the corresponding population mean. An alternative and preferred form of this equation is obtained by substituting $\mu_i = \mu + \alpha_i$, where μ is defined to be the mean of all the μ_i; that is,

Table 11.1

k Random Samples

			Population				
	1	2	$\cdots$	i	$\cdots$	k	
	x_{11}	x_{21}	$\cdots$	x_{i1}	$\cdots$	x_{k1}	
	x_{12}	x_{22}	$\cdots$	x_{i2}	$\cdots$	x_{k2}	
	.	.		.		.	
	.	.		.		.	
	.	.		.		.	
	x_{1n}	x_{2n}	$\cdots$	x_{in}	$\cdots$	x_{kn}	
Total	$T_{1.}$	$T_{2.}$	$\cdots$	$T_{i.}$	$\cdots$	$T_{k.}$	$T_{..}$
Mean	$\bar{x}_{1.}$	$\bar{x}_{2.}$	$\cdots$	$\bar{x}_{i.}$	$\cdots$	$\bar{x}_{k.}$	$\bar{x}_{..}$

$$\mu = \frac{\sum_{i=1}^{k} \mu_i}{k}.$$

Hence we may write

$$x_{ij} = \mu + \alpha_i + \varepsilon_{ij},$$

subject to the restriction that $\sum_{i=1}^{k} \alpha_i = 0$. It is customary to refer to α_i as the *effect* of the ith population.

The null hypothesis that the k population means are equal against the alternative that at least two of the means are unequal may now be replaced by the equivalent hypothesis

$$H_0: \alpha_1 = \alpha_2 = \cdots = \alpha_k = 0,$$

H_1: At least one of the α_i is not equal to zero.

Our test will be based on a comparison of two independent estimates of the common population variance σ^2. These estimates will be obtained by splitting the total variability of our data into two components.

The variance of all the observations grouped into a single sample of size nk is given by the formula

$$s^2 = \frac{\sum_{i=1}^{k} \sum_{j=1}^{n} (x_{ij} - \bar{x}..)^2}{nk - 1}.$$

The double summation means that we sum all possible terms that are obtained by allowing i to assume values from 1 to k for each value of j from 1 to n. The numerator of s^2, called the *total sum of squares*, measures the total variability of our data. It may be partitioned by means of the following identity.

THEOREM 11.1 (SUM-OF-SQUARES IDENTITY)

$$\sum_{i=1}^{k} \sum_{j=1}^{n} (x_{ij} - \bar{x}..)^2 = n \sum_{i=1}^{k} (\bar{x}_{i.} - \bar{x}..)^2 + \sum_{i=1}^{k} \sum_{j=1}^{n} (x_{ij} - \bar{x}_{i.})^2.$$

Proof.
$$\sum_{i=1}^{k} \sum_{j=1}^{n} (x_{ij} - \bar{x}..)^2 = \sum_{i=1}^{k} \sum_{j=1}^{n} [(\bar{x}_{i.} - \bar{x}..) + (x_{ij} - \bar{x}_{i.})]^2$$

$$= \sum_{i=1}^{k} \sum_{j=1}^{n} [(\bar{x}_{i.} - \bar{x}..)^2$$
$$+ 2(\bar{x}_{i.} - \bar{x}..)(x_{ij} - \bar{x}_{i.})$$
$$+ (x_{ij} - \bar{x}_{i.})^2]$$

$$= \sum_{i=1}^{k} \sum_{j=1}^{n} (\bar{x}_{i.} - \bar{x}..)^2$$
$$+ 2 \sum_{i=1}^{k} \sum_{j=1}^{n} (\bar{x}_{i.} - \bar{x}..)(x_{ij} - \bar{x}_{i.})$$
$$+ \sum_{i=1}^{k} \sum_{j=1}^{n} (x_{ij} - \bar{x}_{i.})^2.$$

The middle term sums to zero, since

$$\sum_{j=1}^{n} (x_{ij} - \bar{x}_{i.}) = \sum_{j=1}^{n} x_{ij} - n\bar{x}_{i.}$$

$$= \sum_{j=1}^{n} x_{ij} - n\left(\frac{\sum_{j=1}^{n} x_{ij}}{n}\right) = 0.$$

The first sum does not have j as a subscript and therefore may be written

$$\sum_{i=1}^{k} \sum_{j=1}^{n} (\bar{x}_{i.} - \bar{x}..)^2 = n \sum_{i=1}^{k} (\bar{x}_{i.} - \bar{x}..)^2.$$

Hence

$$\sum_{i=1}^{k} \sum_{j=1}^{n} (x_{ij} - \bar{x}..)^2 = n \sum_{i=1}^{k} (\bar{x}_{i.} - \bar{x}..)^2$$
$$+ \sum_{i=1}^{k} \sum_{j=1}^{n} (x_{ij} - \bar{x}_{i.})^2.$$

It will be convenient in what follows to identify the terms of the sum-of-squares identity by the following notation:

$$\text{SST} = \sum_{i=1}^{k} \sum_{j=1}^{n} (x_{ij} - \bar{x}..)^2 = \text{total sum of squares,}$$

$$\text{SSC} = n \sum_{i=1}^{k} (\bar{x}_{i.} - \bar{x}..)^2 = \text{sum of squares for column means,}$$

$$\text{SSE} = \sum_{i=1}^{k} \sum_{j=1}^{n} (x_{ij} - \bar{x}_{i.})^2 = \text{error sum of squares.}$$

The sum-of-squares identity can then be represented symbolically by the equation

$$\text{SST} = \text{SSC} + \text{SSE.}$$

Many authors refer to the sum of squares for column means as the treatment sum of squares. This terminology is derived from the fact that the k different populations are often classified according to different treatments. Thus the observations x_{ij} ($j = 1, 2, \ldots, n$), represent the n measurements corresponding to the ith treatment. Today the term "treatment" is used more generally to refer to the various classifications, whether they be different analysts, different fertilizers, different manufacturers, or different regions of the country.

One estimate of σ^2, based on $k - 1$ degrees of freedom, is given by

$$s_1^2 = \frac{\text{SSC}}{k - 1}.$$

If H_0 is true, s_1^2 is an unbiased estimate of σ^2. However, if H_1 is true, SSC will have a larger numerical value and s_1^2 overestimates σ^2. A second and independent estimate of σ^2, based on $k(n - 1)$ degrees of freedom is given by

$$s_2^2 = \frac{\text{SSE}}{k(n - 1)}.$$

The estimate s_2^2 is unbiased regardless of the truth or falsity of the null hypothesis. We have already seen that the variance of our grouped data, with $nk - 1$ degrees of freedom, is

$$s^2 = \frac{\text{SST}}{nk - 1},$$

which is an unbiased estimate of σ^2 when H_0 is true. It is important to note that the sum-of-squares identity has not only partitioned the total variability of the data but also the total number of degrees of freedom; that is,

$$nk - 1 = k - 1 + k(n - 1).$$

When H_0 is true, the ratio

$$f = \frac{s_1^2}{s_2^2}$$

is a value of the random variable F having the F distribution with $k - 1$ and $k(n - 1)$ degrees of freedom. Since s_1^2 overestimates σ^2 when H_0 is false, we have a one-tailed test with the critical region entirely in the right tail of the distribution. The null hypothesis H_0 is rejected at the α level of significance when

$$f > f_\alpha[k - 1, k(n - 1)].$$

In practice one usually computes SST and SSC first and then, making use of the sum-of-squares identity, obtains

$$SSE = SST - SSC.$$

The previously defined formulas for SST and SSC are not in the best computational form. Equivalent and preferred formulas are given by

$$SST = \sum_{i=1}^{k} \sum_{j=1}^{n} x_{ij}^2 - \frac{T_{..}^2}{nk},$$

$$SSC = \frac{\sum_{i=1}^{k} T_{i.}^2}{n} - \frac{T_{..}^2}{nk}.$$

The computations in an analysis-of-variance problem are usually summarized in tabular form, as shown in Table 11.2.

Table 11.2

Analysis of Variance for the One-Way Classification

Source of Variation	Sum of Squares	Degrees of Freedom	Mean Square	Computed f
Column means	SSC	$k - 1$	$s_1^2 = \dfrac{SSC}{k - 1}$	$\dfrac{s_1^2}{s_2^2}$
Error	SSE	$k(n - 1)$	$s_2^2 = \dfrac{SSE}{k(n - 1)}$	
Total	SST	$nk - 1$		

Example 11.1 The hypothetical data in Table 11.3 represent 5 random samples, each of size 5, from independent normal distributions with means $\mu_1, \mu_2, \mu_3, \mu_4, \mu_5$ and common variance σ^2. Test the hypothesis at the 0.05 level of significance that $\mu_1 = \mu_2 = \mu_3 = \mu_4 = \mu_5$.

Table 11.3
Hypothetical Samples

	Sample					
	A	*B*	*C*	*D*	*E*	
	5	9	3	2	7	
	4	7	5	3	6	
	8	8	2	4	9	
	6	6	3	1	4	
	3	9	7	4	7	
Total	26	39	20	14	33	132
Mean	5.2	7.8	4.0	2.8	6.6	5.28

Solution.
1. $H_0: \mu_1 = \mu_2 = \mu_3 = \mu_4 = \mu_5$.
2. H_1: At least two of the means are not equal.
3. $\alpha = 0.05$.
4. Critical region: $F > 2.87$.
5. Computations:

$$\text{SST} = 5^2 + 4^2 + \cdots + 7^2 - \frac{132^2}{25}$$

$$= 834 - 696.960 = 137.040,$$

$$\text{SSC} = \frac{26^2 + 39^2 + \cdots + 33^2}{5} - \frac{132^2}{25}$$

$$= 776.400 - 696.960 = 79.440,$$

$$\text{SSE} = 137.040 - 79.440 = 57.60.$$

These results and the remaining computations are exhibited in Table 11.4.

Table 11.4
Analysis of Variance for the Data for Table 11.3

Source of Variation	Sum of Squares	Degrees of Freedom	Mean Square	Computed *f*
Column means	79.440	4	19.860	6.90
Error	57.600	20	2.880	
Total	137.040	24		

6. Conclusion: Reject the hypothesis and conclude that the samples came from different normal populations.

In experimental work one often loses some of the desired observations. For example an experiment might be conducted to determine if college students obtain different grades on the average for classes meeting at different times of the day. Because of dropouts during the semester, it is entirely possible to conclude the experiment with unequal numbers of students in the various sections. The previous analysis for equal sample size will still be valid by slightly modifying the sum-of-squares formulas. We now assume the k random samples to be of size $n_1, n_2, \ldots, n_k$, respectively, with $N = \sum\limits_{i=1}^{k} n_i$. The computational formulas for SST and SSC are now given by

$$\text{SST} = \sum_{i=1}^{k} \sum_{j=1}^{n_i} x_{ij}^2 - \frac{T_{..}^2}{N},$$

$$\text{SSC} = \sum_{i=1}^{k} \frac{T_{i.}^2}{n_i} - \frac{T_{..}^2}{N}.$$

As before, we find SSE by subtraction. The degrees of freedom are partitioned in the same way: $N - 1$ for SST, $k - 1$ for SSC, and $N - 1 - (k - 1) = N - k$ for SSE.

Example 11.2 Test the hypothesis at the 0.05 level of significance that $\mu_1 = \mu_2 = \mu_3$ for the data in Table 11.5.

Table 11.5

Hypothetical Samples

	Sample			
	A	B	C	
	4	5	8	
	7	1	6	
	6	3	8	
	6	5	9	
		3	5	
		4		
Total	23	21	36	80

Solution.   1. $H_0: \mu_1 = \mu_2 = \mu_3$.
2. H_1: At least two of the means are not equal.
3. $\alpha = 0.05$.
4. Critical region: $F > 3.89$.
5. Computations:

$$\text{SST} = 4^2 + 7^2 + \cdots + 5^2 - \frac{80^2}{15} = 65.333,$$

$$\text{SSC} = \frac{23^2}{4} + \frac{21^2}{6} + \frac{36^2}{5} - \frac{80^2}{15} = 38.283,$$

$$\text{SSE} = 65.333 - 38.283 = 27.050.$$

These results and the remaining computations are exhibited in Table 11.6.

Table 11.6

Analysis of Variance for the Data of Table 11.5

Source of Variation	Sum of Squares	Degrees of Freedom	Mean Square	Computed f
Column means	38.283	2	19.142	8.49
Error	27.050	12	2.254	
Total	65.333	14		

6. Conclusion: Reject H_0 and conclude that the population means are not all equal.

In concluding our discussion on the analysis of variance for the one-way classification, we state the advantages in choosing equal sample sizes over the choice of unequal sample sizes. The first advantage is that the f ratio is insensitive to departures from the assumption of equal variances for the k populations when the samples are of equal size. Second, the choice of equal sample size minimizes the probability of committing a type II error. Finally the computation of SSC is simplified if the sample sizes are equal.

11.3 Test for the Equality of Several Variances

We have already stated in Section 11.2 that the f ratio obtained from the analysis-of-variance procedure is insensitive to departures from the assumption of equal variances for the k populations when the samples are of equal size. This is not the case, however, for unequal sample sizes. Consequently, when an experiment results in unequal numbers of observations in the various samples, one may wish to test the hypothesis

$$H_0: \sigma_1^2 = \sigma_2^2 = \cdots = \sigma_k^2$$

against the alternative

$$H_1: \text{The variances are not all equal.}$$

The test most often used, called *Bartlett's test*, is based on a statistic whose sampling distribution is approximated very closely by the chi-square distribution when the k random samples are drawn from independent normal populations.

First we compute the k sample variances $s_1^2, s_2^2, \ldots, s_k^2$ from samples of size $n_1, n_2, \ldots, n_k$, with $\sum_{i=1}^{k} n_i = N$. Second, combine the sample variances to give the pooled estimate

$$s_p^2 = \frac{\sum_{i=1}^{k} (n_i - 1)s_i^2}{N - k}.$$

Now

$$b = 2.3026 \frac{q}{h},$$

where

$$q = (N - k)\log s_p^2 - \sum_{i=1}^{k} (n_i - 1)\log s_i^2,$$

$$h = 1 + \frac{1}{3(k - 1)} \left[\sum_{i=1}^{k} \frac{1}{n_i - 1} - \frac{1}{N - k} \right],$$

is a value of the random variable B having approximately the chi-square distribution with $k - 1$ degrees of freedom.

The quantity q is large when the sample variances differ greatly and is equal to zero when all the sample variances are equal. Hence we reject H_0 at the α level of significance only when $b > \chi_\alpha^2$.

Example 11.3 Use Bartlett's test to test the hypothesis that the variances of the three populations in Example 11.2 are equal.

Solution.
1. $H_0: \sigma_1^2 = \sigma_2^2 = \sigma_3^2$.
2. $H_1:$ The variances are not all equal.
3. $\alpha = 0.05$.
4. Critical region: $B > 5.991$.
5. Computations: Referring to Example 11.2, we have $n_1 = 4, n_2 = 6, n_3 = 5, N = 15, k = 3$. First compute

$$s_1^2 = 1.583, \qquad s_2^2 = 2.300, \qquad s_3^2 = 2.700,$$

and then

$$s_p^2 = \frac{(3)(1.583) + (5)(2.300) + (4)(2.700)}{12} = 2.254.$$

Now

$$q = 12 \log 2.254 - (3 \log 1.583 + 5 \log 2.300$$
$$+ 4 \log 2.700)$$
$$= (12)(0.3530) - [(3)(0.1995) + (5)(0.3617)$$
$$+ (4)(0.4314)]$$
$$= 0.1034,$$
$$h = 1 + \tfrac{1}{6}(\tfrac{1}{3} + \tfrac{1}{5} + \tfrac{1}{4} - \tfrac{1}{12}) = 1.1167.$$

Hence

$$b = \frac{(2.3026)(0.1034)}{1.1167} = 0.213.$$

6. Conclusion: Accept the hypothesis and conclude that the variances of the three populations are equal.

11.4 Multiple-Range Test

The analysis of variance is a powerful procedure for testing the homogeneity of a set of means. However, if we reject the null hypothesis and accept the stated alternative—that the means are not all equal—we still do not know which of the population means are equal and which are different. Several tests are available that separate a set of significantly different means into subsets of homogeneous means. The test that we shall study in this section is called *Duncan's multiple-range test.*

Let us assume that the analysis-of-variance procedure has led to a rejection of the null hypothesis of equal population means. It is also assumed that the k random samples are all of equal size n. The range of any subset of p sample means must exceed a certain value before we consider any of the p population means to be different. This value is called the *least significant range* for the p means, and is denoted by R_p, where

$$R_p = r_p \cdot s_{\bar{x}} = r_p \sqrt{\frac{s^2}{n}}.$$

The sample variance s^2, which is an estimate of the common variance σ^2, is obtained from the error mean square in the analysis of variance. The values of the quantity r_p, called the *least significant studentized range*, depend on the desired level of significance and the number of degrees of freedom of the error mean square. These values may be obtained from Table A.11 for $p = 2, 3, \ldots,$ 10 means.

To illustrate the multiple-range test procedure, let us consider the data in Table 11.3. First we arrange the sample means in increasing order of magnitude:

$\bar{x}_4$	$\bar{x}_3$	$\bar{x}_1$	$\bar{x}_5$	$\bar{x}_2$
2.8	4.0	5.2	6.6	7.8

Next we obtain $s^2 = 2.880$ with 20 degrees of freedom from the error mean square of the analysis of variance in Table 11.4. Let $\alpha = 0.05$. Then the values of r_p are obtained from Table A.11, with $v = 20$ degrees of freedom, for $p = 2, 3, 4,$ and 5. Finally we obtain R_p by multiplying each r_p by $\sqrt{s^2/n} = \sqrt{2.880/5} = 0.76$. The results of these computations are summarized as follows:

p	2	3	4	5
r_p	2.950	3.097	3.190	3.255
R_p	2.24	2.35	2.42	2.47

Comparing these least significant ranges with the differences in ordered means, we arrive at the following conclusions:

1. Since $\bar{x}_2 - \bar{x}_5 = 1.2 < R_2 = 2.24$, we conclude that $\bar{x}_2$ and $\bar{x}_5$ are not significantly different.
2. Since $\bar{x}_2 - \bar{x}_1 = 2.6 > R_3 = 2.35$, we conclude that $\bar{x}_2$ is significantly larger than $\bar{x}_1$ and therefore $\mu_2 > \mu_1$. It also follows that $\mu_2 > \mu_3$ and $\mu_2 > \mu_4$.
3. Since $\bar{x}_5 - \bar{x}_1 = 1.4 < R_2 = 2.24$, we conclude that $\bar{x}_5$ and $\bar{x}_1$ are not significantly different.
4. Since $\bar{x}_5 - \bar{x}_3 = 2.6 > R_3 = 2.35$, we conclude that $\bar{x}_5$ is significantly larger than $\bar{x}_3$ and therefore $\mu_5 > \mu_3$. Also, $\mu_5 > \mu_4$.
5. Since $\bar{x}_1 - \bar{x}_3 = 1.2 < R_2 = 2.24$, we conclude that $\bar{x}_1$ and $\bar{x}_3$ are not significantly different.
6. Since $\bar{x}_1 - \bar{x}_4 = 2.4 > R_3 = 2.35$, we conclude that $\bar{x}_1$ is significantly larger than $\bar{x}_4$ and therefore $\mu_1 > \mu_4$.
7. Since $\bar{x}_3 - \bar{x}_4 = 1.2 < R_2 = 2.24$, we conclude that $\bar{x}_3$ and $\bar{x}_4$ are not significantly different.

It is customary to summarize the above conclusions by drawing a line under any subset of adjacent means that are not significantly different. Thus we have

$\bar{x}_4$	$\bar{x}_3$	$\bar{x}_1$	$\bar{x}_5$	$\bar{x}_2$
2.8	4.0	5.2	6.6	7.8

One can immediately observe from this manner of presentation that $\mu_2 > \mu_1$, $\mu_2 > \mu_3, \mu_2 > \mu_4, \mu_5 > \mu_3, \mu_5 > \mu_4, \mu_1 > \mu_4$, while all other pairs of population means are not considered significantly different.

11.5 Two-Way Classification, Single Observation per Cell

A set of observations may be classified according to two criteria at once by means of a rectangular array in which the columns represent one criterion of classification and the rows represent a second criterion of classification. For example the rectangular array of observations might be the yields of the three varieties of wheat, discussed in Section 11.1, using four different kinds of fertilizer. The yields are given in Table 11.7. Each treatment combination defines a cell in our array for which we have obtained a single observation.

Table 11.7

Yields of Wheat in Bushels per Acre

Fertilizer Treatment	v_1	v_2	v_3	Total
	\multicolumn{3}{Varieties of Wheat}	
t_1	64	72	74	210
t_2	55	57	47	159
t_3	59	66	58	183
t_4	58	57	53	168
Total	236	252	232	720

Table 11.8

Two-Way Classification with One Observation per Cell

Row	\multicolumn{6}{Columns}	Total	Mean
	1	2	$\cdots$	j	$\cdots$	c		
1	x_{11}	x_{12}	$\cdots$	x_{1j}	$\cdots$	x_{1c}	$T_{1.}$	$\bar{x}_{1.}$
2	x_{21}	x_{22}	$\cdots$	x_{2j}	$\cdots$	x_{2c}	$T_{2.}$	$\bar{x}_{2.}$
.	.	.		.		.	.	.
.	.	.		.		.	.	.
.	.	.		.		.	.	.
i	x_{i1}	x_{i2}	$\cdots$	x_{ij}	$\cdots$	x_{ic}	$T_{i.}$	$\bar{x}_{i.}$
.	.	.		.		.	.	.
.	.	.		.		.	.	.
.	.	.		.		.	.	.
r	x_{r1}	x_{r2}	$\cdots$	x_{rj}	$\cdots$	x_{rc}	$T_{r.}$	$\bar{x}_{r.}$
Total	$T_{.1}$	$T_{.2}$	$\cdots$	$T_{.j}$	$\cdots$	$T_{.c}$	$T_{..}$	
Mean	$\bar{x}_{.1}$	$\bar{x}_{.2}$	$\cdots$	$\bar{x}_{.j}$	$\cdots$	$\bar{x}_{.c}$		$\bar{x}_{..}$

In this section we shall derive formulas that will enable us to test whether the variation in our yields is caused by the different varieties of wheat, different kinds of fertilizers, or differences in both.

We shall now generalize and consider a rectangular array consisting of r rows and c columns as in Table 11.8, where x_{ij} denotes an observation in the ith row and jth column. It will be assumed that the x_{ij} are values of independent random variables having normal distributions with means μ_{ij} and the common variance σ^2. In Table 11.8 $T_{i.}$ and $\bar{x}_{i.}$ are the total and mean, respectively, of all observations in the ith row, $T_{.j}$ and $\bar{x}_{.j}$ the total and mean of all observations in the jth column, and $T_{..}$ and $\bar{x}_{..}$ the total and mean of all rc observations.

The average of the population means for the ith row, $\mu_{i.}$, is defined by

$$\mu_{i.} = \frac{\sum\limits_{j=1}^{c} \mu_{ij}}{c}.$$

Similarly the average of the population means for the jth column, $\mu_{.j}$, is defined by

$$\mu_{.j} = \frac{\sum\limits_{i=1}^{r} \mu_{ij}}{r},$$

and the average of the rc population means, μ, is defined by

$$\mu = \frac{\sum\limits_{i=1}^{r} \sum\limits_{j=1}^{c} \mu_{ij}}{rc}.$$

To determine if part of the variation in our observations is due to differences among the rows, we consider the test

$$H_0': \mu_{1.} = \mu_{2.} = \cdots = \mu_{r.} = \mu,$$
$$H_1': \text{The } \mu_{i.} \text{ are not all equal.}$$

Also, to determine if part of the variation is due to differences among the columns, we consider the test

$$H_0'': \mu_{.1} = \mu_{.2} = \cdots = \mu_{.c} = \mu,$$
$$H_1'': \text{The } \mu_{.j} \text{ are not all equal.}$$

Each observation may be written in the form

$$x_{ij} = \mu_{ij} + \varepsilon_{ij},$$

where ε_{ij} measures the deviation of the observed value x_{ij} from the population mean μ_{ij}. The preferred form of this equation is obtained by substituting

$$\mu_{ij} = \mu + \alpha_i + \beta_j,$$

where α_i is the effect of the ith row and β_j is the effect of the jth column. It is assumed that the row and column effects are additive. Hence we may write

$$x_{ij} = \mu + \alpha_i + \beta_j + \varepsilon_{ij}.$$

If we now impose the restrictions that

$$\sum_{i=1}^{r} \alpha_i = 0 \qquad \text{and} \qquad \sum_{j=1}^{c} \beta_j = 0,$$

then

$$\mu_{i.} = \frac{\sum_{j=1}^{c} (\mu + \alpha_i + \beta_j)}{c} = \mu + \alpha_i,$$

$$\mu_{.j} = \frac{\sum_{i=1}^{r} (\mu + \alpha_i + \beta_j)}{r} = \mu + \beta_j.$$

The null hypothesis that the r row means $\mu_{i.}$ are equal, and therefore equal to μ, is now equivalent to testing the hypothesis

$$H_0': \alpha_1 = \alpha_2 = \cdots = \alpha_r = 0,$$

H_1' : At least one of the α_i is not equal to zero.

Similarly the null hypothesis that the c column means $\mu_{.j}$ are equal is equivalent to testing the hypothesis

$$H_0'': \beta_1 = \beta_2 = \cdots = \beta_c = 0,$$

H_1'' : At least one of the β_j is not equal to zero.

Each of these tests will be based on a comparison of independent estimates of the common population variance σ^2. These estimates will be obtained by splitting the total sum of squares of our data into three components by means of the following identity.

THEOREM 11.2 (SUM-OF-SQUARES IDENTITY)

$$\sum_{i=1}^{r} \sum_{j=1}^{c} (x_{ij} - \bar{x}..)^2 = c \sum_{i=1}^{r} (\bar{x}_{i.} - \bar{x}..)^2 + r \sum_{j=1}^{c} (\bar{x}_{.j} - \bar{x}..)^2$$

$$+ \sum_{i=1}^{r} \sum_{j=1}^{c} (x_{ij} - \bar{x}_{i.} - \bar{x}_{.j} + \bar{x}..)^2.$$

Proof. $\displaystyle\sum_{i=1}^{r} \sum_{j=1}^{c} (x_{ij} - \bar{x}..)^2$

$$= \sum_{i=1}^{r} \sum_{j=1}^{c} [(\bar{x}_{i.} - \bar{x}..) + (\bar{x}_{.j} - \bar{x}..)$$

$$+ (x_{ij} - \bar{x}_{i.} - \bar{x}_{.j} + \bar{x}..)]^2$$

$$= \sum_{i=1}^{r} \sum_{j=1}^{c} (\bar{x}_{i.} - \bar{x}..)^2 + \sum_{i=1}^{r} \sum_{j=1}^{c} (\bar{x}_{.j} - \bar{x}..)^2$$

$$+ \sum_{i=1}^{r} \sum_{j=1}^{c} (x_{ij} - \bar{x}_{i.} - \bar{x}_{.j} + \bar{x}..)^2$$

$$+ 2 \sum_{i=1}^{r} \sum_{j=1}^{c} (\bar{x}_{i.} - \bar{x}..)(\bar{x}_{.j} - \bar{x}..)$$

$$+ 2 \sum_{i=1}^{r} \sum_{j=1}^{c} (\bar{x}_{i.} - \bar{x}..)(x_{ij} - \bar{x}_{i.} - \bar{x}_{.j} + \bar{x}..)$$

$$+ 2 \sum_{i=1}^{r} \sum_{j=1}^{c} (\bar{x}_{.j} - \bar{x}..)(x_{ij} - \bar{x}_{i.} - \bar{x}_{.j} + \bar{x}..).$$

The cross-product terms are all equal to zero. Hence

$$\sum_{i=1}^{r} \sum_{j=1}^{c} (x_{ij} - \bar{x}..)^2 = c \sum_{i=1}^{r} (\bar{x}_{i.} - \bar{x}..)^2$$

$$+ r \sum_{j=1}^{c} (\bar{x}_{.j} - \bar{x}..)^2$$

$$+ \sum_{i=1}^{r} \sum_{j=1}^{c} (x_{ij} - \bar{x}_{i.} - \bar{x}_{.j} + \bar{x}..)^2.$$

The sum-of-squares identity may be represented symbolically by the equation

$$\text{SST} = \text{SSR} + \text{SSC} + \text{SSE},$$

where

$$\text{SST} = \sum_{i=1}^{r} \sum_{j=1}^{c} (x_{ij} - \bar{x}..)^2 = \text{total sum of squares},$$

$$\text{SSR} = c \sum_{i=1}^{r} (\bar{x}_{i.} - \bar{x}_{..})^2 = \text{sum of squares for row means,}$$

$$\text{SSC} = r \sum_{j=1}^{c} (\bar{x}_{.j} - \bar{x}_{..})^2 = \text{sum of squares for column means,}$$

$$\text{SSE} = \sum_{i=1}^{r} \sum_{j=1}^{c} (x_{ij} - \bar{x}_{i.} - \bar{x}_{.j} + \bar{x}_{..})^2 = \text{error sum of squares.}$$

One estimate of σ^2, based on $r - 1$ degrees of freedom, is given by

$$s_1^2 = \frac{\text{SSR}}{r - 1}.$$

If the row effects $\alpha_1 = \alpha_2 = \cdots = \alpha_r = 0$, s_1^2 is an unbiased estimate of σ^2. However, if the row effects are not all zero, SSR will have an inflated numerical value and s_1^2 overestimates σ^2. A second estimate of σ^2, based on $c - 1$ degrees of freedom, is given by

$$s_2^2 = \frac{\text{SSC}}{c - 1}.$$

The estimate s_2^2 is an unbiased estimate of σ^2 when the column effects $\beta_1 = \beta_2 = \cdots = \beta_c = 0$. If the column effects are not all zero, SSC will also be inflated and s_2^2 will overestimate σ^2. A third estimate of σ^2, based on $(r - 1)(c - 1)$ degrees of freedom and independent of s_1^2 and s_2^2, is given by

$$s_3^2 = \frac{\text{SSE}}{(r - 1)(c - 1)},$$

which is unbiased regardless of the truth or falsity of either null hypothesis.

To test the null hypothesis that the row effects are all equal to zero, we compute the ratio

$$f_1 = \frac{s_1^2}{s_3^2},$$

which is a value of the random variable F_1 having the F distribution with $r - 1$ and $(r - 1)(c - 1)$ degrees of freedom when the null hypothesis is true. The null hypothesis is rejected at the α level of significance when $f_1 > f_\alpha[r - 1, (r - 1)(c - 1)]$.

Similarly, to test the null hypothesis that the column effects are all equal to zero, we compute the ratio

$$f_2 = \frac{s_2^2}{s_3^2},$$

which is a value of the random variable F_2 having the F distribution with $c - 1$ and $(r - 1)(c - 1)$ degrees of freedom when the null hypothesis is true. In this case the null hypothesis is rejected at the α level of significance when $f_2 > f_\alpha[c - 1, (r - 1)(c - 1)]$.

In practice we first compute SST, SSR, and SSC and then obtain SSE by subtraction by the formula

$$\text{SSE} = \text{SST} - \text{SSR} - \text{SSC}.$$

The degrees of freedom associated with SSE are also usually obtained by subtraction. It is not difficult to verify the identity

$$(r - 1)(c - 1) = (rc - 1) - (r - 1) - (c - 1).$$

Preferred computational formulas for the sums of squares are given as follows:

$$\text{SST} = \sum_{i=1}^{r} \sum_{j=1}^{c} x_{ij}^2 - \frac{T_{..}^2}{rc},$$

$$\text{SSR} = \frac{\sum_{i=1}^{r} T_{i.}^2}{c} - \frac{T_{..}^2}{rc},$$

$$\text{SSC} = \frac{\sum_{j=1}^{c} T_{.j}^2}{r} - \frac{T_{..}^2}{rc}.$$

The computations in an analysis-of-variance problem, for a two-way classification with a single observation per cell, may be summarized as in Table 11.9.

Table 11.9

Analysis of Variance for the Two-Way Classification with a Single Observation per Cell

Source of Variation	Sum of Squares	Degrees of Freedom	Mean Square	Computed f
Row means	SSR	$r - 1$	$s_1^2 = \dfrac{\text{SSR}}{r - 1}$	$f_1 = \dfrac{s_1^2}{s_3^2}$
Column means	SSC	$c - 1$	$s_2^2 = \dfrac{\text{SSC}}{c - 1}$	$f_2 = \dfrac{s_2^2}{s_3^2}$
Error	SSE	$(r - 1)(c - 1)$	$s_3^2 = \dfrac{\text{SSE}}{(r - 1)(c - 1)}$	
Total	SST	$rc - 1$		

Example 11.4 For the data in Table 11.7, test the hypothesis H_0', at the 0.05 level of significance, that there is no difference in the average yield of wheat when different kinds of fertilizer are used. Also test the hypothesis H_0'' that there is no difference in the average yield of the three varieties of wheat.

Solution. 1. (a) $H_0': \alpha_1 = \alpha_2 = \alpha_3 = \alpha_4 = 0$ (row effects are zero),
    (b) $H_0'': \beta_1 = \beta_2 = \beta_3 = 0$ (column effects are zero).
  2. (a) H_1': At least one of the α_i is not equal to zero,
    (b) H_1'': At least one of the β_j is not equal to zero.
  3. $\alpha = 0.05$.
  4. Critical regions: (a) $F_1 > 4.76$, (b) $F_2 > 5.14$.
  5. Computations:

$$\text{SST} = 64^2 + 55^2 + \cdots + 53^2 - \frac{720^2}{12} = 662,$$

$$\text{SSR} = \frac{210^2 + 159^2 + 183^2 + 168^2}{3} - \frac{720^2}{12} = 498,$$

$$\text{SSC} = \frac{236^2 + 252^2 + 232^2}{4} - \frac{720^2}{12} = 56,$$

$$\text{SSE} = 662 - 498 - 56 = 108.$$

These results and the remaining computations are exhibited in Table 11.10.

Table 11.10

Analysis of Variance for the Data of Table 11.7

Source of Variation	Sum of Squares	Degrees of Freedom	Mean Square	Computed f
Row means	498	3	166	9.22
Column means	56	2	28	1.56
Error	108	6	18	
Total	662	11		

6. Conclusions:
   (a) Reject H_0' and conclude that there is a difference in the average yield of wheat when different kinds of fertilizer are used.
   (b) Accept H_0'' and conclude that there is no difference in the average yield of the three varieties of wheat.

11.6 Two-Way Classification, Several Observations per Cell

In Section 11.5 it was assumed that the row and column effects were additive. This is equivalent to stating that $\mu_{ij} - \mu_{ij'} = \mu_{i'j} - \mu_{i'j'}$ or $\mu_{ij} - \mu_{i'j} = \mu_{ij'} - \mu_{i'j'}$, for every value of i, i', j, j'. That is, the difference between the population means for columns j and j' is the same for every row and the difference between the population means for rows i and i' is the same for every column. Referring to Table 11.7, this implies that if variety v_2 produces on the average 5 more bushels of wheat per acre than variety v_1 when fertilizer treatment t_1 is used, then v_2 will still produce on the average 5 more bushels than v_1 when t_2, t_3, or t_4 is used. Likewise, if v_1 produces an average of 3 more bushels per acre using fertilizer treatment t_2 rather than t_4, then v_2 or v_3 will also produce an average of 3 more bushels per acre using fertilizer treatment t_2 instead of t_4.

Table 11.11
Yields of Wheat for Three
Replications

Fertilizer Treatment	Varieties of Wheat		
	v_1	v_2	v_3
t_1	64	72	74
	66	81	51
	70	64	65
t_2	65	57	47
	63	43	58
	58	52	67
t_3	59	66	58
	68	71	39
	65	59	42
t_4	58	57	53
	41	61	59
	46	53	38

In many experiments the assumption of additivity does not hold and the analysis of Section 11.5 leads to erroneous conclusions. Suppose, for instance, that variety v_2 produces on the average 5 more bushels of wheat per acre than variety v_1 when fertilizer treatment t_1 is used but produces an average of 2 bushels per acre less than v_1 when fertilizer treatment t_2 is used. The varieties of wheat and the kinds of fertilizer are now said to *interact*.

An inspection of Table 11.7 suggests the presence of interaction. This apparent interaction may be real or it may be due to experimental error. The analysis of Example 11.4 was based on the assumption that the apparent interaction was due entirely to experimental error. If the total variability of our data was in part the result of an interaction effect, this source of variation remained a part of the error sum of squares, causing the error mean square to overestimate σ^2, and thereby increased the probability of committing a type II error.

Table 11.12

Two-Way Classification with Several Observations per Cell

Rows	Columns 1	2	$\cdots$	c	Total	Mean
1	x_{111} x_{112} . . . x_{11n}	x_{121} x_{122} . . . x_{12n}	$\cdots$	x_{1c1} x_{1c2} . . . x_{1cn}	$T_{1..}$	$\bar{x}_{1..}$
2	x_{211} x_{212} . . . x_{21n}	x_{221} x_{222} . . . x_{22n}	$\cdots$	x_{2c1} x_{2c2} . . . x_{2cn}	$T_{2..}$	$\bar{x}_{2..}$
. . .	. . .	. . .	$\cdots$	. . .	.	.
r	x_{r11} x_{r12} . . . x_{r1n}	x_{r21} x_{r22} . . . x_{r2n}	$\cdots$	x_{rc1} x_{rc2} . . . x_{rcn}	$T_{r..}$	$\bar{x}_{r..}$
Total Mean	$T_{.1.}$ $\bar{x}_{.1.}$	$T_{.2.}$ $\bar{x}_{.2.}$	$\cdots$ $\cdots$	$T_{.c.}$ $\bar{x}_{.c.}$	$T...$	$\bar{x}...$

To test for differences among the row and column means when the interaction is a significant factor, we must obtain an unbiased and independent estimate of σ^2. This is best accomplished by considering the variation of repeated measurements obtained under similar conditions. Suppose that we have reason to believe that the varieties of wheat and kinds of fertilizer in Table 11.7 interact. We repeat the experiment twice, using 36 one-acre plots, rather than 12, and record the results in Table 11.11. It is customary to say that the experiment has been *replicated* three times.

To present general formulas for the analysis of variance using repeated observations, we shall consider the case of n replications and then demonstrate the use of these formulas by applying them to the data of Table 11.11. As before, we shall consider a rectangular array consisting of r rows and c columns. Thus we still have rc cells, but now each cell contains n observations. We denote the kth observation in the ith row and jth column by x_{ijk}; the rcn observations are shown in Table 11.12.

The observations in the ijth cell constitute a random sample of size n from a population that is assumed to be normally distributed with mean μ_{ij} and variance σ^2. All rc populations are assumed to have the same variance σ^2. Let us define the following useful symbols, some of which are used in Table 11.12:

$$T_{ij.} = \text{sum of the observations in the } ij\text{th cell,}$$
$$T_{i..} = \text{sum of the observations in the } i\text{th row,}$$
$$T_{.j.} = \text{sum of the observations in the } j\text{th column,}$$
$$T_{...} = \text{sum of all } rcn \text{ observations,}$$
$$\bar{x}_{ij.} = \text{mean of the observations in the } ij\text{th cell,}$$
$$\bar{x}_{i..} = \text{mean of the observations in the } i\text{th row,}$$
$$\bar{x}_{.j.} = \text{mean of the observations in the } j\text{th column,}$$
$$\bar{x}_{...} = \text{mean of all } rcn \text{ observations.}$$

Each observation in Table 11.12 may be written in the form

$$x_{ijk} = \mu_{ij} + \varepsilon_{ijk},$$

where ε_{ijk} measures the deviations of the observed x_{ijk} values in the ijth cell from the population mean μ_{ij}. If we let $(\alpha\beta)_{ij}$ denote the interaction effect of the ith row and the jth column, α_i the effect of the ith row, β_j the effect of the jth column, and μ the over-all mean, we can write

$$\mu_{ij} = \mu + \alpha_i + \beta_j + (\alpha\beta)_{ij},$$

and then

$$x_{ijk} = \mu + \alpha_i + \beta_j + (\alpha\beta)_{ij} + \varepsilon_{ijk},$$

on which we impose the restrictions

$$\sum_{i=1}^{r} \alpha_i = 0, \qquad \sum_{j=1}^{c} \beta_j = 0, \qquad \sum_{i=1}^{r} (\alpha\beta)_{ij} = 0, \qquad \sum_{j=1}^{c} (\alpha\beta)_{ij} = 0.$$

The three hypotheses to be tested are as follows:

1. $H_0': \alpha_1 = \alpha_2 = \cdots = \alpha_r = 0$,
   H_1': At least one of the α_i is not equal to zero.
2. $H_0'': \beta_1 = \beta_2 = \cdots = \beta_c = 0$,
   H_1'': At least one of the β_j is not equal to zero.

3. $H_0''' : (\alpha\beta)_{11} = (\alpha\beta)_{12} = \cdots = (\alpha\beta)_{rc} = 0$,
   $H_1''' :$ At least one of the $(\alpha\beta)_{ij}$ is not equal to zero.

Each of these tests will be based on a comparison of independent estimates of σ^2 provided by the splitting of the total sum of squares of our data into four components by means of the following identity.

THEOREM 11.3 (SUM-OF-SQUARES IDENTITY)

$$\sum_{i=1}^{r} \sum_{j=1}^{c} \sum_{k=1}^{n} (x_{ijk} - \bar{x}...)^2 = cn \sum_{i=1}^{r} (\bar{x}_{i..} - \bar{x}...)^2 + rn \sum_{j=1}^{c} (\bar{x}_{.j.} - \bar{x}...)^2$$

$$+ n \sum_{i=1}^{r} \sum_{j=1}^{c} (\bar{x}_{ij.} - \bar{x}_{i..} - \bar{x}_{.j.} + \bar{x}...)^2$$

$$+ \sum_{i=1}^{r} \sum_{j=1}^{c} \sum_{k=1}^{n} (x_{ijk} - \bar{x}_{ij.})^2.$$

Proof. $\displaystyle\sum_{i=1}^{r} \sum_{j=1}^{c} \sum_{k=1}^{n} (x_{ijk} - \bar{x}...)^2$

$$= \sum_{i=1}^{r} \sum_{j=1}^{c} \sum_{k=1}^{n} [(\bar{x}_{i..} - \bar{x}...) + (\bar{x}_{.j.} - \bar{x}...)$$

$$+ (\bar{x}_{ij.} - \bar{x}_{i..} - \bar{x}_{.j.} + \bar{x}...) + (x_{ijk} - \bar{x}_{ij.})]^2$$

$$= \sum_{i=1}^{r} \sum_{j=1}^{c} \sum_{k=1}^{n} (\bar{x}_{i..} - \bar{x}...)^2$$

$$+ \sum_{i=1}^{r} \sum_{j=1}^{c} \sum_{k=1}^{n} (\bar{x}_{.j.} - \bar{x}...)^2$$

$$+ \sum_{i=1}^{r} \sum_{j=1}^{c} \sum_{k=1}^{n} (\bar{x}_{ij.} - \bar{x}_{i..} - \bar{x}_{.j.} + \bar{x}...)^2$$

$$+ \sum_{i=1}^{r} \sum_{j=1}^{c} \sum_{k=1}^{n} (x_{ijk} - \bar{x}_{ij.})^2$$

$+$ 6 cross-product terms.

The cross-product terms are all equal to zero. Hence

$$\sum_{i=1}^{r} \sum_{j=1}^{c} \sum_{k=1}^{n} (x_{ijk} - \bar{x}...)^2$$

$$= cn \sum_{i=1}^{r} (\bar{x}_{i..} - \bar{x}...)^2 + rn \sum_{j=1}^{c} (\bar{x}_{.j.} - \bar{x}...)^2$$

$$+ n \sum_{i=1}^{r} \sum_{j=1}^{c} (\bar{x}_{ij.} - \bar{x}_{i..} - \bar{x}_{.j.} + \bar{x}...)^2$$

$$+ \sum_{i=1}^{r} \sum_{j=1}^{c} \sum_{k=1}^{n} (x_{ijk} - \bar{x}_{ij.})^2.$$

Symbolically we write the sum-of-squares identity

$$SST = SSR + SSC + SS(RC) + SSE,$$

where SSE, SSR, and SSC are the sum of squares for error, row means, and column means, respectively, and SS(RC) is the sum of squares for interaction of rows and columns. The degrees of freedom are partitioned according to the identity

$$rcn - 1 = (r - 1) + (c - 1) + (r - 1)(c - 1) + rc(n - 1).$$

Dividing each of the sum of squares on the right side of the sum of squares identity by their corresponding number of degrees of freedom, we obtain the four estimates

$$s_1^2 = \frac{SSR}{r - 1}, \quad s_2^2 = \frac{SSC}{c - 1}, \quad s_3^2 = \frac{SS(RC)}{(r - 1)(c - 1)}, \quad s_4^2 = \frac{SSE}{rc(n - 1)},$$

of σ^2, which are all unbiased when H_0', H_0'', and H_0''' are true.

To test the hypothesis H_0' that the row effects are all equal to zero, we compute the ratio

$$f_1 = \frac{s_1^2}{s_4^2},$$

which is a value of the random variable F_1 having the F distribution with $r - 1$ and $rc(n - 1)$ degrees of freedom when H_0' is true. The null hypothesis is rejected at the α level of significance when $f_1 > f_\alpha[r - 1, rc(n - 1)]$. Similarly, to test the hypothesis H_0'' that the column effects are all equal to zero, we compute the ratio

$$f_2 = \frac{s_2^2}{s_4^2},$$

which is a value of the random variable F_2 having the F distribution with $c - 1$ and $rc(n - 1)$ degrees of freedom when H_0'' is true. This hypothesis is rejected at the α level of significance when $f_2 > f_\alpha[c - 1, rc(n - 1)]$. Finally, to test the hypothesis H_0''' that the interaction effects are all equal to zero, we compute the ratio

$$f_3 = \frac{s_3^2}{s_4^2},$$

which is a value of the random variable F_3 having the F distribution with $(r - 1)(c - 1)$ and $rc(n - 1)$ degrees of freedom.

The computations in an analysis-of-variance problem, for a two-way classification with several observations per cell, are usually summarized as in Table 11.13.

Table 11.13

Analysis of Variance for the Two-Way Classification with
Several Observations per Cell

Source of Variation	Sum of Squares	Degrees of Freedom	Mean Square	Computed f
Row means	SSR	$r - 1$	$s_1^2 = \dfrac{SSR}{r - 1}$	$f_1 = \dfrac{s_1^2}{s_4^2}$
Column means	SSC	$c - 1$	$s_2^2 = \dfrac{SSC}{c - 1}$	$f_2 = \dfrac{s_2^2}{s_4^2}$
Interaction	SS(RC)	$(r - 1)(c - 1)$	$s_3^2 = \dfrac{SS(RC)}{(r - 1)(c - 1)}$	$f_3 = \dfrac{s_3^2}{s_4^2}$
Error	SSE	$rc(n - 1)$	$s_4^2 = \dfrac{SSE}{rc(n - 1)}$	
Total	SST	$rcn - 1$		

The sums of squares are usually obtained by means of the following computational formulas:

$$SST = \sum_{i=1}^{r} \sum_{j=1}^{c} \sum_{k=1}^{n} x_{ijk}^2 - \frac{T_{...}^2}{rcn},$$

$$SSR = \frac{\sum_{i=1}^{r} T_{i..}^2}{cn} - \frac{T_{...}^2}{rcn},$$

$$SSC = \frac{\sum_{j=1}^{c} T_{.j.}^2}{rn} - \frac{T_{...}^2}{rcn},$$

$$SS(RC) = \frac{\sum_{i=1}^{r} \sum_{j=1}^{c} T_{ij.}^2}{n} - \frac{\sum_{i=1}^{r} T_{i..}^2}{cn} - \frac{\sum_{j=1}^{c} T_{.j.}^2}{rn} + \frac{T_{...}^2}{rcn}.$$

As before, SSE is obtained by subtraction using the formula

$$SSE = SST - SSR - SSC - SS(RC).$$

Example 11.5 For the data in Table 11.11, use a 0.05 level of significance to test the following hypotheses: (a) H_0': There is no difference in the average yield of wheat when different kinds of fertilizer are used; (b) H_0'': there is no difference in the average yield of the three varieties of wheat; and (c) H_0''': there is no interaction between the different kinds of fertilizer and the different varieties of wheat.

Solution.
1. (a) H_0': $\alpha_1 = \alpha_2 = \alpha_3 = \alpha_4 = 0$,
   (b) H_0'': $\beta_1 = \beta_2 = \beta_3 = 0$,
   (c) H_0''': $(\alpha\beta)_{11} = (\alpha\beta)_{12} = \cdots = (\alpha\beta)_{43} = 0$.
2. (a) H_1': At least one of the α_i is not equal to zero,
   (b) H_1'': At least one of the β_j is not equal to zero,
   (c) H_1''': At least one of the $(\alpha\beta)_{ij}$ is not equal to zero.
3. $\alpha = 0.05$.
4. Critical regions: (a) $F_1 > 3.01$, (b) $F_2 > 3.40$, (c) $F_3 > 2.51$.
5. Computations: From Table 11.11 we first construct the following table of totals:

	v_1	v_2	v_3	*Total*
t_1	200	217	190	607
t_2	186	152	172	510
t_3	192	196	139	527
t_4	145	171	150	466
Total	723	736	651	2110

Now

$$SST = 64^2 + 66^2 + \cdots + 38^2 - \frac{2110^2}{36}$$

$$= 127,448 - 123,669 = 3779,$$

$$SSR = \frac{607^2 + 510^2 + 527^2 + 466^2}{9} - \frac{2110^2}{36}$$

$$= 124,826 - 123,669 = 1157,$$

$$SSC = \frac{723^2 + 736^2 + 651^2}{12} - \frac{2110^2}{36}$$

$$= 124,019 - 123,669 = 350,$$

$$SS(RC) = \frac{200^2 + 186^2 + \cdots + 150^2}{3}$$

$$- 124,826 - 124,019 + 123,669$$

$$= 771,$$

$$SSE = 3779 - 1157 - 350 - 771 = 1501.$$

These results, with the remaining computations, are given in Table 11.14.

Table 11.14

Analysis of Variance for the Data of Table 11.11

Source of Variation	Sum of Squares	Degrees of Freedom	Mean Square	Computed f
Row means	1157	3	385.667	6.17
Column means	350	2	175.000	2.80
Interaction	771	6	128.500	2.05
Error	1501	24	62.542	
Total	3779	35		

6. Conclusions:
  (a) Reject H_0' and conclude that a difference in the average yields of wheat exists when different kinds of fertilizer are used.
  (b) Accept H_0'' and conclude that there is no difference in the average yield of the three varieties of wheat.
  (c) Accept H_0''' and conclude that there is no interaction between the different kinds of fertilizer and the different varieties of wheat.

11.7 Brief Discussion of Experimental Designs

The analysis-of-variance technique has been described as a process for splitting the total variation of a set of experimental data into meaningful components that measure different sources of variation. The precise steps in carrying out the analysis will depend on the experimental design used to generate the data. In most cases the scientist or statistician may solve his problem by planning his experiment to fit one of several catalogued designs.

The simplest of all the experimental designs is called the *completely randomized design*. This design may be defined as one in which the treatments are randomly arranged over the whole of the experimental material. If the treatments in question represent three different varieties of wheat planted at random on several 1-acre plots, the design is said to be completely randomized and the data are recorded and analyzed using the methods of Section 11.2 for the one-way classification. When the treatments represent all possible combinations of two criteria, such as the 12 possible combinations using 3 varieties of wheat and 4 kinds of fertilizer, and the experimental material is large enough for one or more replications of the 12 treatments, then the design is analyzed by the methods described in Section 11.5 or in Section 11.6 for the two-way classification. The completely randomized design is very easy to lay out and the analysis is simple to perform. It should be used only when the number of treatments is small and the experimental material is homogeneous. If, in the above illustration, the plots are not uniform in their soil composition, it is possible to select a more efficient design.

Perhaps the next simplest experimental design is called the *randomized block design*. The experimental material is divided into groups or blocks such that the units making up a particular block are homogeneous. Each block constitutes a replication of the treatments. The treatments are assigned at random to the units in each block. If the treatments represent a single criterion of classification, the analysis-of-variance procedure is exactly the same as for the two-way classification, where the rows and columns represent blocks and treatments, respectively. Analysis-of-variance procedures for multiway classifications are required to analyze a randomized block design where the treatments are all possible combinations of two or more criteria of classification. The chief disadvantage of the randomized design is that it is not suitable for large numbers of treatments, owing to the difficulty in obtaining the homogeneous blocks. This disadvantage may be overcome by choosing a design from the catalog of *incomplete block designs*. These designs allow one to investigate differences among v treatments arranged in b blocks containing k experimental units, where $k < v$.

The randomized block design is very effective in reducing the experimental error by removing one source of variation. Another design that is particularly useful in controlling two sources of variation, and at the same time reduces the required number of treatment combinations, is called the *Latin square*. Suppose that we are interested in the yields of four varieties of wheat using four different fertilizers over a period of 4 years. The total number of treatment combinations for a completely randomized design would be 64. By selecting the same number of categories for all three criteria of classification, we may select a Latin-square design and perform the analysis of variance using the results of only 16 treatment combinations. A typical Latin square, selected at random from all possible 4 × 4 squares, might be the following:

	Column			
Row	1	2	3	4
1	*A*	*B*	*C*	*D*
2	*D*	*A*	*B*	*C*
3	*C*	*D*	*A*	*B*
4	*B*	*C*	*D*	*A*

The four letters *A*, *B*, *C*, and *D* represent the 4 varieties of wheat, the rows represent the 4 fertilizers, and the columns account for the 4 years. We now see that each treatment occurs exactly once in each row and each column. With such a balanced arrangement the analysis of variance enables one to separate the variation due to the different fertilizers and different years from the error sum of squares and thereby obtain a more accurate test for differences in the yielding capabilities of the 4 varieties of wheat.

We have considered here only a few of the simplest and often used experimental designs. It would take a separate book to give a complete and detailed account of the numerous designs that are now available. Lacking knowledge in this field of specialization, the experimenter would be wise to consult with a competent statistician prior to conducting his investigation. Only a well-planned experiment can guarantee reliable conclusions.

EXERCISES

1. Consider the following 5 random samples:

		Sample		
A	B	C	D	E
21	35	45	32	45
35	12	60	53	29
32	27	33	29	31
28	41	36	42	22
14	19	31	40	36
47	23	40	23	29
25	31	43	35	42
38	20	48	42	30

Perform the analysis of variance, and test the hypothesis, at the 0.05 level of significance, that the samples come from populations having the same mean.

2. Six different machines are being considered for use in manufacturing rubber seals. The machines are being compared with respect to tensile strength of the product. A random sample of 4 seals from each machine is used to determine whether or not the mean tensile strength varies from machine to machine. The following are the tensile-strength measurements in pounds per square inch $\times 10^{-2}$:

			Machine		
1	2	3	4	5	6
17.5	16.4	20.3	14.6	17.5	18.3
16.9	19.2	15.7	16.7	19.2	16.2
15.8	17.7	17.8	20.8	16.5	17.5
18.6	15.4	18.9	18.9	20.5	20.1

Perform the analysis of variance at the 0.05 level of significance and indicate whether or not the treatment means differ significantly.

3. Show that the computing formula for SSC, in the analysis of variance of the one-way classification, is equivalent to the corresponding term in the identity of Theorem 11.1.

4. Three sections of the same elementary mathematics course are taught by 3 teachers. The final grades were recorded as follows:

Teacher		
A	B	C
73	88	68
89	78	79
82	48	56
43	91	91
80	51	71
73	85	71
66	74	87
60	77	41
45	31	59
93	78	68
36	62	53
77	76	79
	96	15
	80	
	56	

Is there a significant difference in the average grades given by the 3 teachers? Use a 0.05 level of significance.

5. Test for homogeneity of variances in Exercise 4.

6. Four laboratories are being used to perform chemical analyses. Samples of the same material are sent to the laboratories for analysis as part of the study to determine whether or not they give, on the average, the same results. The analytical results for the 4 laboratories are as follows:

Laboratory			
A	B	C	D
58.7	62.7	55.9	60.7
61.4	64.5	56.1	60.3
60.9	63.1	57.3	60.9
59.1	59.2	55.2	61.4
58.2	60.3	58.1	62.3

(a) Use Bartlett's test to determine if the within-laboratory variances are significantly different at the $\alpha = 0.05$ level of significance.

(b) If the hypothesis in **(a)** is accepted, perform the analysis of variance and give conclusions concerning the laboratories.

7. Use the multiple-range test, with a 0.05 level of significance, to analyze the means of the 5 samples in Exercise 1.

8. An investigation was conducted to determine the source of reduction in yield of a certain chemical product. It was known that the loss in yield occurred in the mother liquor; that is, the material removed at the filtration stage. It was felt that different blends of the original material may result in different yield reductions at the mother liquor stage. The following are results of the percentage reduction for 3 batches at each of 4 preselected blends:

	Blend		
1	2	3	4
25.6	25.2	20.8	31.6
24.3	28.6	26.7	29.8
27.9	24.7	22.2	34.3

(a) Is there a significant difference in the average percentage reduction in yield for the different blends? Use a 0.05 level of significance.

(b) Use Duncan's multiple-range test to determine which blends differ.

9. The following data represent the final grades obtained by 5 students in mathematics, English, French, and biology:

Student	Mathematics	English	French	Biology
1	68	57	73	61
2	83	94	91	86
3	72	81	63	59
4	55	73	77	66
5	92	68	75	87

Use a 0.05 level of significance to test the hypothesis that:

(a) The courses are of equal difficulty.

(b) The students have equal ability.

10. Referring to the proof of Theorem 11.2, show that the cross-product term

$$\sum_{i=1}^{r} \sum_{j=1}^{c} (x_{ij} - \bar{x}_{i.} - \bar{x}_{.j} + \bar{x}_{..})(\bar{x}_{.j} - \bar{x}_{..}) = 0.$$

11. Show that the computing formula for SSR, in the analysis of variance of the two-way classification with a single observation per cell, is equivalent to the corresponding term in the identity of Theorem 11.2.

12. Use the multiple-range test, with a 0.05 level of significance, to analyze the mean grades obtained by the 5 students in Exercise 9.

13. Three varieties of potatoes are being compared for yield. The experiment was conducted by assigning each variety at random to 3 equal-size plots at each of 4 different locations. The following yields, in bushels per plot, were recorded:

	Variety of Potato		
Location	A	B	C
1	18	13	12
2	20	23	21
3	14	12	9
4	11	17	10

Use a 0.05 level of significance to test the hypothesis that there is no difference in the yielding capabilities of the 3 varieties of potatoes. Was it necessary to plant each variety at each location to reach a valid conclusion concerning the 3 varieties?

14. An experiment is conducted in which 4 treatments are to be compared using 5 subjects. The following data are generated:

Treatment	Subject				
	1	2	3	4	5
1	12.8	10.6	11.7	10.7	11.0
2	11.7	14.2	11.8	9.9	13.8
3	11.5	14.7	13.6	10.7	15.9
4	12.6	16.5	15.4	9.6	17.1

Perform the analysis of variance, separating out the treatment, subject, and error sums of squares. Use a 0.05 level of significance to test the hypothesis that there is no difference between the treatment means.

15. The following data represent the results of 4 quizzes obtained by 5 students in mathematics, English, French, and biology:

Student	Mathematics		English		French		Biology	
1	88	63	51	58	73	81	87	81
	79	80	72	65	77	77	92	76
2	79	96	85	95	82	36	80	93
	56	68	67	88	80	68	62	67
3	67	66	74	47	91	95	77	70
	51	89	59	82	59	92	84	73
4	35	60	76	49	43	52	55	49
	64	70	26	76	42	32	53	56
5	99	77	84	94	95	81	83	76
	87	95	83	76	98	96	87	80

Use a 0.05 level of significance to test the hypotheses that:
(a) The courses are of equal difficulty.
(b) The students have equal ability.
(c) The students and subjects do not interact.

16. Suppose, in Exercise 13, that the experiment was conducted using 9 uniform plots at each location. The 3 varieties of potatoes are each planted on 3 plots selected at random for each location. The yields, in bushels per plot, were as follows:

Location	Variety of Potatoes		
	A	B	C
1	15	20	22
	19	24	17
	12	18	14

16. (*continued*)

2	17	24	26
	10	18	19
	13	22	21
3	9	12	10
	12	15	5
	6	10	8
4	14	21	19
	8	16	15
	11	14	12

Use a 0.05 level of significance to test the hypotheses that:
(a) There is no difference in the yielding capabilities of the 3 varieties of potatoes.
(b) Different locations have no effect on the yields.
(c) The locations and varieties of potatoes do not interact.

17. Show that

$$\sum_{i=1}^{r} \sum_{j=1}^{c} \sum_{k=1}^{n} (x_{ijk} - \bar{x}_{ij.})(\bar{x}_{i..} - \bar{x}...) = 0.$$

18. Show that the computational formulas for the sum of squares in a two-way classification, with several observations per cell, are equivalent to the corresponding terms in the identity of Theorem 11.3.

19. In an experiment conducted to determine which of 3 missile systems is preferable, the propellant burning rate for 24 static firings was measured. Four propellant types were used. The experiment yielded duplicate observations of burning rates at each combination of the treatments. The data, after coding, were recorded as follows:

Missile System	Propellant Type			
	b_1	b_2	b_3	b_4
a_1	34.0	30.1	29.8	29.0
	32.7	32.8	26.7	28.9
a_2	32.0	30.2	28.7	27.6
	33.2	29.8	28.1	27.8
a_3	28.4	27.3	29.7	28.8
	29.3	28.9	27.3	29.1

Use a 0.05 level of significance to test the following hypotheses:
(a) H'_0: There is no difference in the mean propellant burning rates when different missile systems are used.
(b) H''_0: There is no difference in the mean propellant burning rates of the 4 propellant types.
(c) H'''_0: There is no interaction between the different missile systems and the different propellant types.

20. Three strains of rats were studied under 2 environmental conditions for their performance in a maze test. The error scores for the 48 rats were recorded as follows:

Environment	Strain					
	Bright		Mixed		Dull	
Free	28	12	33	83	101	94
	22	23	36	14	33	56
	25	10	41	76	122	83
	36	86	22	58	35	23
Restricted	72	32	60	89	136	120
	48	93	35	126	38	153
	25	31	83	110	64	128
	91	19	99	118	87	140

Use a 0.01 level of significance to test the hypotheses that:

(a) There is no difference in error scores for different environments.

(b) There is no difference in error scores for different strains.

(c) The environments and strains of rats do not interact.

REFERENCES

Alder, H. L., and E. B. Roessler. *Introduction to Probability and Statistics*, 4th ed. San Francisco: W. H. Freeman & Co., Publishers, 1968.

Brunk, H. D. *An Introduction to Mathematical Statistics*, 2nd ed. Waltham, Mass.: Ginn/Blaisdell, 1965.

Dixon, W. J., and F. J. Massey, Jr. *Introduction to Statistical Analysis*, 3rd ed. New York: McGraw-Hill Book Company, 1969.

Ezekiel, M. *Methods of Correlation Analysis*. New York: John Wiley & Sons, Inc., 1930.

Fehr, H. F., L. N. H. Bunt, and G. Grossman. *An Introduction to Sets*, *Probability and Hypothesis Testing*. Lexington, Mass.: D. C. Heath & Company, 1964.

Freund, J. E. *Mathematical Statistics*, 2nd ed. Englewood Cliffs, N.J.: Prentice-Hall, Inc., 1971.

Guenther, W. C. *Analysis of Variance*. Englewood Cliffs, N.J.: Prentice-Hall, Inc., 1964.

Guenther, W. C. *Concepts of Statistical Inference*. New York: McGraw-Hill Book Company, 1965.

Hicks, C. R. *Fundamental Concepts in the Design of Experiments*. New York: Holt, Rinehart and Winston, Inc., 1964.

Hodges, J. L., Jr., and E. L. Lehmann. *Basic Concepts of Probability and Statistics*, 2nd ed. San Francisco: Holden-Day, Inc., 1970.

Hoel, P. G. *Elementary Statistics*, 3rd ed. New York: John Wiley & Sons, Inc., 1971.

Hogg, R. V., and A. T. Craig. *Introduction to Mathematical Statistics*, 3rd ed. New York: Macmillan Publishing Co., Inc., 1970.

Kemeny, J. G., J. L. Snell, and G. L. Thompson. *Introduction to Finite Mathematics*, 2nd ed. Englewood Cliffs, N.J.: Prentice-Hall, Inc., 1966.

Kurtz, T. E. *Basic Statistics*. Englewood Cliffs, N.J.: Prentice-Hall, Inc., 1963.

Li, J. C. R. *Introduction to Statistical Inference*. Ann Arbor, Mich.: J. W. Edwards, Publisher, Incorporated, 1961.

Lindgren, B. W., and G. W. McElrath. *Introduction to Probability and Statistics*, 3rd ed. New York: Macmillan Publishing Co., Inc., 1969.

Mendenhall, W. *Introduction to Probability and Statistics*, 3rd ed. Belmont, Calif.: Wadsworth Publishing Co., Inc., 1971.

Miller, I., and J. E. Freund. *Probability and Statistics for Engineers*. Englewood Cliffs, N.J.: Prentice-Hall, Inc., 1965.

Mode, E. B. *Elements of Probability and Statistics*. Englewood Cliffs, N.J.: Prentice-Hall, Inc., 1966.

Mosteller, F., R. E. K. Rourke, and G. B. Thomas, Jr. *Probability with Statistical Applications*. Reading, Mass.: Addison-Wesley Publishing Co., Inc., 1961.

Scheffe, H. *Analysis of Variance*. New York: John Wiley & Sons, Inc., 1959.

Siegel, S. *Nonparametric Statistics for the Behavioral Sciences*. New York: McGraw-Hill Book Company, 1956.

Ullman, N. R. *Statistics: An Applied Approach*. Lexington, Mass.: Xerox College Publishing, 1972.

Walpole, R. E., and R. H. Myers. *Probability and Statistics for Engineers and Scientists*. New York: Macmillan Publishing Co., Inc., 1972.

Weinberg, G. H., and J. A. Schumaker. *Statistics: An Intuitive Approach*. 2nd ed. Belmont, Calif.: Wadsworth Publishing Co., Inc., 1969.

Wine, R. L. *Statistics for Scientists and Engineers*. Englewood Cliffs, N.J.: Prentice-Hall, Inc., 1964.

Wonnacott, T. H., and R. J. Wonnacott. *Introductory Statistics*, 2nd ed. New York: John Wiley & Sons, Inc., 1972.

APPENDIX
STATISTICAL TABLES

Table A.1
Squares and Square Roots

n	n^2	$\sqrt{n}$	$\sqrt{10n}$	n	n^2	$\sqrt{n}$	$\sqrt{10n}$
1.0	1.00	1.000	3.162	5.5	30.25	2.345	7.416
1.1	1.21	1.049	3.317	5.6	31.36	2.366	7.483
1.2	1.44	1.095	3.464	5.7	32.49	2.387	7.550
1.3	1.69	1.140	3.606	5.8	33.64	2.408	7.616
1.4	1.96	1.183	3.742	5.9	34.81	2.429	7.681
1.5	2.25	1.225	3.873	6.0	36.00	2.449	7.746
1.6	2.56	1.265	4.000	6.1	37.21	2.470	7.810
1.7	2.89	1.304	4.123	6.2	38.44	2.490	7.874
1.8	3.24	1.342	4.243	6.3	39.69	2.510	7.937
1.9	3.61	1.378	4.359	6.4	40.96	2.530	8.000
2.0	4.00	1.414	4.472	6.5	42.25	2.550	8.062
2.1	4.41	1.449	4.583	6.6	43.56	2.569	8.124
2.2	4.84	1.483	4.690	6.7	44.89	2.588	8.185
2.3	5.29	1.517	4.796	6.8	46.24	2.608	8.246
2.4	5.76	1.549	4.899	6.9	47.61	2.627	8.307
2.5	6.25	1.581	5.000	7.0	49.00	2.646	8.367
2.6	6.76	1.612	5.099	7.1	50.41	2.665	8.426
2.7	7.29	1.643	5.196	7.2	51.84	2.683	8.485
2.8	7.84	1.673	5.292	7.3	53.29	2.702	8.544
2.9	8.41	1.703	5.385	7.4	54.76	2.720	8.602
3.0	9.00	1.732	5.477	7.5	56.25	2.739	8.660
3.1	9.61	1.761	5.568	7.6	57.76	2.757	8.718
3.2	10.24	1.789	5.657	7.7	59.29	2.775	8.775
3.3	10.89	1.817	5.745	7.8	60.84	2.793	8.832
3.4	11.56	1.844	5.831	7.9	62.41	2.811	8.888
3.5	12.25	1.871	5.916	8.0	64.00	2.828	8.944
3.6	12.96	1.897	6.000	8.1	65.61	2.846	9.000
3.7	13.69	1.924	6.083	8.2	67.24	2.864	9.055
3.8	14.44	1.949	6.164	8.3	68.89	2.881	9.110
3.9	15.21	1.975	6.245	8.4	70.56	2.898	9.165
4.0	16.00	2.000	6.325	8.5	72.25	2.915	9.220
4.1	16.81	2.025	6.403	8.6	73.96	2.933	9.274
4.2	17.64	2.049	6.481	8.7	75.69	2.950	9.327
4.3	18.49	2.074	6.557	8.8	77.44	2.966	9.381
4.4	19.36	2.098	6.633	8.9	79.21	2.983	9.434
4.5	20.25	2.121	6.708	9.0	81.00	3.000	9.487
4.6	21.16	2.145	6.782	9.1	82.81	3.017	9.539
4.7	22.09	2.168	6.856	9.2	84.64	3.033	9.592
4.8	23.04	2.191	6.928	9.3	86.49	3.050	9.644
4.9	24.01	2.214	7.000	9.4	88.36	3.066	9.695
5.0	25.00	2.236	7.071	9.5	90.25	3.082	9.747
5.1	26.01	2.258	7.141	9.6	92.16	3.098	9.798
5.2	27.04	2.280	7.211	9.7	94.09	3.114	9.849
5.3	28.09	2.302	7.280	9.8	96.04	3.130	9.899
5.4	29.16	2.324	7.348	9.9	98.01	3.146	9.950

Table A.2

Binomial Probability Sums $\sum_{x=0}^{r} b(x; n, p)$

						p					
n	*r*	0.10	0.20	0.25	0.30	0.40	0.50	0.60	0.70	0.80	0.90
5	0	0.5905	0.3277	0.2373	0.1681	0.0778	0.0312	0.0102	0.0024	0.0003	0.0000
	1	0.9185	0.7373	0.6328	0.5282	0.3370	0.1875	0.0870	0.0308	0.0067	0.0005
	2	0.9914	0.9421	0.8965	0.8369	0.6826	0.5000	0.3174	0.1631	0.0579	0.0086
	3	0.9995	0.9933	0.9844	0.9692	0.9130	0.8125	0.6630	0.4718	0.2627	0.0815
	4	1.0000	0.9997	0.9990	0.9976	0.9898	0.9688	0.9222	0.8319	0.6723	0.4095
	5	1.0000	1.0000	1.0000	1.0000	1.0000	1.0000	1.0000	1.0000	1.0000	1.0000
10	0	0.3487	0.1074	0.0563	0.0282	0.0060	0.0010	0.0001	0.0000	0.0000	0.0000
	1	0.7361	0.3758	0.2440	0.1493	0.0464	0.0107	0.0017	0.0001	0.0000	0.0000
	2	0.9298	0.6778	0.5256	0.3828	0.1673	0.0547	0.0123	0.0016	0.0001	0.0000
	3	0.9872	0.8791	0.7759	0.6496	0.3823	0.1719	0.0548	0.0106	0.0009	0.0000
	4	0.9984	0.9672	0.9219	0.8497	0.6331	0.3770	0.1662	0.0474	0.0064	0.0002
	5	0.9999	0.9936	0.9803	0.9527	0.8338	0.6230	0.3669	0.1503	0.0328	0.0016
	6	1.0000	0.9991	0.9965	0.9894	0.9452	0.8281	0.6177	0.3504	0.1209	0.0128
	7	1.0000	0.9999	0.9996	0.9984	0.9877	0.9453	0.8327	0.6172	0.3222	0.0702
	8	1.0000	1.0000	1.0000	0.9999	0.9983	0.9893	0.9536	0.8507	0.6242	0.2639
	9	1.0000	1.0000	1.0000	1.0000	0.9999	0.9990	0.9940	0.9718	0.8926	0.6513
	10	1.0000	1.0000	1.0000	1.0000	1.0000	1.0000	1.0000	1.0000	1.0000	1.0000
15	0	0.2059	0.0352	0.0134	0.0047	0.0005	0.0000	0.0000	0.0000	0.0000	0.0000
	1	0.5490	0.1671	0.0802	0.0353	0.0052	0.0005	0.0000	0.0000	0.0000	0.0000
	2	0.8159	0.3980	0.2361	0.1268	0.0271	0.0037	0.0003	0.0000	0.0000	0.0000
	3	0.9444	0.6482	0.4613	0.2969	0.0905	0.0176	0.0019	0.0001	0.0000	0.0000
	4	0.9873	0.8358	0.6865	0.5155	0.2173	0.0592	0.0094	0.0007	0.0000	0.0000
	5	0.9978	0.9389	0.8516	0.7216	0.4032	0.1509	0.0338	0.0037	0.0001	0.0000
	6	0.9997	0.9819	0.9434	0.8689	0.6098	0.3036	0.0951	0.0152	0.0008	0.0000
	7	1.0000	0.9958	0.9827	0.9500	0.7869	0.5000	0.2131	0.0500	0.0042	0.0000
	8	1.0000	0.9992	0.9958	0.9848	0.9050	0.6964	0.3902	0.1311	0.0181	0.0003
	9	1.0000	0.9999	0.9992	0.9963	0.9662	0.8491	0.5968	0.2784	0.0611	0.0023
	10	1.0000	1.0000	0.9999	0.9993	0.9907	0.9408	0.7827	0.4845	0.1642	0.0127
	11	1.0000	1.0000	1.0000	0.9999	0.9981	0.9824	0.9095	0.7031	0.3518	0.0556
	12	1.0000	1.0000	1.0000	1.0000	0.9997	0.9963	0.9729	0.8732	0.6020	0.1841
	13	1.0000	1.0000	1.0000	1.0000	1.0000	0.9995	0.9948	0.9647	0.8329	0.4510
	14	1.0000	1.0000	1.0000	1.0000	1.0000	1.0000	0.9995	0.9953	0.9648	0.7941
	15	1.0000	1.0000	1.0000	1.0000	1.0000	1.0000	1.0000	1.0000	1.0000	1.0000
20	0	0.1216	0.0115	0.0032	0.0008	0.0000	0.0000	0.0000	0.0000	0.0000	0.0000
	1	0.3917	0.0692	0.0243	0.0076	0.0005	0.0000	0.0000	0.0000	0.0000	0.0000
	2	0.6769	0.2061	0.0913	0.0355	0.0036	0.0002	0.0000	0.0000	0.0000	0.0000
	3	0.8670	0.4114	0.2252	0.1071	0.0160	0.0013	0.0001	0.0000	0.0000	0.0000
	4	0.9568	0.6296	0.4148	0.2375	0.0510	0.0059	0.0003	0.0000	0.0000	0.0000
	5	0.9887	0.8042	0.6172	0.4164	0.1256	0.0207	0.0016	0.0000	0.0000	0.0000
	6	0.9976	0.9133	0.7858	0.6080	0.2500	0.0577	0.0065	0.0003	0.0000	0.0000
	7	0.9996	0.9679	0.8982	0.7723	0.4159	0.1316	0.0210	0.0013	0.0000	0.0000
	8	0.9999	0.9900	0.9591	0.8867	0.5956	0.2517	0.0565	0.0051	0.0001	0.0000
	9	1.0000	0.9974	0.9861	0.9520	0.7553	0.4119	0.1275	0.0171	0.0006	0.0000
	10	1.0000	0.9994	0.9961	0.9829	0.8725	0.5881	0.2447	0.0480	0.0026	0.0000
	11	1.0000	0.9999	0.9991	0.9949	0.9435	0.7483	0.4044	0.1133	0.0100	0.0001
	12	1.0000	1.0000	0.9998	0.9987	0.9790	0.8684	0.5841	0.2277	0.0321	0.0004
	13	1.0000	1.0000	1.0000	0.9997	0.9935	0.9423	0.7500	0.3920	0.0867	0.0024
	14	1.0000	1.0000	1.0000	1.0000	0.9984	0.9793	0.8744	0.5836	0.1958	0.0113
	15	1.0000	1.0000	1.0000	1.0000	0.9997	0.9941	0.9490	0.7625	0.3704	0.0432
	16	1.0000	1.0000	1.0000	1.0000	1.0000	0.9987	0.9840	0.8929	0.5886	0.1330
	17	1.0000	1.0000	1.0000	1.0000	1.0000	0.9998	0.9964	0.9645	0.7939	0.3231
	18	1.0000	1.0000	1.0000	1.0000	1.0000	1.0000	0.9995	0.9924	0.9308	0.6083
	19	1.0000	1.0000	1.0000	1.0000	1.0000	1.0000	1.0000	0.9992	0.9885	0.8784
	20	1.0000	1.0000	1.0000	1.0000	1.0000	1.0000	1.0000	1.0000	1.0000	1.0000

Table A.3*

Poisson Probability Sums $\sum\limits_{x=0}^{r} p(x;\mu)$

					μ				
r	0.1	0.2	0.3	0.4	0.5	0.6	0.7	0.8	0.9
0	0.9048	0.8187	0.7408	0.6730	0.6065	0.5488	0.4966	0.4493	0.4066
1	0.9953	0.9825	0.9631	0.9384	0.9098	0.8781	0.8442	0.8088	0.7725
2	0.9998	0.9989	0.9964	0.9921	0.9856	0.9769	0.9659	0.9526	0.9371
3	1.0000	0.9999	0.9997	0.9992	0.9982	0.9966	0.9942	0.9909	0.9865
4		1.0000	1.0000	0.9999	0.9998	0.9996	0.9992	0.9986	0.9977
5				1.0000	1.0000	1.0000	0.9999	0.9998	0.9997
6							1.0000	1.0000	1.0000

					μ				
r	1.0	1.5	2.0	2.5	3.0	3.5	4.0	4.5	5.0
0	0.3679	0.2231	0.1353	0.0821	0.0498	0.0302	0.0183	0.0111	0.0067
1	0.7358	0.5578	0.4060	0.2873	0.1991	0.1359	0.0916	0.0611	0.0404
2	0.9197	0.8088	0.6767	0.5438	0.4232	0.3208	0.2381	0.1736	0.1247
3	0.9810	0.9344	0.8571	0.7576	0.6472	0.5366	0.4335	0.3423	0.2650
4	0.9963	0.9814	0.9473	0.8912	0.8153	0.7254	0.6288	0.5321	0.4405
5	0.9994	0.9955	0.9834	0.9580	0.9161	0.8576	0.7851	0.7029	0.6160
6	0.9999	0.9991	0.9955	0.9858	0.9665	0.9347	0.8893	0.8311	0.7622
7	1.0000	0.9998	0.9989	0.9958	0.9881	0.9733	0.9489	0.9134	0.8666
8		1.0000	0.9998	0.9989	0.9962	0.9901	0.9786	0.9597	0.9319
9			1.0000	0.9997	0.9989	0.9967	0.9919	0.9829	0.9682
10				0.9999	0.9997	0.9990	0.9972	0.9933	0.9863
11				1.0000	0.9999	0.9997	0.9991	0.9976	0.9945
12					1.0000	0.9999	0.9997	0.9992	0.9980
13						1.0000	0.9999	0.9997	0.9993
14							1.0000	0.9999	0.9998
15								1.0000	0.9999
16									1.0000

Table A.3*

Poisson Probability Sums $\sum\limits_{x=0}^{r} p(x;\mu)$ (*continued*)

					μ				
r	5.5	6.0	6.5	7.0	7.5	8.0	8.5	9.0	9.5
0	0.0041	0.0025	0.0015	0.0009	0.0006	0.0003	0.0002	0.0001	0.0001
1	0.0266	0.0174	0.0113	0.0073	0.0047	0.0030	0.0019	0.0012	0.0008
2	0.0884	0.0620	0.0430	0.0296	0.0203	0.0138	0.0093	0.0062	0.0042
3	0.2017	0.1512	0.1118	0.0818	0.0591	0.0424	0.0301	0.0212	0.0149
4	0.3575	0.2851	0.2237	0.1730	0.1321	0.0996	0.0744	0.0550	0.0403
5	0.5289	0.4457	0.3690	0.3007	0.2414	0.1912	0.1496	0.1157	0.0885
6	0.6860	0.6063	0.5265	0.4497	0.3782	0.3134	0.2562	0.2068	0.1649
7	0.8095	0.7440	0.6728	0.5987	0.5246	0.4530	0.3856	0.3239	0.2687
8	0.8944	0.8472	0.7916	0.7291	0.6620	0.5925	0.5231	0.4557	0.3918
9	0.9462	0.9161	0.8774	0.8305	0.7764	0.7166	0.6530	0.5874	0.5218
10	0.9747	0.9574	0.9332	0.9015	0.8622	0.8159	0.7634	0.7060	0.6453
11	0.9890	0.9799	0.9661	0.9466	0.9208	0.8881	0.8487	0.8030	0.7520
12	0.9955	0.9912	0.9840	0.9730	0.9573	0.9362	0.9091	0.8758	0.8364
13	0.9983	0.9964	0.9929	0.9872	0.9784	0.9658	0.9486	0.9261	0.8981
14	0.9994	0.9986	0.9970	0.9943	0.9897	0.9827	0.9726	0.9585	0.9400
15	0.9998	0.9995	0.9988	0.9976	0.9954	0.9918	0.9862	0.9780	0.9665
16	0.9999	0.9998	0.9996	0.9990	0.9980	0.9963	0.9934	0.9889	0.9823
17	1.0000	0.9999	0.9998	0.9996	0.9992	0.9984	0.9970	0.9947	0.9911
18		1.0000	0.9999	0.9999	0.9997	0.9994	0.9987	0.9976	0.9957
19			1.0000	1.0000	0.9999	0.9997	0.9995	0.9989	0.9980
20					1.0000	0.9999	0.9998	0.9996	0.9991
21						1.0000	0.9999	0.9998	0.9996
22							1.0000	0.9999	0.9999
23								1.0000	0.9999
24									1.0000

Table A.3*

Poisson Probability Sums $\sum_{x=0}^{r} p(x; \mu)$ (continued)

r	μ								
	10.0	11.0	12.0	13.0	14.0	15.0	16.0	17.0	18.0
0	0.0000	0.0000	0.0000						
1	0.0005	0.0002	0.0001	0.0000	0.0000				
2	0.0028	0.0012	0.0005	0.0002	0.0001	0.0000	0.0000		
3	0.0103	0.0049	0.0023	0.0010	0.0005	0.0002	0.0001	0.0000	0.0000
4	0.0293	0.0151	0.0076	0.0037	0.0018	0.0009	0.0004	0.0002	0.0001
5	0.0671	0.0375	0.0203	0.0107	0.0055	0.0028	0.0014	0.0007	0.0003
6	0.1301	0.0786	0.0458	0.0259	0.0142	0.0076	0.0040	0.0021	0.0010
7	0.2202	0.1432	0.0895	0.0540	0.0316	0.0180	0.0100	0.0054	0.0029
8	0.3328	0.2320	0.1550	0.0998	0.0621	0.0374	0.0220	0.0126	0.0071
9	0.4579	0.3405	0.2424	0.1658	0.1094	0.0699	0.0433	0.0261	0.0154
10	0.5830	0.4599	0.3472	0.2517	0.1757	0.1185	0.0774	0.0491	0.0304
11	0.6968	0.5793	0.4616	0.3532	0.2600	0.1848	0.1270	0.0847	0.0549
12	0.7916	0.6887	0.5760	0.4631	0.3585	0.2676	0.1931	0.1350	0.0917
13	0.8645	0.7813	0.6815	0.5730	0.4644	0.3632	0.2745	0.2009	0.1426
14	0.9165	0.8540	0.7720	0.6751	0.5704	0.4657	0.3675	0.2808	0.2081
15	0.9513	0.9074	0.8444	0.7636	0.6694	0.5681	0.4667	0.3715	0.2867
16	0.9730	0.9441	0.8987	0.8355	0.7559	0.6641	0.5660	0.4677	0.3750
17	0.9857	0.9678	0.9370	0.8905	0.8272	0.7489	0.6593	0.5640	0.4686
18	0.9928	0.9823	0.9626	0.9302	0.8826	0.8195	0.7423	0.6550	0.5622
19	0.9965	0.9907	0.9787	0.9573	0.9235	0.8752	0.8122	0.7363	0.6509
20	0.9984	0.9953	0.9884	0.9750	0.9521	0.9170	0.8682	0.8055	0.7307
21	0.9993	0.9977	0.9939	0.9859	0.9712	0.9469	0.9108	0.8615	0.7991
22	0.9997	0.9990	0.9970	0.9924	0.9833	0.9673	0.9418	0.9047	0.8551
23	0.9999	0.9995	0.9985	0.9960	0.9907	0.9805	0.9633	0.9367	0.8989
24	1.0000	0.9998	0.9993	0.9980	0.9950	0.9888	0.9777	0.9594	0.9317
25		0.9999	0.9997	0.9990	0.9974	0.9938	0.9869	0.9748	0.9554
26		1.0000	0.9999	0.9995	0.9987	0.9967	0.9925	0.9848	0.9718
27			0.9999	0.9998	0.9994	0.9983	0.9959	0.9912	0.9827
28			1.0000	0.9999	0.9997	0.9991	0.9978	0.9950	0.9897
29				1.0000	0.9999	0.9996	0.9989	0.9973	0.9941
30					0.9999	0.9998	0.9994	0.9986	0.9967
31					1.0000	0.9999	0.9997	0.9993	0.9982
32						1.0000	0.9999	0.9996	0.9990
33							0.9999	0.9998	0.9995
34							1.0000	0.9999	0.9998
35								1.0000	0.9999
36									0.9999
37									1.0000

* From E. C. Molina's *Poisson's Exponential Binomial Limit*, Copyright 1942, Princeton, N.J.: D. Van Nostrand Company, Inc., by permission of the publisher.

Table A.4
Areas Under the Normal Curve

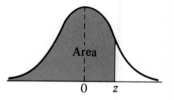

z	0.00	0.01	0.02	0.03	0.04	0.05	0.06	0.07	0.08	0.09
−3.4	0.0003	0.0003	0.0003	0.0003	0.0003	0.0003	0.0003	0.0003	0.0003	0.0002
−3.3	0.0005	0.0005	0.0005	0.0004	0.0004	0.0004	0.0004	0.0004	0.0004	0.0003
−3.2	0.0007	0.0007	0.0006	0.0006	0.0006	0.0006	0.0006	0.0005	0.0005	0.0005
−3.1	0.0010	0.0009	0.0009	0.0009	0.0008	0.0008	0.0008	0.0008	0.0007	0.0007
−3.0	0.0013	0.0013	0.0013	0.0012	0.0012	0.0011	0.0011	0.0011	0.0010	0.0010
−2.9	0.0019	0.0018	0.0017	0.0017	0.0016	0.0016	0.0015	0.0015	0.0014	0.0014
−2.8	0.0026	0.0025	0.0024	0.0023	0.0023	0.0022	0.0021	0.0021	0.0020	0.0019
−2.7	0.0035	0.0034	0.0033	0.0032	0.0031	0.0030	0.0029	0.0028	0.0027	0.0026
−2.6	0.0047	0.0045	0.0044	0.0043	0.0041	0.0040	0.0039	0.0038	0.0037	0.0036
−2.5	0.0062	0.0060	0.0059	0.0057	0.0055	0.0054	0.0052	0.0051	0.0049	0.0048
−2.4	0.0082	0.0080	0.0078	0.0075	0.0073	0.0071	0.0069	0.0068	0.0066	0.0064
−2.3	0.0107	0.0104	0.0102	0.0099	0.0096	0.0094	0.0091	0.0089	0.0087	0.0084
−2.2	0.0139	0.0136	0.0132	0.0129	0.0125	0.0122	0.0119	0.0116	0.0113	0.0110
−2.1	0.0179	0.0174	0.0170	0.0166	0.0162	0.0158	0.0154	0.0150	0.0146	0.0143
−2.0	0.0228	0.0222	0.0217	0.0212	0.0207	0.0202	0.0197	0.0192	0.0188	0.0183
−1.9	0.0287	0.0281	0.0274	0.0268	0.0262	0.0256	0.0250	0.0244	0.0239	0.0233
−1.8	0.0359	0.0352	0.0344	0.0336	0.0329	0.0322	0.0314	0.0307	0.0301	0.0294
−1.7	0.0446	0.0436	0.0427	0.0418	0.0409	0.0401	0.0392	0.0384	0.0375	0.0367
−1.6	0.0548	0.0537	0.0526	0.0516	0.0505	0.0495	0.0485	0.0475	0.0465	0.0455
−1.5	0.0668	0.0655	0.0643	0.0630	0.0618	0.0606	0.0594	0.0582	0.0571	0.0559
−1.4	0.0808	0.0793	0.0778	0.0764	0.0749	0.0735	0.0722	0.0708	0.0694	0.0681
−1.3	0.0968	0.0951	0.0934	0.0918	0.0901	0.0885	0.0869	0.0853	0.0838	0.0823
−1.2	0.1151	0.1131	0.1112	0.1093	0.1075	0.1056	0.1038	0.1020	0.1003	0.0985
−1.1	0.1357	0.1335	0.1314	0.1292	0.1271	0.1251	0.1230	0.1210	0.1190	0.1170
−1.0	0.1587	0.1562	0.1539	0.1515	0.1492	0.1469	0.1446	0.1423	0.1401	0.1379
−0.9	0.1841	0.1814	0.1788	0.1762	0.1736	0.1711	0.1685	0.1660	0.1635	0.1611
−0.8	0.2119	0.2090	0.2061	0.2033	0.2005	0.1977	0.1949	0.1922	0.1894	0.1867
−0.7	0.2420	0.2389	0.2358	0.2327	0.2296	0.2266	0.2236	0.2206	0.2177	0.2148
−0.6	0.2743	0.2709	0.2676	0.2643	0.2611	0.2578	0.2546	0.2514	0.2483	0.2451
−0.5	0.3085	0.3050	0.3015	0.2981	0.2946	0.2912	0.2877	0.2843	0.2810	0.2776
−0.4	0.3446	0.3409	0.3372	0.3336	0.3300	0.3264	0.3228	0.3192	0.3156	0.3121
−0.3	0.3821	0.3783	0.3745	0.3707	0.3669	0.3632	0.3594	0.3557	0.3520	0.3483
−0.2	0.4207	0.4168	0.4129	0.4090	0.4052	0.4013	0.3974	0.3936	0.3897	0.3859
−0.1	0.4602	0.4562	0.4522	0.4483	0.4443	0.4404	0.4364	0.4325	0.4286	0.4247
−0.0	0.5000	0.4960	0.4920	0.4880	0.4840	0.4801	0.4761	0.4721	0.4681	0.4641
0.0	0.5000	0.5040	0.5080	0.5120	0.5160	0.5199	0.5239	0.5279	0.5319	0.5359
0.1	0.5398	0.5438	0.5478	0.5517	0.5557	0.5596	0.5636	0.5675	0.5714	0.5753
0.2	0.5793	0.5832	0.5871	0.5910	0.5948	0.5987	0.6026	0.6064	0.6103	0.6141
0.3	0.6179	0.6217	0.6255	0.6293	0.6331	0.6368	0.6406	0.6443	0.6480	0.6517
0.4	0.6554	0.6591	0.6628	0.6664	0.6700	0.6736	0.6772	0.6808	0.6844	0.6879
0.5	0.6915	0.6950	0.6985	0.7019	0.7054	0.7088	0.7123	0.7157	0.7190	0.7224
0.6	0.7257	0.7291	0.7324	0.7357	0.7389	0.7422	0.7454	0.7486	0.7517	0.7549
0.7	0.7580	0.7611	0.7642	0.7673	0.7704	0.7734	0.7764	0.7794	0.7823	0.7852
0.8	0.7881	0.7910	0.7939	0.7967	0.7995	0.8023	0.8051	0.8078	0.8106	0.8133
0.9	0.8159	0.8186	0.8212	0.8238	0.8264	0.8289	0.8315	0.8340	0.8365	0.8389
1.0	0.8413	0.8438	0.8461	0.8485	0.8508	0.8531	0.8554	0.8577	0.8599	0.8621
1.1	0.8643	0.8665	0.8686	0.8708	0.8729	0.8749	0.8770	0.8790	0.8810	0.8830
1.2	0.8849	0.8869	0.8888	0.8907	0.8925	0.8944	0.8962	0.8980	0.8997	0.9015
1.3	0.9032	0.9049	0.9066	0.9082	0.9099	0.9115	0.9131	0.9147	0.9162	0.9177
1.4	0.9192	0.9207	0.9222	0.9236	0.9251	0.9265	0.9278	0.9292	0.9306	0.9319
1.5	0.9332	0.9345	0.9357	0.9370	0.9382	0.9394	0.9406	0.9418	0.9429	0.9441
1.6	0.9452	0.9463	0.9474	0.9484	0.9495	0.9505	0.9515	0.9525	0.9535	0.9545
1.7	0.9554	0.9564	0.9573	0.9582	0.9591	0.9599	0.9608	0.9616	0.9625	0.9633
1.8	0.9641	0.9649	0.9656	0.9664	0.9671	0.9678	0.9686	0.9693	0.9699	0.9706
1.9	0.9713	0.9719	0.9726	0.9732	0.9738	0.9744	0.9750	0.9756	0.9761	0.9767
2.0	0.9772	0.9778	0.9783	0.9788	0.9793	0.9798	0.9803	0.9808	0.9812	0.9817
2.1	0.9821	0.9826	0.9830	0.9834	0.9838	0.9842	0.9846	0.9850	0.9854	0.9857
2.2	0.9861	0.9864	0.9868	0.9871	0.9875	0.9878	0.9881	0.9884	0.9887	0.9890
2.3	0.9893	0.9896	0.9898	0.9901	0.9904	0.9906	0.9909	0.9911	0.9913	0.9916
2.4	0.9918	0.9920	0.9922	0.9925	0.9927	0.9929	0.9931	0.9932	0.9934	0.9936
2.5	0.9938	0.9940	0.9941	0.9943	0.9945	0.9946	0.9948	0.9949	0.9951	0.9952
2.6	0.9953	0.9955	0.9956	0.9957	0.9959	0.9960	0.9961	0.9962	0.9963	0.9964
2.7	0.9965	0.9966	0.9967	0.9968	0.9969	0.9970	0.9971	0.9972	0.9973	0.9974
2.8	0.9974	0.9975	0.9976	0.9977	0.9977	0.9978	0.9979	0.9979	0.9980	0.9981
2.9	0.9981	0.9982	0.9982	0.9983	0.9984	0.9984	0.9985	0.9985	0.9986	0.9986
3.0	0.9987	0.9987	0.9987	0.9988	0.9988	0.9989	0.9989	0.9989	0.9990	0.9990
3.1	0.9990	0.9991	0.9991	0.9991	0.9992	0.9992	0.9992	0.9992	0.9993	0.9993
3.2	0.9993	0.9993	0.9994	0.9994	0.9994	0.9994	0.9994	0.9995	0.9995	0.9995
3.3	0.9995	0.9995	0.9995	0.9996	0.9996	0.9996	0.9996	0.9996	0.9996	0.9997
3.4	0.9997	0.9997	0.9997	0.9997	0.9997	0.9997	0.9997	0.9997	0.9997	0.9998

Table A.5*
Critical Values of the *t* Distribution

degrees of freedom

.025 for 95% C.I.

	α				
ν	0.10	0.05	0.025	0.01	0.005
1	3.078	6.314	12.706	31.821	63.657
2	1.886	2.920	4.303	6.965	9.925
3	1.638	2.353	3.182	4.541	5.841
4	1.533	2.132	2.776	3.747	4.604
5	1.476	2.015	2.571	3.365	4.032
6	1.440	1.943	2.447	3.143	3.707
7	1.415	1.895	2.365	2.998	3.499
8	1.397	1.860	2.306	2.896	3.355
9	1.383	1.833	2.262	2.821	3.250
10	1.372	1.812	2.228	2.764	3.169
11	1.363	1.796	2.201	2.718	3.106
12	1.356	1.782	2.179	2.681	3.055
13	1.350	1.771	2.160	2.650	3.012
14	1.345	1.761	2.145	2.624	2.977
15	1.341	1.753	2.131	2.602	2.947
16	1.337	1.746	2.120	2.583	2.921
17	1.333	1.740	2.110	2.567	2.898
18	1.330	1.734	2.101	2.552	2.878
19	1.328	1.729	2.093	2.539	2.861
20	1.325	1.725	2.086	2.528	2.845
21	1.323	1.721	2.080	2.518	2.831
22	1.321	1.717	2.074	2.508	2.819
23	1.319	1.714	2.069	2.500	2.807
24	1.318	1.711	2.064	2.492	2.797
25	1.316	1.708	2.060	2.485	2.787
26	1.315	1.706	2.056	2.479	2.779
27	1.314	1.703	2.052	2.473	2.771
28	1.313	1.701	2.048	2.467	2.763
29	1.311	1.699	2.045	2.462	2.756
inf.	1.282	1.645	1.960	2.326	2.576

* Table A.5 is taken from Table IV of R. A. Fisher: *Statistical Methods for Research Workers*, published by Oliver & Boyd Ltd., Edinburgh, by permission of the author and publishers.

Table A.6*
Critical Values of the Chi-square Distribution

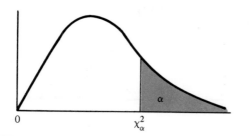

ν	α							
	0.995	0.99	0.975	0.95	0.05	0.025	0.01	0.005
1	0.0⁴393	0.0³157	0.0³982	0.0²393	3.841	5.024	6.635	7.879
2	0.0100	0.0201	0.0506	0.103	5.991	7.378	9.210	10.597
3	0.0717	0.115	0.216	0.352	7.815	9.348	11.345	12.838
4	0.207	0.297	0.484	0.711	9.488	11.143	13.277	14.860
5	0.412	0.554	0.831	1.145	11.070	12.832	15.086	16.750
6	0.676	0.872	1.237	1.635	12.592	14.449	16.812	18.548
7	0.989	1.239	1.690	2.167	14.067	16.013	18.475	20.278
8	1.344	1.646	2.180	2.733	15.507	17.535	20.090	21.955
9	1.735	2.088	2.700	3.325	16.919	19.023	21.666	23.589
10	2.156	2.558	3.247	3.940	18.307	20.483	23.209	25.188
11	2.603	3.053	3.816	4.575	19.675	21.920	24.725	26.757
12	3.074	3.571	4.404	5.226	21.026	23.337	26.217	28.300
13	3.565	4.107	5.009	5.892	22.362	24.736	27.688	29.819
14	4.075	4.660	5.629	6.571	23.685	26.119	29.141	31.319
15	4.601	5.229	6.262	7.261	24.996	27.488	30.578	32.801
16	5.142	5.812	6.908	7.962	26.296	28.845	32.000	34.267
17	5.697	6.408	7.564	8.672	27.587	30.191	33.409	35.718
18	6.265	7.015	8.231	9.390	28.869	31.526	34.805	37.156
19	6.844	7.633	8.907	10.117	30.144	32.852	36.191	38.582
20	7.434	8.260	9.591	10.851	31.410	34.170	37.566	39.997
21	8.034	8.897	10.283	11.591	32.671	35.479	38.932	41.401
22	8.643	9.542	10.982	12.338	33.924	36.781	40.289	42.796
23	9.260	10.196	11.689	13.091	35.172	38.076	41.638	44.181
24	9.886	10.856	12.401	13.848	36.415	39.364	42.980	45.558
25	10.520	11.524	13.120	14.611	37.652	40.646	44.314	46.928
26	11.160	12.198	13.844	15.379	38.885	41.923	45.642	48.290
27	11.808	12.879	14.573	16.151	40.113	43.194	46.963	49.645
28	12.461	13.565	15.308	16.928	41.337	44.461	48.278	50.993
29	13.121	14.256	16.047	17.708	42.557	45.722	49.588	52.336
30	13.787	14.953	16.791	18.493	43.773	46.979	50.892	53.672

* Abridged from Table 8 of *Biometrika Tables for Statisticians*, Vol. I, by permission of E. S. Pearson and the Biometrika Trustees.

Table A.7*

Critical Values of the F Distribution

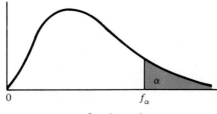

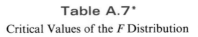

$$f_{0.05}(\nu_1, \nu_2)$$

ν_2	ν_1								
	1	2	3	4	5	6	7	8	9
1	161.4	199.5	215.7	224.6	230.2	234.0	236.8	238.9	240.5
2	18.51	19.00	19.16	19.25	19.30	19.33	19.35	19.37	19.38
3	10.13	9.55	9.28	9.12	9.01	8.94	8.89	8.85	8.81
4	7.71	6.94	6.59	6.39	6.26	6.16	6.09	6.04	6.00
5	6.61	5.79	5.41	5.19	5.05	4.95	4.88	4.82	4.77
6	5.99	5.14	4.76	4.53	4.39	4.28	4.21	4.15	4.10
7	5.59	4.74	4.35	4.12	3.97	3.87	3.79	3.73	3.68
8	5.32	4.46	4.07	3.84	3.69	3.58	3.50	3.44	3.39
9	5.12	4.26	3.86	3.63	3.48	3.37	3.29	3.23	3.18
10	4.96	4.10	3.71	3.48	3.33	3.22	3.14	3.07	3.02
11	4.84	3.98	3.59	3.36	3.20	3.09	3.01	2.95	2.90
12	4.75	3.89	3.49	3.26	3.11	3.00	2.91	2.85	2.80
13	4.67	3.81	3.41	3.18	3.03	2.92	2.83	2.77	2.71
14	4.60	3.74	3.34	3.11	2.96	2.85	2.76	2.70	2.65
15	4.54	3.68	3.29	3.06	2.90	2.79	2.71	2.64	2.59
16	4.49	3.63	3.24	3.01	2.85	2.74	2.66	2.59	2.54
17	4.45	3.59	3.20	2.96	2.81	2.70	2.61	2.55	2.49
18	4.41	3.55	3.16	2.93	2.77	2.66	2.58	2.51	2.46
19	4.38	3.52	3.13	2.90	2.74	2.63	2.54	2.48	2.42
20	4.35	3.49	3.10	2.87	2.71	2.60	2.51	2.45	2.39
21	4.32	3.47	3.07	2.84	2.68	2.57	2.49	2.42	2.37
22	4.30	3.44	3.05	2.82	2.66	2.55	2.46	2.40	2.34
23	4.28	3.42	3.03	2.80	2.64	2.53	2.44	2.37	2.32
24	4.26	3.40	3.01	2.78	2.62	2.51	2.42	2.36	2.30
25	4.24	3.39	2.99	2.76	2.60	2.49	2.40	2.34	2.28
26	4.23	3.37	2.98	2.74	2.59	2.47	2.39	2.32	2.27
27	4.21	3.35	2.96	2.73	2.57	2.46	2.37	2.31	2.25
28	4.20	3.34	2.95	2.71	2.56	2.45	2.36	2.29	2.24
29	4.18	3.33	2.93	2.70	2.55	2.43	2.35	2.28	2.22
30	4.17	3.32	2.92	2.69	2.53	2.42	2.33	2.27	2.21
40	4.08	3.23	2.84	2.61	2.45	2.34	2.25	2.18	2.12
60	4.00	3.15	2.76	2.53	2.37	2.25	2.17	2.10	2.04
120	3.92	3.07	2.68	2.45	2.29	2.17	2.09	2.02	1.96
∞	3.84	3.00	2.60	2.37	2.21	2.10	2.01	1.94	1.88

Table A.7*
Critical Values of the F Distribution (*continued*)

$$f_{0.05}(v_1, v_2)$$

v_2	\multicolumn{10}{c}{v_1}
	10	12	15	20	24	30	40	60	120	∞
1	241.9	243.9	245.9	248.0	249.1	250.1	251.1	252.2	253.3	254.3
2	19.40	19.41	19.43	19.45	19.45	19.46	19.47	19.48	19.49	19.50
3	8.79	8.74	8.70	8.66	8.64	8.62	8.59	8.57	8.55	8.53
4	5.96	5.91	5.86	5.80	5.77	5.75	5.72	5.69	5.66	5.63
5	4.74	4.68	4.62	4.56	4.53	4.50	4.46	4.43	4.40	4.36
6	4.06	4.00	3.94	3.87	3.84	3.81	3.77	3.74	3.70	3.67
7	3.64	3.57	3.51	3.44	3.41	3.38	3.34	3.30	3.27	3.23
8	3.35	3.28	3.22	3.15	3.12	3.08	3.04	3.01	2.97	2.93
9	3.14	3.07	3.01	2.94	2.90	2.86	2.83	2.79	2.75	2.71
10	2.98	2.91	2.85	2.77	2.74	2.70	2.66	2.62	2.58	2.54
11	2.85	2.79	2.72	2.65	2.61	2.57	2.53	2.49	2.45	2.40
12	2.75	2.69	2.62	2.54	2.51	2.47	2.43	2.38	2.34	2.30
13	2.67	2.60	2.53	2.46	2.42	2.38	2.34	2.30	2.25	2.21
14	2.60	2.53	2.46	2.39	2.35	2.31	2.27	2.22	2.18	2.13
15	2.54	2.48	2.40	2.33	2.29	2.25	2.20	2.16	2.11	2.07
16	2.49	2.42	2.35	2.28	2.24	2.19	2.15	2.11	2.06	2.01
17	2.45	2.38	2.31	2.23	2.19	2.15	2.10	2.06	2.01	1.96
18	2.41	2.34	2.27	2.19	2.15	2.11	2.06	2.02	1.97	1.92
19	2.38	2.31	2.23	2.16	2.11	2.07	2.03	1.98	1.93	1.88
20	2.35	2.28	2.20	2.12	2.08	2.04	1.99	1.95	1.90	1.84
21	2.32	2.25	2.18	2.10	2.05	2.01	1.96	1.92	1.87	1.81
22	2.30	2.23	2.15	2.07	2.03	1.98	1.94	1.89	1.84	1.78
23	2.27	2.20	2.13	2.05	2.01	1.96	1.91	1.86	1.81	1.76
24	2.25	2.18	2.11	2.03	1.98	1.94	1.89	1.84	1.79	1.73
25	2.24	2.16	2.09	2.01	1.96	1.92	1.87	1.82	1.77	1.71
26	2.22	2.15	2.07	1.99	1.95	1.90	1.85	1.80	1.75	1.69
27	2.20	2.13	2.06	1.97	1.93	1.88	1.84	1.79	1.73	1.67
28	2.19	2.12	2.04	1.96	1.91	1.87	1.82	1.77	1.71	1.65
29	2.18	2.10	2.03	1.94	1.90	1.85	1.81	1.75	1.70	1.64
30	2.16	2.09	2.01	1.93	1.89	1.84	1.79	1.74	1.68	1.62
40	2.08	2.00	1.92	1.84	1.79	1.74	1.69	1.64	1.58	1.51
60	1.99	1.92	1.84	1.75	1.70	1.65	1.59	1.53	1.47	1.39
120	1.91	1.83	1.75	1.66	1.61	1.55	1.50	1.43	1.35	1.25
∞	1.83	1.75	1.67	1.57	1.52	1.46	1.39	1.32	1.22	1.00

Table A.7*

Critical Values of the F Distribution (*continued*)

$$f_{0.01}(v_1, v_2)$$

v_2	\multicolumn{9}{c}{v_1}
	1	2	3	4	5	6	7	8	9
1	4052	4999.5	5403	5625	5764	5859	5928	5981	6022
2	98.50	99.00	99.17	99.25	99.30	99.33	99.36	99.37	99.39
3	34.12	30.82	29.46	28.71	28.24	27.91	27.67	27.49	27.35
4	21.20	18.00	16.69	15.98	15.52	15.21	14.98	14.80	14.66
5	16.26	13.27	12.06	11.39	10.97	10.67	10.46	10.29	10.16
6	13.75	10.92	9.78	9.15	8.75	8.47	8.26	8.10	7.98
7	12.25	9.55	8.45	7.85	7.46	7.19	6.99	6.84	6.72
8	11.26	8.65	7.59	7.01	6.63	6.37	6.18	6.03	5.91
9	10.56	8.02	6.99	6.42	6.06	5.80	5.61	5.47	5.35
10	10.04	7.56	6.55	5.99	5.64	5.39	5.20	5.06	4.94
11	9.65	7.21	6.22	5.67	5.32	5.07	4.89	4.74	4.63
12	9.33	6.93	5.95	5.41	5.06	4.82	4.64	4.50	4.39
13	9.07	6.70	5.74	5.21	4.86	4.62	4.44	4.30	4.19
14	8.86	6.51	5.56	5.04	4.69	4.46	4.28	4.14	4.03
15	8.68	6.36	5.42	4.89	4.56	4.32	4.14	4.00	3.89
16	8.53	6.23	5.29	4.77	4.44	4.20	4.03	3.89	3.78
17	8.40	6.11	5.18	4.67	4.34	4.10	3.93	3.79	3.68
18	8.29	6.01	5.09	4.58	4.25	4.01	3.84	3.71	3.60
19	8.18	5.93	5.01	4.50	4.17	3.94	3.77	3.63	3.52
20	8.10	5.85	4.94	4.43	4.10	3.87	3.70	3.56	3.46
21	8.02	5.78	4.87	4.37	4.04	3.81	3.64	3.51	3.40
22	7.95	5.72	4.82	4.31	3.99	3.76	3.59	3.45	3.35
23	7.88	5.66	4.76	4.26	3.94	3.71	3.54	3.41	3.30
24	7.82	5.61	4.72	4.22	3.90	3.67	3.50	3.36	3.26
25	7.77	5.57	4.68	4.18	3.85	3.63	3.46	3.32	3.22
26	7.72	5.53	4.64	4.14	3.82	3.59	3.42	3.29	3.18
27	7.68	5.49	4.60	4.11	3.78	3.56	3.39	3.26	3.15
28	7.64	5.45	4.57	4.07	3.75	3.53	3.36	3.23	3.12
29	7.60	5.42	4.54	4.04	3.73	3.50	3.33	3.20	3.09
30	7.56	5.39	4.51	4.02	3.70	3.47	3.30	3.17	3.07
40	7.31	5.18	4.31	3.83	3.51	3.29	3.12	2.99	2.89
60	7.08	4.98	4.13	3.65	3.34	3.12	2.95	2.82	2.72
120	6.85	4.79	3.95	3.48	3.17	2.96	2.79	2.66	2.56
∞	6.63	4.61	3.78	3.32	3.02	2.80	2.64	2.51	2.41

Table A.7*

Critical Values of the F Distribution (*continued*)

$$f_{0.01}(v_1, v_2)$$

v_2	\multicolumn{10}{c}{v_1}
	10	12	15	20	24	30	40	60	120	∞
1	6056	6106	6157	6209	6235	6261	6287	6313	6339	6366
2	99.40	99.42	99.43	99.45	99.46	99.47	99.47	99.48	99.49	99.50
3	27.23	27.05	26.87	26.69	26.60	26.50	26.41	26.32	26.22	26.13
4	14.55	14.37	14.20	14.02	13.93	13.84	13.75	13.65	13.56	13.46
5	10.05	9.89	9.72	9.55	9.47	9.38	9.29	9.20	9.11	9.02
6	7.87	7.72	7.56	7.40	7.31	7.23	7.14	7.06	6.97	6.88
7	6.62	6.47	6.31	6.16	6.07	5.99	5.91	5.82	5.74	5.65
8	5.81	5.67	5.52	5.36	5.28	5.20	5.12	5.03	4.95	4.86
9	5.26	5.11	4.96	4.81	4.73	4.65	4.57	4.48	4.40	4.31
10	4.85	4.71	4.56	4.41	4.33	4.25	4.17	4.08	4.00	3.91
11	4.54	4.40	4.25	4.10	4.02	3.94	3.86	3.78	3.69	3.60
12	4.30	4.16	4.01	3.86	3.78	3.70	3.62	3:54	3.45	3.36
13	4.10	3.96	3.82	3.66	3.59	3.51	3.43	3.34	3.25	3.17
14	3.94	3.80	3.66	3.51	3.43	3.35	3.27	3.18	3.09	3.00
15	3.80	3.67	3.52	3.37	3.29	3.21	3.13	3.05	2.96	2.87
16	3.69	3.55	3.41	3.26	3.18	3.10	3.02	2.93	2.84	2.75
17	3.59	3.46	3.31	3.16	3.08	3.00	2.92	2.83	2.75	2.65
18	3.51	3.37	3.23	3.08	3.00	2.92	2.84	2.75	2.66	2.57
19	3.43	3.30	3.15	3.00	2.92	2.84	2.76	2.67	2.58	2.49
20	3.37	3.23	3.09	2.94	2.86	2.78	2.69	2.61	2.52	2.42
21	3.31	3.17	3.03	2.88	2.80	2.72	2.64	2.55	2.46	2.36
22	3.26	3.12	2.98	2.83	2.75	2.67	2.58	2.50	2.40	2.31
23	3.21	3.07	2.93	2.78	2.70	2.62	2.54	2.45	2.35	2.26
24	3.17	3.03	2.89	2.74	2.66	2.58	2.49	2.40	2.31	2.21
25	3.13	2.99	2.85	2.70	2.62	2.54	2.45	2.36	2.27	2.17
26	3.09	2.96	2.81	2.66	2.58	2.50	2.42	2.33	2.23	2.13
27	3.06	2.93	2.78	2.63	2.55	2.47	2.38	2.29	2.20	2.10
28	3.03	2.90	2.75	2.60	2.52	2.44	2.35	2.26	2.17	2.06
29	3.00	2.87	2.73	2.57	2.49	2.41	2.33	2.23	2.14	2.03
30	2.98	2.84	2.70	2.55	2.47	2.39	2.30	2.21	2.11	2.01
40	2.80	2.66	2.52	2.37	2.29	2.20	2.11	2.02	1.92	1.80
60	2.63	2.50	2.35	2.20	2.12	2.03	1.94	1.84	1.73	1.60
120	2.47	2.34	2.19	2.03	1.95	1.86	1.76	1.66	1.53	1.38
∞	2.32	2.18	2.04	1.88	1.79	1.70	1.59	1.47	1.32	1.00

Table A.8*

$\Pr(U \leq u | H_0$ is true) in the Wilcoxon Two-Sample Test

$n_2 = 3$

u	n_1		
	1	2	3
0	0.250	0.100	0.050
1	0.500	0.200	0.100
2	0.750	0.400	0.200
3		0.600	0.350
4			0.500
5			0.650

$n_2 = 4$

u	n_1			
	1	2	3	4
0	0.200	0.067	0.028	0.014
1	0.400	0.133	0.057	0.029
2	0.600	0.267	0.114	0.057
3		0.400	0.200	0.100
4		0.600	0.314	0.171
5			0.429	0.243
6			0.571	0.343
7				0.443
8				0.557

$n_2 = 5$

u	n_1				
	1	2	3	4	5
0	0.167	0.047	0.018	0.008	0.004
1	0.333	0.095	0.036	0.016	0.008
2	0.500	0.190	0.071	0.032	0.016
3	0.667	0.286	0.125	0.056	0.028
4		0.429	0.196	0.095	0.048
5		0.571	0.286	0.143	0.075
6			0.393	0.206	0.111
7			0.500	0.278	0.155
8			0.607	0.365	0.210
9				0.452	0.274
10				0.548	0.345
11					0.421
12					0.500
13					0.579

$Pr(U \le u|H_0$ is true) in the Wilcoxon Two-Sample Test (*continued*)

$n_2 = 6$

u	1	2	3	4	5	6
				n_1		
0	0.143	0.036	0.012	0.005	0.002	0.001
1	0.286	0.071	0.024	0.010	0.004	0.002
2	0.428	0.143	0.048	0.019	0.009	0.004
3	0.571	0.214	0.083	0.033	0.015	0.008
4		0.321	0.131	0.057	0.026	0.013
5		0.429	0.190	0.086	0.041	0.021
6		0.571	0.274	0.129	0.063	0.032
7			0.357	0.176	0.089	0.047
8			0.452	0.238	0.123	0.066
9			0.548	0.305	0.165	0.090
10				0.381	0.214	0.120
11				0.457	0.268	0.155
12				0.545	0.331	0.197
13					0.396	0.242
14					0.465	0.294
15					0.535	0.350
16						0.409
17						0.469
18						0.531

$n_2 = 7$

u	1	2	3	4	5	6	7
				n_1			
0	0.125	0.028	0.008	0.003	0.001	0.001	0.000
1	0.250	0.056	0.017	0.006	0.003	0.001	0.001
2	0.375	0.111	0.033	0.012	0.005	0.002	0.001
3	0.500	0.167	0.058	0.021	0.009	0.004	0.002
4	0.625	0.250	0.092	0.036	0.015	0.007	0.003
5		0.333	0.133	0.055	0.024	0.011	0.006
6		0.444	0.192	0.082	0.037	0.017	0.009
7		0.556	0.258	0.115	0.053	0.026	0.013
8			0.333	0.158	0.074	0.037	0.019
9			0.417	0.206	0.101	0.051	0.027
10			0.500	0.264	0.134	0.069	0.036
11			0.583	0.324	0.172	0.090	0.049
12				0.394	0.216	0.117	0.064
13				0.464	0.265	0.147	0.082
14				0.538	0.319	0.183	0.104
15					0.378	0.223	0.130
16					0.438	0.267	0.159
17					0.500	0.314	0.191
18					0.562	0.365	0.228
19						0.418	0.267
20						0.473	0.310
21						0.527	0.355
22							0.402
23							0.451
24							0.500
25							0.549

Table A.8*

$\Pr(U \le u | H_0$ is true) in the Wilcoxon Two-Sample Test (*continued*)

$$n_2 = 8$$

	n_1							
u	1	2	3	4	5	6	7	8
0	0.111	0.022	0.006	0.002	0.001	0.000	0.000	0.000
1	0.222	0.044	0.012	0.004	0.002	0.001	0.000	0.000
2	0.333	0.089	0.024	0.008	0.003	0.001	0.001	0.000
3	0.444	0.133	0.042	0.014	0.005	0.002	0.001	0.001
4	0.556	0.200	0.067	0.024	0.009	0.004	0.002	0.001
5		0.267	0.097	0.036	0.015	0.006	0.003	0.001
6		0.356	0.139	0.055	0.023	0.010	0.005	0.002
7		0.444	0.188	0.077	0.033	0.015	0.007	0.003
8		0.556	0.248	0.107	0.047	0.021	0.010	0.005
9			0.315	0.141	0.064	0.030	0.014	0.007
10			0.387	0.184	0.085	0.041	0.020	0.010
11			0.461	0.230	0.111	0.054	0.027	0.014
12			0.539	0.285	0.142	0.071	0.036	0.019
13				0.341	0.177	0.091	0.047	0.025
14				0.404	0.217	0.114	0.060	0.032
15				0.467	0.262	0.141	0.076	0.041
16				0.533	0.311	0.172	0.095	0.052
17					0.362	0.207	0.116	0.065
18					0.416	0.245	0.140	0.080
19					0.472	0.286	0.168	0.097
20					0.528	0.331	0.198	0.117
21						0.377	0.232	0.139
22						0.426	0.268	0.164
23						0.475	0.306	0.191
24						0.525	0.347	0.221
25							0.389	0.253
26							0.433	0.287
27							0.478	0.323
28							0.522	0.360
29								0.399
30								0.439
31								0.480
32								0.520

* Reproduced from H. B. Mann and D. R. Whitney, "On a test of whether one of two random variables is stochastically larger than the other," *Ann. Math. Statist.*, vol. 18, pp. 52–54 (1947), by permission of the authors and the publisher.

Table A.9*

Critical Values of U in the Wilcoxon Two-Sample Test

One-Tailed Test at $\alpha = 0.001$ or Two-Tailed Test at $\alpha = 0.002$

n_1	9	10	11	12	13	14	15	16	17	18	19	20
1												
2												
3									0	0	0	0
4		0	0	0	1	1	1	2	2	3	3	3
5	1	1	2	2	3	3	4	5	5	6	7	7
6	2	3	4	4	5	6	7	8	9	10	11	12
7	3	5	6	7	8	9	10	11	13	14	15	16
8	5	6	8	9	11	12	14	15	17	18	20	21
9	7	8	10	12	14	15	17	19	21	23	25	26
10	8	10	12	14	17	19	21	23	25	27	29	32
11	10	12	15	17	20	22	24	27	29	32	34	37
12	12	14	17	20	23	25	28	31	34	37	40	42
13	14	17	20	23	26	29	32	35	38	42	45	48
14	15	19	22	25	29	32	36	39	43	46	50	54
15	17	21	24	28	32	36	40	43	47	51	55	59
16	19	23	27	31	35	39	43	48	52	56	60	65
17	21	25	29	34	38	43	47	52	57	61	66	70
18	23	27	32	37	42	46	51	56	61	66	71	76
19	25	29	34	40	45	50	55	60	66	71	77	82
20	26	32	37	42	48	54	59	65	70	76	82	88

One-Tailed Test at $\alpha = 0.01$ or Two-Tailed Test at $\alpha = 0.02$

n_1	9	10	11	12	13	14	15	16	17	18	19	20
1												
2					0	0	0	0	0	0	1	1
3	1	1	1	2	2	2	3	3	4	4	4	5
4	3	3	4	5	5	6	7	7	8	9	9	10
5	5	6	7	8	9	10	11	12	13	14	15	16
6	7	8	9	11	12	13	15	16	18	19	20	22
7	9	11	12	14	16	17	19	21	23	24	26	28
8	11	13	15	17	20	22	24	26	28	30	32	34
9	14	16	18	21	23	26	28	31	33	36	38	40
10	16	19	22	24	27	30	33	36	38	41	44	47
11	18	22	25	28	31	34	37	41	44	47	50	53
12	21	24	28	31	35	38	42	46	49	53	56	60
13	23	27	31	35	39	43	47	51	55	59	63	67
14	26	30	34	38	43	47	51	56	60	65	69	73
15	28	33	37	42	47	51	56	61	66	70	75	80
16	31	36	41	46	51	56	61	66	71	76	82	87
17	33	38	44	49	55	60	66	71	77	82	88	93
18	36	41	47	53	59	65	70	76	82	88	94	100
19	38	44	50	56	63	69	75	82	88	94	101	107
20	40	47	53	60	67	73	80	87	93	100	107	114

Table A.9* (*continued*)

One-Tailed Test at α = 0.025 or Two-Tailed Test at α = 0.05

n_1	n_2											
	9	10	11	12	13	14	15	16	17	18	19	20
1												
2	0	0	0	1	1	1	1	1	2	2	2	2
3	2	3	3	4	4	5	5	6	6	7	7	8
4	4	5	6	7	8	9	10	11	11	12	13	13
5	7	8	9	11	12	13	14	15	17	18	19	20
6	10	11	13	14	16	17	19	21	22	24	25	27
7	12	14	16	18	20	22	24	26	28	30	32	34
8	15	17	19	22	24	26	29	31	34	36	38	41
9	17	20	23	26	28	31	34	37	39	42	45	48
10	20	23	26	29	33	36	39	42	45	48	52	55
11	23	26	30	33	37	40	44	47	51	55	58	62
12	26	29	33	37	41	45	49	53	57	61	65	69
13	28	33	37	41	45	50	54	59	63	67	72	76
14	31	36	40	45	50	55	59	64	67	74	78	83
15	34	39	44	49	54	59	64	70	75	80	85	90
16	37	42	47	53	59	64	70	75	81	86	92	98
17	39	45	51	57	63	67	75	81	87	93	99	105
18	42	48	55	61	67	74	80	86	93	99	106	112
19	45	52	58	65	72	78	85	92	99	106	113	119
20	48	55	62	69	76	83	90	98	105	112	119	127

One-Tailed Test at α = 0.05 or Two-Tailed Test at α = 0.10

n_1	n_2											
	9	10	11	12	13	14	15	16	17	18	19	20
1											0	0
2	1	1	1	2	2	2	3	3	3	4	4	4
3	3	4	5	5	6	7	7	8	9	9	10	11
4	6	7	8	9	10	11	12	14	15	16	17	18
5	9	11	12	13	15	16	18	19	20	22	23	25
6	12	14	16	17	19	21	23	25	26	28	30	32
7	15	17	19	21	24	26	28	30	33	35	37	39
8	18	20	23	26	28	31	33	36	39	41	44	47
9	21	24	27	30	33	36	39	42	45	48	51	54
10	24	27	31	34	37	41	44	48	51	55	58	62
11	27	31	34	38	42	46	50	54	57	61	65	69
12	30	34	38	42	47	51	55	60	64	68	72	77
13	33	37	42	47	51	56	61	65	70	75	80	84
14	36	41	46	51	56	61	66	71	77	82	87	92
15	39	44	50	55	61	66	72	77	83	88	94	100
16	42	48	54	60	65	71	77	83	89	95	101	107
17	45	51	57	64	70	77	83	89	96	102	109	115
18	48	55	61	68	75	82	88	95	102	109	116	123
19	51	58	65	72	80	87	94	101	109	116	123	130
20	54	62	69	77	84	92	100	107	115	123	130	138

* Adapted and abridged from Tables 1, 3, 5, and 7 of D. Auble, "Extended tables for the Mann–Whitney statistic," *Bulletin of the Institute of Educational Research at Indiana University*, vol. 1, no. 2 (1953), by permission of the director.

Table A.10*

Critical Values of W in the Wilcoxon Test for Paired Observations

n	One-sided $\alpha = 0.01$ Two-sided $\alpha = 0.02$	One-sided $\alpha = 0.025$ Two-sided $\alpha = 0.05$	One-sided $\alpha = 0.05$ Two-sided $\alpha = 0.10$
5			1
6		1	2
7	0	2	4
8	2	4	6
9	3	6	8
10	5	8	11
11	7	11	14
12	10	14	17
13	13	17	21
14	16	21	26
15	20	25	30
16	24	30	36
17	28	35	41
18	33	40	47
19	38	46	54
20	43	52	60
21	49	59	68
22	56	66	75
23	62	73	83
24	69	81	92
25	77	90	101
26	85	98	110
27	93	107	120
28	102	117	130
29	111	127	141
30	120	137	152

* Reproduced from F. Wilcoxon and R. A. Wilcox, *Some Rapid Approximate Statistical Procedures*, American Cyanamid Company, Pearl River, N.Y., 1964, by permission of the American Cyanamid Company.

Table A.11*

Least Significant Studentized Ranges r_p

$\alpha = 0.05$

ν	p								
	2	3	4	5	6	7	8	9	10
1	17.97	17.97	17.97	17.97	17.97	17.97	17.97	17.97	17.97
2	6.085	6.085	6.085	6.085	6.085	6.085	6.085	6.085	6.085
3	4.501	4.516	4.516	4.516	4.516	4.516	4.516	4.516	4.516
4	3.927	4.013	4.033	4.033	4.033	4.033	4.033	4.033	4.033
5	3.635	3.749	3.797	3.814	3.814	3.814	3.814	3.814	3.814
6	3.461	3.587	3.649	3.680	3.694	3.697	3.697	3.697	3.697
7	3.344	3.477	3.548	3.588	3.611	3.622	3.626	3.626	3.626
8	3.261	3.399	3.475	3.521	3.549	3.566	3.575	3.579	3.579
9	3.199	3.339	3.420	3.470	3.502	3.523	3.536	3.544	3.547
10	3.151	3.293	3.376	3.430	3.465	3.489	3.505	3.516	3.522
11	3.113	3.256	3.342	3.397	3.435	3.462	3.480	3.493	3.501
12	3.082	3.225	3.313	3.370	3.410	3.439	3.459	3.474	3.484
13	3.055	3.200	3.289	3.348	3.389	3.419	3.442	3.458	3.470
14	3.033	3.178	3.268	3.329	3.372	3.403	3.426	3.444	3.457
15	3.014	3.160	3.250	3.312	3.356	3.389	3.413	3.432	3.446
16	2.998	3.144	3.235	3.298	3.343	3.376	3.402	3.422	3.437
17	2.984	3.130	3.222	3.285	3.331	3.366	3.392	3.412	3.429
18	2.971	3.118	3.210	3.274	3.321	3.356	3.383	3.405	3.421
19	2.960	3.107	3.199	3.264	3.311	3.347	3.375	3.397	3.415
20	2.950	3.097	3.190	3.255	3.303	3.339	3.368	3.391	3.409
24	2.919	3.066	3.160	3.226	3.276	3.315	3.345	3.370	3.390
30	2.888	3.035	3.131	3.199	3.250	3.290	3.322	3.349	3.371
40	2.858	3.006	3.102	3.171	3.224	3.266	3.300	3.328	3.352
60	2.829	2.976	3.073	3.143	3.198	3.241	3.277	3.307	3.333
120	2.800	2.947	3.045	3.116	3.172	3.217	3.254	3.287	3.314
∞	2.772	2.918	3.017	3.089	3.146	3.193	3.232	3.265	3.294

Table A.11*

Least Significant Studentized Ranges r_p (*continued*)

$\alpha = 0.01$

ν					p				
	2	3	4	5	6	7	8	9	10
1	90.03	90.03	90.03	90.03	90.03	90.03	90.03	90.03	90.03
2	14.04	14.04	14.04	14.04	14.04	14.04	14.04	14.04	14.04
3	8.261	8.321	8.321	8.321	8.321	8.321	8.321	8.321	8.321
4	6.512	6.677	6.740	6.756	6.756	6.756	6.756	6.756	6.756
5	5.702	5.893	5.989	6.040	6.065	6.074	6.074	6.074	6.074
6	5.243	5.439	5.549	5.614	5.655	5.680	5.694	5.701	5.703
7	4.949	5.145	5.260	5.334	5.383	5.416	5.439	5.454	5.464
8	4.746	4.939	5.057	5.135	5.189	5.227	5.256	5.276	5.291
9	4.596	4.787	4.906	4.986	5.043	5.086	5.118	5.142	5.160
10	4.482	4.671	4.790	4.871	4.931	4.975	5.010	5.037	5.058
11	4.392	4.579	4.697	4.780	4.841	4.887	4.924	4.952	4.975
12	4.320	4.504	4.622	4.706	4.767	4.815	4.852	4.883	4.907
13	4.260	4.442	4.560	4.644	4.706	4.755	4.793	4.824	4.850
14	4.210	4.391	4.508	4.591	4.654	4.704	4.743	4.775	4.802
15	4.168	4.347	4.463	4.547	4.610	4.660	4.700	4.733	4.760
16	4.131	4.309	4.425	4.509	4.572	4.622	4.663	4.696	4.724
17	4.099	4.275	4.391	4.475	4.539	4.589	4.630	4.664	4.693
18	4.071	4.246	4.362	4.445	4.509	4.560	4.601	4.635	4.664
19	4.046	4.220	4.335	4.419	4.483	4.534	4.575	4.610	4.639
20	4.024	4.197	4.312	4.395	4.459	4.510	4.552	4.587	4.617
24	3.956	4.126	4.239	4.322	4.386	4.437	4.480	4.516	4.546
30	3.889	4.056	4.168	4.250	4.314	4.366	4.409	4.445	4.477
40	3.825	3.988	4.098	4.180	4.244	4.296	4.339	4.376	4.408
60	3.762	3.922	4.031	4.111	4.174	4.226	4.270	4.307	4.340
120	3.702	3.858	3.965	4.044	4.107	4.158	4.202	4.239	4.272
∞	3.643	3.796	3.900	3.978	4.040	4.091	4.135	4.172	4.205

* Abridged from H. L. Harter, "Critical values for Duncan's new multiple range test," *Biometrics*, vol. 16, no. 4 (1960), by permission of the author and the editor.

ANSWERS TO EXERCISES

Chapter 2

1. (a) $\{7, 14, 21, 28, 35, 42, 49\}$.
  (b) $\{-3, 2\}$.
  (c) $\{H1, H2, H3, H4, H5, H6, T1, T2, T3, T4, T5, T6\}$.
  (d) {N. America, S. America, Europe, Asia, Africa, Australia, Antarctica}.
  (e) $\varnothing$.

2. (a) T. **(b)** F. **(c)** F. **(d)** T. **(e)** T. **(f)** T.

3. $A = C$.

4. $16; \{p, q, r\}, \{p, r, s\}, \{p, q, s\}, \{q, r, s\}$.

5. {air, land}, {air, sea}, {land, sea}, {air}, {land}, {sea}, $\varnothing$.

8. (a) $\{0, 2, 3, 4, 5, 6, 8\}$. **(b)** $\varnothing$. **(c)** $\{0, 1, 6, 7, 8, 9\}$. **(d)** $\{1, 3, 5, 6, 7, 9\}$.
  (e) $\{0, 1, 6, 7, 8, 9\}$. **(f)** $\{2, 4\}$.

9. (a) {nitrogen, potassium, uranium, oxygen}. **(b)** {copper, sodium, zinc, oxygen}.
  (c) {copper, sodium, nitrogen, potassium, uranium, zinc}.
  (d) {copper, uranium, zinc}. **(e)** $\varnothing$. **(f)** {oxygen}.

10. $P \cup Q = \{z \mid z < 9\}; P \cap Q = \{z \mid 1 < z < 5\}$.

12. (a)

Green	Red					
	1	2	3	4	5	6
1	(1, 1)	(1, 2)	(1, 3)	(1, 4)	(1, 5)	(1, 6)
2	(2, 1)	(2, 2)	(2, 3)	(2, 4)	(2, 5)	(2, 6)

12. (a) (*continued*)

3	(3, 1)	(3, 2)	(3, 3)	(3, 4)	(3, 5)	(3, 6)
4	(4, 1)	(4, 2)	(4, 3)	(4, 4)	(4, 5)	(4, 6)
5	(5, 1)	(5, 2)	(5, 3)	(5, 4)	(5, 5)	(5, 6)
6	(6, 1)	(6, 2)	(6, 3)	(6, 4)	(6, 5)	(6, 6)

(b) $A = \{(1, 1), (1, 2), (1, 3), (2, 1), (3, 1), (2, 2)\}$.
(c) $B = \{(1, 6), (2, 6), (3, 6), (4, 6), (5, 6), (6, 6), (6, 1), (6, 2), (6, 3), (6, 4), (6, 5)\}$.
(d) $C = \{(2, 1), (2, 2), (2, 3), (2, 4), (2, 5), (2, 6)\}$.

13. (a) $S = \{HH, HT, T1, T2, T3, T4, T5, T6\}$.
   (b) $A = \{T1, T2, T3\}$.
   (c) $B = \varnothing$.

14. (a) $S = \{YYY, YYN, YNY, NYY, YNN, NYN, NNY, NNN\}$.
   (b) $E = \{YYY, YYN, YNY, NYY\}$.
   (c) One possible event: "The second woman interviewed uses brand X."

15. (a) $S = \{M_1M_2, M_1F_1, M_1F_2, M_2M_1, M_2F_1, M_2F_2, F_1M_1, F_1M_2, F_1F_2, F_2M_1,$
$F_2M_2, F_2F_1\}$.
   (b) $A = \{M_1M_2, M_1F_1, M_1F_2, M_2M_1, M_2F_1, M_2F_2\}$.
   (c) $B = \{M_1F_1, M_1F_2, M_2F_1, M_2F_2, F_1M_1, F_1M_2, F_2M_1, F_2M_2\}$.

16. (a) 120. **(b)** 72.   **17.** 24.

18. 256.   **19.** 50,400.

20. 15,120.   **21. (a)** 100. **(b)** 48. **(c)** 48.

22. 120.   **23.** 144.

24. (a) 40,320. **(b)** 384. **(c)** 576.   **25.** 120.

26. 210.   **27.** 280.

28. 6930.   **29. (a)** 56. **(b)** 30. **(c)** 10.

30. 8,211,173,256.   **31.** 180.

32. 70.   **33.** $\Pr(C) = \frac{1}{2}$; $\Pr(A') = \frac{3}{4}$.

34. $\frac{1}{6}$.   **35.** $S = \{\$10, \$25, \$100\}$; $\frac{9}{10}$.

36. $\frac{8}{15}$.   **37.** $\frac{1}{9}$; $\frac{1}{6}$.

38. $\frac{33}{54,145}$; $\frac{33}{66,640}$.   **39. (a)** $\frac{3}{8}$; **(b)** $\frac{9}{28}$.

40. $\frac{46}{221}$.   **41.** A and B.

42. (a) 0.9. **(b)** 0.6. **(c)** 0.5.   **43. (a)** 0.7. **(b)** 0.5. **(c)** 0.4.

44. (a) 0.0001. **(b)** 0.9999.   **45.** $\frac{13}{120}$.

46. (a) $\frac{2}{11}$. **(b)** $\frac{5}{11}$.   **47.** 0.72.

48. (a) $\frac{1}{4}$. **(b)** $\frac{5}{27}$. **(c)** $\frac{1}{20}$.   **49.** $\frac{38}{63}$.

50. (a) 0.35. **(b)** 0.875. **(c)** 0.55.   **51.** 0.04.

52. $\frac{1}{4}$.   **53.** $\frac{2}{9}$.

54. $\frac{19}{64}$; $\frac{45}{64}$.   **55.** $\frac{5}{8}$.

56. $\frac{3}{10}$.   **57.** $\frac{15}{23}$.

58. 0.174.

Chapter 3

1. Discrete; continuous; continuous; discrete; discrete; continuous.

2.

x	0	1	2	3
$\Pr(X = x)$	$\frac{8}{27}$	$\frac{4}{9}$	$\frac{2}{9}$	$\frac{1}{27}$

3. $f(x) = \dfrac{\dbinom{5}{x}\dbinom{5}{4-x}}{\dbinom{10}{4}}$, $x = 0, 1, 2, 3, 4$.

4. $f(x) = \frac{1}{6}$, $x = 1, 2, \ldots, 6$.

5.

x	0	1	2
$\Pr(X = x)$	$\frac{1}{5}$	$\frac{3}{5}$	$\frac{1}{5}$

6.

x	0	1	2	3
$\Pr(X = x)$	$\frac{1}{27}$	$\frac{2}{9}$	$\frac{4}{9}$	$\frac{8}{27}$

7. (b) $\frac{1}{4}$. **(c)** 0.3.

8. (a) $\frac{16}{27}$. **(b)** $\frac{1}{3}$.

15. Assuming the lower class interval to be 9–17, $P_{14} = 38.65$, $Q_1 = 54.79$, $D_7 = 78.7$.

16. Assuming the lower class interval to be \$3.21–\$6.41, $P_{56} = \$12.478$, $Q_3 = \$15.564$, $D_9 = \$19.255$.

17. Assuming the lower class interval to be 0.1–1.1, $P_{37} = 1.535$, $Q_3 = 5.138$, $D_2 = 0.783$.

18. (a)   **(b)** $\frac{1}{2}$.

$f(x, y)$		0	1	2	3
	0		$\frac{3}{70}$	$\frac{9}{70}$	$\frac{3}{70}$
y	1	$\frac{2}{70}$	$\frac{18}{70}$	$\frac{18}{70}$	$\frac{2}{70}$
	2	$\frac{3}{70}$	$\frac{9}{70}$	$\frac{3}{70}$	

19. (a)   **(b)** $\frac{11}{12}$.

$f(x, y)$		0	1	2
	0	$\frac{16}{36}$	$\frac{8}{36}$	$\frac{1}{36}$
y	1	$\frac{8}{36}$	$\frac{2}{36}$	
	2	$\frac{1}{36}$		

20. (a)

y	0	1	2
$f(y\|2)$	$\frac{3}{10}$	$\frac{3}{5}$	$\frac{1}{10}$

(b) $\frac{3}{10}$.

21. (a)

x	1	2	3
$g(x)$	$\frac{1}{3}$	$\frac{19}{36}$	$\frac{5}{36}$

y	1	2	3
$h(y)$	$\frac{1}{4}$	$\frac{14}{45}$	$\frac{79}{180}$

(b) $\frac{9}{19}$.

22.

x	2	4
$g(x)$	0.40	0.60

y	1	3	5
$h(y)$	0.25	0.50	0.25

; X and Y are independent.

23. Not independent.

Chapter 4

1. (a) $w_6^2 + w_7^2 + w_8^2 + w_9^2 + w_{10}^2$. **(b)** $x_2 + x_3 + x_4 + 9$.
  (c) $3(v_1 + v_2 + v_3 + v_4 + v_5) - 30$.

2. (a) $12x^2 + 36x + 29$. **(b)** $4x^3 + 18x^2 + 42x + 36$.

3. (a) 66. **(b)** 53. **(c)** $\frac{5}{3}$.

4. (a) -7. **(b)** -3. **(c)** -14.

9. 1. **10.** \$200.

11. \$1.08. **12.** 2.

13. $\frac{3}{4}$. **14.** $\frac{76}{3}$.

15. \$420. **16.** 35.2.

17. (a) 209. **(b)** $\frac{65}{4}$. **18.** $-\frac{3}{7}$.

19. (a) 7. **(b)** 0. **(c)** 12.25. **20. (a)** -2.60. **(b)** 9.60.

21. $\frac{2}{5}$. **22.** 3.041.

23. $\mu = 1.125$; $\sigma^2 = 0.502$. **24.** $-\frac{3}{14}$.

25. -0.1244. **27. (a)** $\frac{175}{12}$. **(b)** $\frac{175}{6}$.

28. 68. **29.** 52.

30. (a) At least $\frac{3}{4}$. **(b)** At least $\frac{8}{9}$.

31. 104–136; less than 96 and greater than 144.

Chapter 5

1. $f(x) = \frac{1}{10}$, $x = 1, 2, \ldots, 10$; $\Pr(X < 4) = \frac{3}{10}$.

2. $f(x) = \frac{1}{25}$, $x = 1, 2, \ldots, 25$.

3. $f(x) = \frac{1}{20}$, $x = 1, 2, \ldots, 20$.

4. 0.4219.

5. Uniform and binomial.

6. 0.3134.

7. 0.1240.

8. 0.1035.

9. 0.0006.

10. $\frac{63}{64}$.

11. $f(x) = \binom{5}{x} (\frac{1}{4})^x (\frac{3}{4})^{5-x}$, $x = 0, 1, 2, \ldots, 5$; $\mu \pm 2\sigma = 1.25 \pm 1.936$.

12. Four-engine plane.

13. Two-engine plane when $q = \frac{1}{2}$; either plane when $q = \frac{1}{3}$.

14. 3.75; $\mu \pm 2\sigma = 3.75 \pm 3.35$.

15. 32; 24–40.

16. $\frac{15}{128}$.

17. 0.0095.

18. 0.0025.

19. $\frac{21}{256}$.

20. (a) 0.2963. (b) 0.3972.   **21.** $\frac{5}{14}$.

22. $f(x) = \dfrac{\binom{4}{x}\binom{2}{3-x}}{\binom{6}{3}}$, $x = 1, 2, 3$. $\Pr(2 \le X \le 3) = \frac{4}{5}$.

23. (a) $\frac{1}{6}$. (b) $\frac{203}{210}$.   **24.** 1.2.

25. 3.25; 0.52–5.98.   **26.** 0.2131.

27. 0.9453.   **28.** 0.0129.

29. $\frac{4}{33}$.   **30.** $\frac{17}{63}$.

31. 0.1008.   **32.** (a) 0.1429. (b) 0.1353.

33. (a) 0.1512. (b) 0.4015.   **34.** 0.3840.

35. 0.6288.   **36.** 0.2657.

37. 0–8.   **38.** 0.0515.

39. 0.0651.   **40.** 0.1172.

41. $\frac{63}{64}$.   **42.** (a) 0.0630. (b) 0.9730.

Chapter 6

1. (a) 0.0913. (b) 0.9849. (c) 0.3362. (d) 39.244. (e) 46.756.

2. (a) 0.9192. (b) 0.9821. (c) 0.6106. (d) 208.42. (e) 188.5; 211.5.

3. (a) 0.1151. (b) 16.375. (c) 0.5403. (d) 20.55.

4. (a) 0.0548. (b) 0.4514. (c) 23. (d) 6.66 ounces.

5. (a) 0.0062. (b) 0.6826. (c) 3.986.

6. (a) 0.0572. (b) 99.11%. (c) 0.3963. (d) 27.941 minutes.

7. (a) 64. (b) 86.

8. (a) 17. (b) 524. (c) 72. (d) 26.

9. 62.

10. (a) 57.11%. (b) $4.23.

11. (a) 0.0401. (b) 0.7734. (c) 0.7888.

12. (a) 0.0401. (b) 0.0244.

13. (a) 19.36%. (b) 39.70%.

14. 26.

15. (a) 0.0023. (b) 0.2160. (c) 0.0520.

16. 6.238 years.

17. 0.0018.

18. (a) 0.7925. (b) 0.0352. (c) 0.0101.

19. (a) 0.8643. (b) 0.2978. (c) 0.0796.

20. 0.9515.

21. Less than 75%.

22. (a) 0.0846. (b) 0.1630.

23. 0.1179.

24. 0.1357.

25. 0.4356.

Chapter 7

1. (a) Responses of all people in Richmond who have telephones.
   (b) Outcomes for a large or infinite number of tosses of a coin.
   (c) Length of life of all such combat boots when worn in Vietnam.
   (d) All possible distances that this golfer can hit the ball with his driver.

2. $\bar{x} = 2$; $\tilde{x} = 1.5$; $m = 1$.

3. $\bar{x} = 5.875$; $\tilde{x} = 6.5$; $m = 4$ and 7.

4. $\bar{x} = 10$; $\tilde{x} = 7$; $m = 7$.

5. $\bar{x} = 25.2$; $\tilde{x} = 17$; $m = 11$.

6. Range $= 5$; $s = 1.57$.

7. Range $= 30$; $s^2 = 76$.

8. $s = 0.55$.

11. (a) 3.367. (b) 13.468. (c) 3.367.

12. 2.

13. (c) 3.18; 4.82.

14. 0.1922.

15. (c) -0.07; 4.07.

16. 0.3159.

17. 100.

18. Adjust the machine.

19. (a) $\mu_{\bar{x}} = 68.5$, $\sigma_{\bar{x}} = 0.54$. (b) 161. (c) 0.

20. (a) $\frac{7}{3}$. (b) $\frac{25}{9}$.

21. 0.7064.

22. (a) 0.0772. (b) 0.2814.

23. (a) 2.110. (b) -2.764. (c) 1.714.

24. No; $\mu > 20$.

25. Yes.

26. (a) 34.805. (b) 16.047. (c) 13.277.

27. (a) 0.05. (b) 0.94.

28. Values are not valid.

29. (a) 2.71. (b) 3.51. (c) 2.92. (d) 0.47. (e) 0.34.

30. 0.05.

31. 0.99.

Chapter 8

1. (a) $\mu = \frac{10}{3}$; $\sigma^2 = \frac{14}{9}$.

3. $765 < \mu < 795$.

4. $7.24 < \mu < 7.56$.

5. (a) $67.61 < \mu < 69.39$.   (b) $e < 0.89$.

6. (a) $13,882 < \mu < 15,118$.   (b) $e < 618$.

7. 68.

8. 11.

9. 28.

10. $9.81 < \mu < 10.31$.

11. $30.69 < \mu < 34.91$.

12. $15.63 < \mu < 21.57$.

13. $\$7.09 < \mu < \$8.91$.

14. $2.9 < \mu_1 - \mu_2 < 7.1$.

15. $6.56 < \mu_1 - \mu_2 < 11.24$.

16. $638 < \mu_1 - \mu_2 < 1362$.

17. $0.3 < \mu_1 - \mu_2 < 9.7$.

18. $1.5 < \mu_1 - \mu_2 < 12.5$.

19. $-4036 < \mu_1 - \mu_2 < 1836$.

20. $-11.9 < \mu_{II} - \mu_{I} < 36.5$.

21. $-0.7 < \mu_D < 6.3$.

22. $-1795 < \mu_D < 445$.

23. $3.2 < \mu_D < 13.1$.

24. (a) $0.498 < p < 0.642$.   (b) $e < 0.072$.

25. (a) $0.1442 < p < 0.1998$.   (b) $e < 0.0278$.

26. $0.1971 < p < 0.2589$.

27. $0.1205 < p < 0.3055$.

28. (a) 0.85.   (b) $0.739 < p < 0.961$.   (c) No.

29. 2593.

30. 382.

31. 9604.

32. 16,577.

33. $0.1179 < p_1 - p_2 < 0.3781$.

34. $0.016 < p_A - p_B < 0.164$.

35. $-0.030 < p_M - p_W < 0.250$.

36. $0.0284 < p_1 - p_2 < 0.1416$.

37. $0.023 < \sigma^2 < 0.313$.

38. $3.430 < \sigma < 6.587$.

39. $1.410 < \sigma < 6.385$.

40. $1.712 < \sigma^2 < 7.363$.

41. $0.600 < \sigma_1/\sigma_2 < 2.819$; yes.

42. $0.236 < \sigma_1^2/\sigma_2^2 < 1.877$.

43. $0.016 < \sigma_1^2/\sigma_2^2 < 0.454$.

44. $R(\hat{P}; p) = pq/n$.

45. $R(\Theta_1; \theta) = \begin{cases} 0 & \text{for } \theta = 0 \\ \frac{2}{3} & \text{for } \theta = 1 \\ \frac{2}{3} & \text{for } \theta = 2 \\ 0 & \text{for } \theta = 3. \end{cases}$

46. $R(\Theta_2; \theta) = \begin{cases} 0 & \text{for } \theta = 0 \\ \frac{1}{3} & \text{for } \theta = 1 \\ 1 & \text{for } \theta = 2 \\ 0 & \text{for } \theta = 3. \end{cases}$

47. $\hat{\Theta}_1$.

48. $\hat{\Theta}_2$.

Chapter 9

1. $\alpha = 0.0853$; $\beta = 0.8287$; $\beta = 0.7817$; not a good test procedure.

2. $\alpha = 0.0548$; $\beta = 0.3504$; $\beta = 0.6177$; $\beta = 0.8281$.

3. $\alpha = 0.0466$; $\beta = 0.0022$.   **4.** $\alpha = 0.0796$; $\beta = 0.0796$; $\beta = 0.5$.

5. $z = -1.643$; accept H_0.   **6.** $z = -1.25$; accept H_0.

7. $z = 3.143$; $\mu > 68.5$.   **8.** $z = 10.417$; $\mu > 12,000$.

9. $t = 0.771$; accept H_0.   **10.** $t = 2.776$; $\mu > 30$.

11. $t = 1.296$; valid claim.   **12.** $t = 1.781$; accept H_0.

13. $z = 4.222$; $\mu_1 > \mu_2$.   **14.** $z = -2.603$; $\mu_A - \mu_B < 12$.

15. $z = 2.448$; $\mu_1 - \mu_2 > \$500$.   **16.** $t = 6.575$; $\mu_1 > \mu_2$.

17. $t = 1.501$; yes.   **18.** $t = -0.922$; yes.

19. $t = -0.850$; accept H_0.   **20.** $t' = 0.215$; accept H_0.

21. $t = 1.821$; accept H_0.   **22.** $t = -2.109$; accept H_0.

23. $t = -0.910$; accept H_0.   **24.** $\chi^2 = 18.120$; accept H_0.

25. $\chi^2 = 10.735$; accept H_0.   **26.** $\chi^2 = 17.530$; reject H_0.

27. $\chi^2 = 17.188$; accept H_0.   **28.** $f = 1.325$; accept H_0.

29. $f = 0.748$; accept H_0.   **30.** $f = 0.086$; reject H_0.

31. $u = 12.5$; accept H_0.   **32.** $u = 43.5$; $\mu_I = \mu_{II}$.

33. $u = 15$; $\mu_1 = \mu_2$.   **34.** $w = 1$; accept H_0.

35. $w = 8.5$; accept H_0.   **36.** $w = 4.5$; accept H_0.

37. $w = 9.5$; accept H_0.   **38.** $z = -1.443$; $p = 0.6$.

39. Valid claim.   **40.** $z = -3.265$; $p \neq \frac{2}{3}$.

41. $z = 1.339$; valid estimate. **42.** $z = 2.395$; yes.

43. $z = 1.878$; yes. **44.** $z = 1.109$; no.

45. $z = 2.090$; yes. **46.** $\chi^2 = 4.467$; yes.

47. $\chi^2 = 6.76$; no. **48.** $\chi^2 = 1.667$; accept H_0.

49. $\chi^2 = 2.326$; accept H_0. **50.** $\chi^2 = 10.000$; reject H_0.

51. $\chi^2 = 2.571$; accept H_0. **54.** $\chi^2 = 9.613$; not normal.

55. $\chi^2 = 4.617$; normal. **56.** $\chi^2 = 14.464$; not independent.

57. $\chi^2 = 16.816$; not independent. **58.** χ^2 (corrected) $= 0.538$; independent.

59. $\chi^2 = 9.048$; independent. **60.** $\chi^2 = 6.239$; accept H_0.

Chapter 10

1. (a) $\bar{y}_x = 5.799 - 0.514x$. **(c)** $\bar{y}_x = 3.743$.

2. (a) $\bar{y}_x = 12.073 + 0.777x$. **(b)** $\bar{y}_{85} = 78$.

3. (a) $\bar{y}_x = 6.414 + 1.809x$. **(b)** $\bar{y}_{1.75} = 9.580$.

4. (a) $\bar{y}_x = 5.811 + 0.568x$. **(c)** $\bar{y}_{50} = 34.211$.

5. (b) $\bar{y}_x = 32.510 + 0.471x$. **(d)** $x = 59$.

6. (b) $\bar{y}_x = 343.699 + 3.221x$. **(d)** $\bar{y}_{35} = \$456$.

7. $t = -1.855$; accept $\beta = 0$.

8. $t = -0.186$; accept $\alpha = 6$.

9. $-69.911 < \alpha < 94.057$; $-0.248 < \beta < 1.802$.

10. $4.323 < \alpha < 8.505$; $0.445 < \beta < 3.173$.

11. $2.747 < \alpha < 8.875$; $0.501 < \beta < 0.635$.

12. $5.954 < \alpha < 59.066$.

13. $0.975 < \beta < 5.467$.

14. $58.805 < \mu_{Y|80} < 89.661$.

15. $7.808 < y_{1.6} < 10.808$.

16. $32.285 < \mu_{Y|50} < 36.137$; $26.630 < y_{50} < 41.792$.

17. $36.931 < \mu_{Y|35} < 61.059$; $12.938 < y_{35} < 85.052$.

18. $452.970 < \mu_{Y|45} < 524.318$; $390.904 < y_{45} < 586.384$.

19. $f = 1.361$; regression is linear.

20. $f = 1.118$; regression is linear.

21. (a) $\bar{y}_x = (2760)(0.804)^x$. **(b)** $\bar{y}_x = \$1151$.

22. (a) $\gamma = 2.660$; $C = 2.63 \times 10^6$. **(b)** $P = 22.9$.

23. (a) $\bar{y}_{x_1, x_2} = 23.6745 - 0.4880x_1 - 0.2934x_2$. **(b)** $\bar{y}_{10.3, 5.8} = 16.9$.

24. (a) $\bar{y}_{x_1, x_2} = 1.2522 + 0.0028x_1 - 0.1837x_2$. **(b)** $\bar{y}_{575, 3} = 2.311$.

25. (a) $\bar{y}_x = 8.697 - 2.341x + 0.288x^2$. **(b)** $\bar{y}_2 = 5.2$.

26. (a) $\bar{d}_v = 38.124 - 1.080v + 0.0877v^2$. **(b)** $\bar{d}_{45} = 167$ ft.

27. $r = -0.526$.

28. $r = 0.244$.

29. $r = 0.240$.

30. $z = 1.028$; accept H_0.

31. $z = 0.424$; accept H_0.

Chapter 11

1. $f = 3.593$; reject H_0.

2. $f = 0.307$; no significant difference.

4. $f = 0.464$; no significant difference.

5. $b = 0.191$; variances are equal.

6. (a) $b = 3.80$; variances are equal.   **(b)** $f = 13.33$; reject H_0.

7. 26   30   <u>33   37</u>   42

8. (a) $f = 7.10$; blends differ significantly.
   (b) Blend 4 differs significantly from all others.

9. (a) $f = 0.151$; accept H'_0.   **(b)** $f = 4.368$; reject H''_0.

12. 64.75   67.75   68.75   <u>80.50   88.50</u>

13. $f(\text{varieties}) = 1.786$; no difference in the yielding capabilities of the different varieties.
   $f(\text{locations}) = 8.143$; yes, it is necessary to plant each variety at each location.

14. $f(\text{subjects}) = 4.86$; significant. $f(\text{treatments}) = 3.33$; no significant difference.

15. (a) $f = 0.323$; accept H'_0.   **(b)** $f = 15.377$; reject H''_0.   **(c)** $f = 1.668$; accept H'''_0.

16. (a) $f = 9.533$; reject H'_0.   **(b)** $f = 15.175$; reject H''_0.   **(c)** $f = 1.281$; accept H'''_0.

19. (a) $f = 5.85$; reject H'_0.   **(b)** $f = 10.77$; reject H''_0.   **(c)** $f = 2.98$; accept H'''_0.

20. (a) $f = 14.81$; reject H'_0.   **(b)** $f = 9.04$; reject H''_0.   **(c)** $f = 0.61$; accept H'''_0.

INDEX